MW00845696

Demystifying Electromagnetic Equations

A Complete Explanation of EM Unit Systems and Equation Transformations

Demystifying Electromagnetic Equations

A Complete Explanation of EM Unit Systems and Equation Transformations

Douglas L. Cohen

SPIE PRESS

A Publication of SPIE—The International Society for Optical Engineering
Bellingham, Washington USA

Library of Congress Cataloging-in-Publication Data

Cohen, Douglas L.
    Demystifying electromagnetic equations : a complete explanation of EM unit systems
    and equation transformations / by Douglas L. Cohen.
        p. cm.
    Includes bibliographical references and index.
    ISBN 0-8194-4234-8
    1.Electromagnetic theory–Mathematics. 2. Electric units. I. Title.

QC670 .C49 2001
530.14'1'0151--dc21

                                                                                    2001032770

Published by

SPIE—The International Society for Optical Engineering
P.O. Box 10
Bellingham, Washington 98227-0010
Phone: 360.676.3290
Fax: 360.647.1445
Email: spie@spie.org
WWW: www.spie.org

Copyright © 2001 The Society of Photo-Optical Instrumentation Engineers

All rights reserved. No part of this publication may be reproduced or distributed
in any form or by any means without written permission of the publisher.

Printed in the United States of America.

Contents

Preface

In classical Newtonian mechanics, equations and formulas never change form. However, the same thing cannot be said about the equations and formulas of electromagnetic theory, which often change form when converted from one system of units to another. For this reason electromagnetic textbooks are almost always written using a single system of units, and the technical professionals who read them end up being comfortable in only that system. When they encounter a new and important formula in unfamiliar units later on, they must either use a conversion table to change the formula to their preferred system of units or try to become familiar with the formula's units. Although conversion tables usually give the correct answer, they turn their users into computers who must push around numbers and variables without any true understanding of what is being done. It is probably unwise to rely blindly on conversion tables if one must be absolutely sure the transformed formula is correct. That leaves the second option: becoming familiar with the formula's units. The drawback here is that even if a textbook can be found that uses the formula's units, it has been written to teach the basic principles of electromagnetism rather than what the technical professional is looking for, i.e., a detailed explanation of how to convert equations from one system of units to another. This book provides exactly that, while at the same time assuming a good—but not necessarily advanced—understanding of electricity and magnetism.

There are five widely recognized systems of electromagnetic units; four are connected to the centimeter-gram-second (cgs) system of mechanical units and one is connected to the meter-kilogram-second (mks) system of mechanical units. The four connected to the cgs mechanical units are the cgs Gaussian system, the Heaviside-Lorentz system, the cgs electrostatic system, and the cgs electromagnetic system. The system connected to the mks mechanical units is the *Système International* or rationalized mks system. The units of the *Système International* or rationalized mks system are often called SI units. The cgs electrostatic and cgs electromagnetic systems of units were developed first. These are the units in which Maxwell's equations—the foundation of classical electromagnetic theory—were first proposed during the middle of the nineteenth century. The Heaviside-Lorentz and cgs Gaussian systems were introduced at the end of the nineteenth century, followed almost immediately at the beginning of the twentieth century by the rationalized mks system (SI units). The rationalized mks system is the most popular electromagnetic system in use today; almost all introductory textbooks use SI units to explain the principles of electricity and magnetism. This book explains all five systems in depth, along with two systems of mostly historical interest; the nineteenth century system of "practical" units and the unrationalized mks system.

One chronic problem found in many articles and books about systems of units is that the customary language of physics and engineering can permit ambiguity while sounding exact. Suppose, for example, we say

"The electric-current unit in the cgs electrostatic system is the statamp and the electric-current unit in the cgs electromagnetic system is the abamp, with

$$1 \text{ abamp} = c \cdot \text{statamp}$$

where c is the speed of light in cgs units."

This seems clear enough, but notice that $c = 2.99792 \cdot 10^{10}$cm/sec in cgs units. In the above equation, should we take "c" to be "$2.99792 \cdot 10^{10}$" or "2.99792×10^{10}cm/sec?" A naive student might assume c was the pure number $2.99792 \cdot 10^{10}$ because obviously all electric current is the same sort of thing and must have the same type of unit; but later on, possibly in another book, that same student might discover the cgs electrostatic unit of current is $\text{gm}^{1/2} \cdot \text{cm}^{3/2} \cdot \text{sec}^{-2}$ and the cgs electromagnetic unit of current is $\text{gm}^{1/2} \cdot \text{cm}^{1/2} \cdot \text{sec}^{-1}$. At this point confusion sets in, because this is not compatible with the equation $1 \text{ abamp} = c \cdot \text{statamp}$, no matter how it is interpreted.

To avoid this sort of ambiguity, we introduce here the idea of U and N operators, with a U operator returning just the units associated with a physical quantity and an N operator returning just the pure number, or numeric component, associated with a physical quantity. In the cgs system, for example, we have

$$\underset{\text{cgs}}{\text{N}}(c) = 2.99792 \cdot 10^{10}$$

and

$$\underset{\text{cgs}}{\text{U}}(c) = \text{cm/sec}.$$

Authors who put the equation $1 \text{ abamp} = c \cdot \text{statamp}$ in their books and articles are using it to say that $\underset{\text{cgs}}{\text{N}}(c)$ is the conversion factor between the numeric component of the current I in cgs electrostatic units, $\underset{\text{esu}}{\text{N}}(I)$, and the numeric component of the current I in cgs electromagnetic units, $\underset{\text{emu}}{\text{N}}(I)$.

$$\underset{\text{esu}}{\text{N}}(I) = \underset{\text{cgs}}{\text{N}}(c) \cdot \underset{\text{emu}}{\text{N}}(I).$$

The U operator can be used to emphasize that the unit of current in the electrostatic system is not the same as the unit of current in the electromagnetic system.

$$\frac{\text{gm}^{1/2} \cdot \text{cm}^{3/2}}{\text{sec}^2} = \underset{\text{esu}}{\text{U}}(I) \neq \underset{\text{emu}}{\text{U}}(I) = \frac{\text{gm}^{1/2} \cdot \text{cm}^{1/2}}{\text{sec}}$$

The U and N operators make it easy to be precise about the mathematical relationships between different systems of units.

The abbreviations of the SI units are, unfortunately, another possible source of confusion when separating equations into numeric components and units. For example, the standard abbreviation for the SI unit of charge, the coulomb, is C. The capacitance of a circuit element is also traditionally represented as C, and we have already seen that c is used to represent the speed of light. If all three quantities—the coulomb, the speed of light, and the capacitance—have to be included in the same equation, there will be problems. To avoid this source of confusion, we have lengthened the standard abbreviations for the electromagnetic units, representing coulomb by coul, ampere by amp, and so on. This makes the notation less confusing, but the reader should note that the abbreviations used here, although easily understandable, are not the official, internationally approved symbols for the SI units. These international symbols are, in any case, of fairly recent vintage and can be found in virtually all modern textbooks on electromagnetic theory.

One final point worth mentioning is how we treat rationalization of electromagnetic equations. During the middle of the twentieth century it became clear that there were two different schools of thought concerning the rationalization of electromagnetic equations: one that it was a rescaling of the electromagnetic units, and the other that it was a rescaling of the electromagnetic quantities themselves. Both views can be used to deduce the same systems of electromagnetic equations, and both views allow engineers and scientists to transform electromagnetic measurements from one system to another correctly. In the end, neither side convinced the other of the correctness of its views and the controversy faded away. For the purposes of this book, we take the position that rationalization is a rescaling of electromagnetic physical quantities rather than a change of units, not only because it is then easier to describe the units of the rationalized and unrationalized electromagnetic systems but also because it makes the transformation of equations to and from rationalized electromagnetic systems a straightforward process. The opposite position, that rationalization just involves rescaled units, is not necessarily incorrect—that is, after all, how the idea of rationalization was first proposed in the nineteenth century—but it can easily become confusing in a book of this sort.

OUTLINE OF NON-ELECTROMAGNETIC SYSTEMS OF UNITS

The units describing the kinematics and dynamics of mechanical objects are straightforward to learn and use. Different units of length are always the same kind of unit, differing only in size; different energy units are always the same kind of unit, differing only in how much energy is specified; and so on. One happy consequence of this pattern is that all the equations of classical Newtonian physics have the same form no matter what system of units is being used. Indeed, it seems intuitively absurd that any other pattern could exist, that two different units for the same physical quantity could be different in kind as well as size. Nevertheless, an idea can be intuitively absurd without being mathematically absurd, and there are times when it is not clear what kind of unit best describes a well-understood physical quantity. This book will show that the equations and formulas of classical electromagnetic theory routinely change form when expressed in different units exactly because the physical intuition of scientists has been different, leading them to create units that are different in kind as well as size to measure the same electromagnetic quantities.

The idea of many different "kinds" of units for the same physical quantity may seem strange at first, but like most ideas involving units it is really rather simple once the right perspective is adopted. Fortunately, there do exist procedures using elementary mechanical units that are analogous to what goes on when we change the kinds of units used in electromagnetic theory; we can use this analogy to introduce the appropriate perspective for understanding electromagnetic units. This chapter begins by presenting material with which the reader is probably already familiar—what a unit is, what a dimension is, the rules for manipulating units inside equations—and then moves on to describe the procedures applied to standard mechanical units in quantum mechanics and relativistic physics to simplify the forms of complicated equations. It should be emphasized that all these equations are presented as "given," with no expectation that the reader will gain or have any particular knowledge of how the equation is derived; we just show how to simplify the equation by changing the units in which it is expressed. By the end of the chapter the reader will have acquired the rules and terminology needed to describe how and why the equations of classical electromagnetism change form when moving from one system of units to another.

1.1 THE BASIC IDEA OF A UNIT

Measurements create numbers, and units give meaning to numbers by connecting them to measurements. For example, to call a length "8.5" is completely ambiguous; but to call it "8.5 centimeters," or "8.5 feet," or "8.5 miles," does have meaning—and the meaning changes when the attached unit changes. In equations, a physical quantity, such as a length L, is treated as if it were the product of a numeric or pure number, like the 8.5 in "8.5 centimeters," and a unit, like the centimeters in "8.5 centimeters." Length L is specified in centimeters (cm) using the equation

$$L = 8.5 \, \text{cm}. \tag{1.1}$$

The "8.5" behaves like any of the real numbers which have been exhaustively defined and studied by mathematicians, but the cm unit is defined empirically by pointing to the length of some physical object and stating, "This object is exactly 1 cm long and L is 8.5 times as long as this object." For this reason units are regarded as having a quality called dimension that ordinary numbers do not have.* The unit cm is said to have the dimension length, and L also has that dimension since it is a dimensionless number multiplied by a quantity—namely 1 cm—which has the dimension length. Another way of saying that all physical quantities have units attached to them is to say that all physical quantities have dimensions.

At this point we could begin worrying about exactly what sort of mathematical object a unit such as cm is, but for the units encountered in physics and engineering all we need is the assurance that units in formulas and equations are always treated like ordinary real parameters.[1,2] If, for example, we want to convert L to a new unit of length, such as the inch, which is equal to 2.54 cm, we can go directly from

$$1 \, \text{in} = 2.54 \, \text{cm} \tag{1.2a}$$

to

$$1 \, \text{cm} = (2.54)^{-1} \, \text{in}. \tag{1.2b}$$

Substituting Eq. (1.2b) into Eq. (1.1) gives

$$L = 8.5 \cdot \left[(2.54)^{-1} \, \text{in} \right] = \left[8.5 \cdot (2.54)^{-1} \right] \cdot \text{in} = 3.35 \, \text{in}. \tag{1.3}$$

When moving from Eq. (1.1) to Eq. (1.3) we just follow the standard rules of algebra; and, no matter how complicated an algebraic expression is, this sort of substitution can always be used to convert a formula from one set of units to

---

* This use of the word dimension should not be confused with the mathematical use of dimension, where a line has one mathematical dimension, a geometric plane has two dimensions, etc.

another. In fact, going from Eq. (1.1) to Eq. (1.3) is a specific instance of what could be called Rule I for handling units.

RULE I

When unit$a$ and unit$b$ have the same dimensions, and when the numeric parts of a physical quantity q are the pure numbers $q_{[unita]}$ and $q_{[unitb]}$ with respect to unit$a$ and unit$b$, so that

$$q = q_{[unita]} \cdot unita$$

and

$$q = q_{[unitb]} \cdot unitb,$$

it follows that the ratio of $q_{[unita]}$ to $q_{[unitb]}$ is α^{-1},

whenever the ratio of unit$a$ to unit$b$ is the real number α such that unit$a = \alpha \cdot unitb$.

An easy generalization of Eq. (1.3) shows where Rule I comes from. We have

$$q = q_{[\,unita]}\, unita = q_{[unita]}\left(\alpha \cdot unitb\right) = \left(\alpha q_{[\,unita]}\right) unitb. \tag{1.4}$$

All that is required is to stop thinking of α as multiplying unit$b$ and to start thinking of it as multiplying $q_{[unita]}$. Since it is also true that

$$q = q_{[unitb]} unitb,$$

we end up with

$$\alpha q_{[unita]} = q_{[unitb]};$$

or,

$$q_{[unita]} = \frac{1}{\alpha} q_{[unitb]}.$$

Equation (1.4) holds true not only for the units of simple physical quantities such as the centimeter, a unit of length, but also for the units of more sophisticated physical quantities such as velocity, energy, electric field, etc.

1.2 FUNDAMENTAL AND DERIVED UNITS

It is customary to regard the dimensions of sophisticated physical quantities as composed of the products and powers of a smaller number of fundamental dimensions. In many fields of engineering and physics, three different dimensions—mass, length, and time—are regarded as fundamental. This choice is by no means the only one possible. Older textbooks may use an English system of units based on the pound-force, foot, and second, where the fundamental dimensions are force, length, and time. Later on we will discuss exotic systems of units used by particle physicists and relativistic cosmologists where mass and length, or even just length, are taken to be the only fundamental dimensions. For now, though, we follow the path taken by the vast majority of physics and engineering disciplines by treating mass, length, and time as fundamental.

To see how more sophisticated physical quantities are reduced to their fundamental dimensions, consider an area. A unit of area is usually given as the square of some unit of length, such as cm^2, $foot^2$, or $meter^2$; and when it is given a name of its own, such as acre or barn, these names are defined using the square of some length unit:

$$1\,\text{acre} = 43{,}560\,\text{feet}^2$$

$$1\,\text{barn} = 1 \times 10^{-28}\,\text{meter}^2.$$

Equations (1.5) and (1.6) are the well-known formulas for the area A_S of a square whose side is length s, and the area A_C of a circle whose radius is r:

$$A_S = s^2, \tag{1.5}$$

$$A_C = \pi r^2. \tag{1.6}$$

In the second formula, $\pi = 3.14159\ldots$ is taken to be a dimensionless quantity or pure number. We can choose any arbitrary unit of length, e.g., ulength, so that

$$s = s_{[\text{ulength}]} \cdot \text{ulength}$$

and

$$r = r_{[\text{ulength}]} \cdot \text{ulength},$$

where $s_{[\text{ulength}]}$ and $r_{[\text{ulength}]}$ are numerics—that is, pure numbers—such that $s_{[\text{ulength}]}$ and $r_{[\text{ulength}]}$ are the numeric parts of s and r measured in units of ulength. Substituting these expressions into Eqs. (1.5) and (1.6) gives

$$A_S = \left(s_{[\text{ulength}]} \cdot \text{ulength}\right)^2 = \left(s_{[\text{ulength}]}^2\right) \cdot \left(\text{ulength}^2\right)$$

and

$$A_C = \pi \left(r_{\text{[ulength]}} \cdot \text{ulength}\right)^2 = \left(\pi r^2_{\text{[ulength]}}\right) \cdot \left(\text{ulength}^2\right).$$

Clearly, the units of both A_S and A_C are always the square of ulength. No matter what formula is used to calculate an area, it always ends up having units that are the product of two length units, so area is given the dimensions of length^2.

To see how mass and time enter the picture, we need to look at some more specialized equations. The average velocity of a physical object, for example, is defined to be the vector distance $\Delta \vec{s}$ that the object travels during time Δt:

$$\vec{v} = \frac{\Delta \vec{s}}{\Delta t}. \tag{1.7a}$$

That $\vec{v}$ and $\Delta \vec{s}$ are vectors does not matter from the viewpoint of dimensional analysis; for now we can simply disregard their vector nature and write

$$v = \frac{\Delta s}{\Delta t}. \tag{1.7b}$$

Defining an arbitrary unit of time, utime, to go with the arbitrary unit of length ulength, we have

$$\Delta s = \left(\Delta s\right)_{\text{[ulength]}} \cdot \text{ulength} \tag{1.8a}$$

and

$$\Delta t = \left(\Delta t\right)_{\text{[utime]}} \cdot \text{utime}, \tag{1.8b}$$

where $(\Delta s)_{\text{[ulength]}}$ and $(\Delta t)_{\text{[utime]}}$ are numerics. Substitution of Eqs. (1.8a, b) into Eq. (1.7b) gives

$$v = \frac{(\Delta s)_{\text{[ulength]}} \cdot \text{ulength}}{(\Delta t)_{\text{[utime]}} \cdot \text{utime}} = \left[\frac{(\Delta s)_{\text{[ulength]}}}{(\Delta t)_{\text{[utime]}}}\right] \cdot \left(\frac{\text{ulength}}{\text{utime}}\right).$$

Units of velocity are always a unit of length divided by a unit of time, ulength/utime, and so the dimension of velocity is length·time^{-1}. The average acceleration is defined to be the vector change in velocity $\Delta \vec{v}$ over some time Δt, so

$$\vec{a} = \frac{\Delta \vec{v}}{\Delta t} \tag{1.9}$$

and, disregarding the vector nature of the equation,

$$a = \frac{(\Delta v)_{\text{[ulength/utime]}} \cdot (\text{ulength/utime})}{(\Delta t)_{\text{[utime]}} \cdot \text{utime}} = \left[\frac{(\Delta v)_{\text{[ulength/utime]}}}{(\Delta t)_{\text{[utime]}}}\right] \cdot \left(\frac{\text{ulength}}{\text{utime}^2}\right).$$

Units of acceleration are always ulength $\cdot$ utime^{-2}, which makes the dimensions of acceleration length $\cdot$ time^{-2}.

Newton's second law is

$$\vec{F} = m\vec{a}, \qquad\qquad (1.10)$$

where $\vec{F}$ is the force acting on an object of mass m undergoing an acceleration $\vec{a}$. Choosing some unit of mass, umass, we have, disregarding the vector nature of the equation,

$$F = m_{[umass]} \cdot umass \cdot a_{[\,ulength/utime^2]} \cdot \frac{ulength}{utime^2}$$

$$= \left[m_{[umass]} a_{[ulength/utime^2]} \right] \cdot \left(\frac{umass \cdot ulength}{utime^2} \right).$$

This is a good time to introduce Rule II, since we have been using it implicitly when examining Eqs. (1.5) through (1.9).

RULE II

In any equation involving physical quantities, both the left-hand side of the equation and the right-hand side of the equation must have the same dimensions. If both the left-hand side and right-hand side are expressed in the same units, then the numeric part of the left-hand side of the equation is equal to the numeric part of the right-hand side of the equation.

We have just seen that the units of ma are always umass$\cdot$ulength/utime^2, which means the dimensions of ma must be mass$\cdot$length/time^2. Rule II applied to Newton's second law now assures us that the dimensions of force must be the same as the dimensions of ma, namely mass$\cdot$length/time^2.

Although equations involving physical quantities are almost always broken down into numeric components and units in such a way that both sides of the equation are expressed using the same units, the equality can still exist even when each side is in different units. We might, for example, encounter an equation

$$F_{[gm\cdot cm/sec^2]} \cdot \left(\frac{gm \cdot cm}{sec^2} \right) = \left(m_{[ton]} ton \right) \cdot \left(a_{[inch/hour^2]} \frac{inch}{hour^2} \right),$$

which is true even though, according to Rule II, we cannot then expect that

$$F_{[gm\cdot cm/sec^2]} \overset{?}{=} \left(m_{[ton]} \right) \cdot \left(a_{[inch/hour^2]} \right)$$

because the two sides of the equation have different units. The first equality is true exactly because the units have been included in the equation, while the second equality is false because they have been left out. This convention allows us to regard equations like (1.2a, b) as just another equality between physical quantities, in this case the inch and centimeter. Just like the above equations, they are not necessarily true when the units are left out because then we end up with

$$1 \overset{?}{=} 2.54 \quad \text{or} \quad 1 \overset{?}{=} (2.54)^{-1}.$$

By now it should be pretty clear that in any product involving physical quantities the dimensions mass, length, and time act exactly like the arbitrary units umass, ulength, and utime. This observation gives us Rule III for handling units and dimensions.

RULE III

To find the dimensions of any product containing physical quantities, replace all numerics, pure numbers, and other dimensionless quantities by one and all physical quantities by their dimensional formulas.

Using Rule III, the right-hand side of Eq. (1.6) becomes

$$1 \cdot \text{length}^2 = \text{length}^2,$$

giving the dimensions of an area; and the right-hand side of Eq. (1.10) becomes

$$\text{mass} \cdot \left(\text{length} \cdot \text{time}^{-2}\right) = \text{mass} \cdot \text{length} \cdot \text{time}^{-2},$$

giving the dimensions of a force.

We have not yet considered what happens when physical quantities are added or subtracted. Rule IV for handling units and dimensions is the following:

RULE IV

When two physical quantities are added or subtracted they must have the same dimensions. The dimensional formula for the sum or difference of physical quantities is identical to the dimensional formula of either term of the sum or difference. When two physical quantities are expressed in the same units, the numeric part of their sum or difference equals the sum or difference of their numeric parts.

This rule matches what we might expect if units behave like real parameters. If, for example, L_1 and L_2 are two different lengths, with L_1 measured in feet (ft) and L_2 measured in meters (m), then $L_1 + L_2$ becomes (using that $1\,\text{m} = 3.281\,\text{ft}$),

$$(L_1)_{[\text{ft}]}\,\text{ft} + (L_2)_{[\text{m}]}\,\text{m} = (L_1)_{[\text{ft}]} \cdot \text{ft} + (L_2)_{[\text{m}]} \cdot 3.281\,\text{ft} = \left[(L_1)_{[\text{ft}]} + (L_2)_{[\text{ft}]}\right] \cdot \text{ft},$$

where in the last step Rule I is used to get $(L_2)_{[ft]} = 3.281 \cdot (L_2)_{[m]}$ and the common factor of "ft" has been moved outside the parentheses. We end up with the sum of two pure numbers, $(L_1)_{[ft]} + (L_2)_{[ft]}$, which can be added together to get a third number, giving us a perfectly good expression for a new physical quantity—a numeric times the unit ft. Note that we got this result by converting meters to feet, and we can convert to a common unit only when both units have the same dimensions. This explains why physical quantities must have the same dimensions before they can be added or subtracted and why, when they have the same units, we add the quantities by adding together their numeric parts. Rule IV, by the way, applies even when the physical quantities involved are vectors or are complex valued. In the vector sum $\vec{a} + \vec{b}$, both $\vec{a}$ and $\vec{b}$ must be the same type of physical quantity so that their units can have the same dimensions; that is, both $\vec{a}$ and $\vec{b}$ must be vector velocities, or vector accelerations, etc. For example, if $\vec{L}_1$ and $\vec{L}_2$ are position vectors, we can rewrite the above sum as

$$\left(\vec{L}_1\right)_{[ft]} ft + \left(\vec{L}_2\right)_{[m]} m = \left(\vec{L}_1\right)_{[ft]} \cdot ft + \left(\vec{L}_2\right)_{[m]} \cdot 3.281\, ft = \left[\left(\vec{L}_1\right)_{[ft]} + \left(\vec{L}_2\right)_{[ft]}\right] \cdot ft,$$

where $(\vec{L}_1)_{[ft]}$ and $(\vec{L}_2)_{[m]}$ are vectors of pure numbers whose x, y, z components are the numeric parts of the x, y, z components of vectors $\vec{L}_1, \vec{L}_2$ measured in feet and meters, respectively. We see that units are still treated as scalar parameters that can be brought outside the vector when all the x, y, z components have the same units:

$$\begin{bmatrix} (L_{1x})_{[ft]} \cdot ft \\ (L_{1y})_{[ft]} \cdot ft \\ (L_{1z})_{[ft]} \cdot ft \end{bmatrix} = \begin{bmatrix} (L_{1x})_{[ft]} \\ (L_{1y})_{[ft]} \\ (L_{1z})_{[ft]} \end{bmatrix} \cdot ft = (\vec{L}_1)_{[ft]} \cdot ft,$$

$$\begin{bmatrix} (L_{2x})_{[m]} \cdot m \\ (L_{2y})_{[m]} \cdot m \\ (L_{2z})_{[m]} \cdot m \end{bmatrix} = \begin{bmatrix} (L_{2x})_{[m]} \\ (L_{2y})_{[m]} \\ (L_{2z})_{[m]} \end{bmatrix} \cdot m = (\vec{L}_2)_{[m]} \cdot m.$$

This is why we can disregard the vector nature of physical quantities when doing dimensional analysis, analyzing the dimensions and units of vector cross products and dot products the same way we would analyze the products of ordinary real variables. Units are also treated as real scalar parameters in physical quantities represented by complex entities such as $Z = u + iv$, with u and v the real and imaginary parts of Z. If, for example, Z is an impedance measured in ohms, we can write

$$Z = u_{[ohms]} \cdot ohms + i \cdot v_{[ohms]} \cdot ohms = \left(u_{[ohms]} + i \cdot v_{[ohms]}\right) \cdot ohms = Z_{[ohms]} \cdot ohms,$$

where

$$Z_{[ohms]} = u_{[ohms]} + i \cdot v_{[ohms]}.$$

The real and imaginary parts of complex entities always have the same dimensions because $i = \sqrt{-1}$ is regarded as being dimensionless.

1.3 ANALYSIS OF EQUATIONS AND FORMULAS

Rules III and IV make it easy to analyze more complicated physical equations. The equation of motion for a mass m suspended from a spring is

$$m\frac{d^2y}{dt^2} + \gamma\frac{dy}{dt} + ky = 0, \tag{1.11}$$

where t is time, γ is the damping coefficient (from air resistance), k is the spring constant, and y is the vertical displacement (that is, distance) of the mass from its equilibrium height under the pull of gravity. The dimensions of the infinitesimal quantities dy and dt are length and time, respectively, so the dimension of dy/dt is length $\cdot$ time^{-1}. An operator such as d/dt has the dimension of time^{-1} (think of "d" by itself as dimensionless), which gives the operator $d^2/dt^2 = (d/dt)^2$ the dimension time^{-2} and d^2y/dt^2 the dimensions length $\cdot$ time^{-2}. Therefore, the first term in Eq. (1.11) has the dimensions of force because its dimensional formula is (mass $\cdot$ length/time^2). By Rule IV, the second term $\gamma(dy/dt)$ then must also have dimensions of mass $\cdot$ length/time^2. Since dy/dt has dimensions of length $\cdot$ time^{-1}, the damping coefficient γ must have dimensions of mass $\cdot$ time^{-1}. The same type of reasoning shows that the third term ky must have dimensions of mass $\cdot$ length/time^2, giving k the dimensions of mass $\cdot$ time^{-2}. Clearly, Rules III and IV act together to specify the dimensions of constants γ and k based on the dimensions of $m(d^2y/dt^2)$.

Another example of how Rules III and IV work comes from the formula for the energy E or work W done when a force F acts through a distance s:

$$E = W = \int_0^s F \cdot dx. \tag{1.12}$$

The integral in Eq. (1.12) can be regarded as the sum of a large number of infinitesimal terms $F \cdot dx$, as x varies from 0 to s; so by Rules III and IV the dimensional formula for the integral is

force $\cdot$ length = (mass $\cdot$ length $\cdot$ time^{-2}) $\cdot$ length = mass $\cdot$ length^2 $\cdot$ time^{-2}.

By Rule II the dimensions of energy or work are then mass $\cdot$ length^2 $\cdot$ time^{-2}.

Note that, as was pointed out in the discussion following Rule IV, treating $F \cdot dx$ in Eq. (1.12) as the dot product, $\vec{F} \cdot d\vec{x}$, of a vector force $\vec{F}$ and a vector displacement $d\vec{x}$ cannot change the units of E because units interact as real scalar parameters.

The equations of physics are self-consistent, which means the dimensions of a physical quantity such as energy can be found from any equation containing that quantity. The equation for the kinetic energy of a particle of mass m moving with velocity v is

$$E = \frac{1}{2}mv^2. \tag{1.13}$$

By Rules II and III this also specifies the dimensions of energy as mass $\cdot$ length$^2 \cdot$ time^{-2}, because m has dimension mass and velocity has dimension length $\cdot$ time^{-1}.

Another important point worth making explicit is that all formal mathematical functions have arguments and values which are pure numerics.

RULE V

The arguments and values of all formal mathematical functions such as $\sin, \cos, \tan, \tan^{-1}, \ln, \log_{10}$, etc., are always pure numbers—that is, they are always dimensionless. The exponents of numerics and physical quantities must also be dimensionless pure numbers.

We note that when applied to the definition $\exp(ax) = e^{ax}$, Rule V gives two reasons why the product ax must be dimensionless—it is argument of the formal mathematical function exp and the exponent of the numeric e.

To show how Rule V works we construct the solution to Eq. (1.11) using the constraint or boundary condition that at time $t = 0$,

$$\text{mass } m \text{ has zero displacement and mass } m \text{ has velocity } V. \tag{1.14}$$

(When physical quantities are set equal to zero the units are omitted, because 0 seconds is the same amount of time as 0 years; 0 centimeters is the same distance as 0 miles, etc.) The solution to Eq. (1.11) is then

$$y = \frac{V}{\sqrt{\dfrac{k}{m} - \dfrac{\gamma^2}{4m^2}}} e^{-t(\gamma/2m)} \sin\left(t\sqrt{\frac{k}{m} - \frac{\gamma^2}{4m^2}} \right). \tag{1.15}$$

From the first part of Rule V the quantity

$$t\sqrt{\frac{k}{m} - \frac{\gamma^2}{4m^2}}$$

should be dimensionless because it is the argument of the sin. Using Rules III and IV and the dimensions of γ and k given in the discussion following Eq. (1.11), we see that

$$\text{dimensions of} \quad t\sqrt{\frac{k}{m} - \frac{\gamma^2}{4m^2}} = \text{time} \cdot \left(\frac{\text{mass}}{\text{time}^2} \cdot \frac{1}{\text{mass}} - \frac{\text{mass}^2}{\text{time}^2} \cdot \frac{1}{\text{mass}^2}\right)^{1/2}$$
$$= \text{time} \cdot \left(\text{time}^{-2}\right)^{1/2} = 1,$$

which shows that

$$t\sqrt{\frac{k}{m} - \frac{\gamma^2}{4m^2}}$$

is indeed dimensionless. By the second part of Rule V the expression $-t\gamma(2m)^{-1}$, being the exponent of e, must be dimensionless. Using Rule III we have

$$\text{dimensions of} \left(-\frac{t\gamma}{2m}\right) = 1 \cdot \text{time} \cdot \frac{\text{mass}}{\text{time}} \cdot \frac{1}{\text{mass}} = 1,$$

which is again the expected result. The dimension and units of the displacement y are set by the dimension and units of V in Eq. (1.15) because, using Rule III on Eq. (1.15),

$$\text{dimensions of} \quad y = \frac{\text{dimension of } V}{\left[\frac{\text{mass}}{\text{time}^2} \cdot \frac{1}{\text{mass}} - \frac{\text{mass}^2}{\text{time}^2 \cdot \text{mass}^2}\right]^{1/2}} \cdot 1 \cdot 1$$
$$= \text{time} \cdot (\text{dimension of } V),$$

where Rule V has been used to recognize

$$e^{-t(\gamma/2m)} \quad \text{and} \quad \sin\left(t\sqrt{\frac{k}{m} - \frac{\gamma^2}{4m^2}}\right)$$

as pure numbers. The dimension of velocity V is length $\cdot$ time^{-1}, so y ends up with the dimension length, which is what it should have. This, by the way, illustrates one of the major uses of dimensional analysis, examining whether or not formulas are correct by checking that their dimensions make sense.

1.4 DIMENSIONLESS PARAMETERS

In Eq. (1.15) the square root in the denominator and inside the sine suggests that we expect

$$\frac{k}{m} > \frac{\gamma^2}{4m^2},$$

which is indeed the case. A size relationship using ">", "≥", "<", or "≤", just like an equality, must have the same dimensions on both sides. In fact, everything said about equations involving physical quantities in Rule II also holds true for size relationships involving physical quantities. We could state a Rule IIa as follows:

In any size relationship involving physical quantities, both the left-hand side of the size relationship and the right-hand side of the size relationship must have the same dimensions. If both sides are expressed in the same units, then the numeric part of the left-hand side has the same size relationship to the numeric part of the right-hand side, as does the left-hand side's physical quantity to the right-hand side's physical quantity.

Here the size relationship makes sense because both k/m and $\gamma^2/(4m^2)$. have dimensions of time^{-2}.

These types of size relationships are often handled by constructing dimensionless parameters. We could, for example, define

$$\xi = \frac{k/m}{\gamma^2/(4m^2)} = \frac{4km}{\gamma^2} \tag{1.16}$$

and say that Eq. (1.15) is a well-defined solution to Eq. (1.11) when $\xi > 1$.

A dimensionless parameter that is important in the study of fluids is the Reynolds number. The Reynolds number for a fluid of density ρ and of viscosity μ is

$$R = \frac{VD\rho}{\mu}, \tag{1.17}$$

when the fluid flows through a pipe of diameter D at an average velocity V. The density ρ has dimensions mass $\cdot$ length^{-3}, the viscosity μ has dimensions mass $\cdot$ time$^{-1} \cdot$ length^{-1}, and we already know the dimensions of velocity V and diameter D. By Rule III the dimension of the Reynolds number is

$$\text{dimension of } R = \frac{\text{length} \cdot \text{time}^{-1} \cdot \text{length} \cdot \text{mass} \cdot \text{length}^{-3}}{\text{mass} \cdot \text{length}^{-1} \cdot \text{time}^{-1}} = 1, \tag{1.18}$$

exactly as expected.

Suppose we create two different triplets of fundamental units—ulength, utime, umass, and Ulength, Utime, Umass—and examine what happens to the Reynolds number when going from one triplet of units to another. The physical quantities V, D, ρ, and μ can be written using both sets of units:

$$V = V_{[\text{ulength}\cdot\text{utime}^{-1}]} \frac{\text{ulength}}{\text{utime}} = V_{[\text{Ulength}\cdot\text{Utime}^{-1}]} \cdot \frac{\text{Ulength}}{\text{Utime}},$$

$$D = D_{[\text{ulength}]} \text{ulength} = D_{[\text{Ulength}]} \text{Ulength},$$

$$\rho = \rho_{[\text{umass}\cdot\text{ulength}^{-3}]} \frac{\text{umass}}{\text{ulength}^3} = \rho_{[\text{Umass}\cdot\text{Ulength}^{-3}]} \frac{\text{Umass}}{\text{Ulength}^3},$$

$$\mu = \mu_{[\text{umass}\cdot\text{utime}^{-1}\cdot\text{ulength}^{-1}]} \frac{\text{umass}}{\text{ulength}\cdot\text{utime}}$$

$$= \mu_{[\text{Umass}\cdot\text{Utime}^{-1}\cdot\text{Ulength}^{-1}]} \frac{\text{Umass}}{\text{Ulength}\cdot\text{Utime}}.$$

To save space, we define

$$V_u = V_{[\text{ulength}\cdot\text{utime}^{-1}]}, \quad V_U = V_{[\text{Ulength}\cdot\text{Utime}^{-1}]},$$

$$D_u = D_{[\text{ulength}]}, \quad D_U = D_{[\text{Ulength}]},$$

$$\rho_u = \rho_{[\text{umass}\cdot\text{ulength}^{-3}]}, \quad \rho_U = \rho_{[\text{Umass}\cdot\text{Ulength}^{-3}]},$$

$$\mu_u = \mu_{[\text{umass}\cdot\text{utime}^{-1}\cdot\text{ulength}^{-1}]}, \quad \mu_U = \mu_{[\text{Umass}\cdot\text{Utime}^{-1}\cdot\text{Ulength}^{-1}]}.$$

The Reynolds number calculated in the "u" system of units is

$$R_u = \frac{V_u \dfrac{\text{ulength}}{\text{utime}} \cdot D_u \text{ulength} \cdot \rho_u \dfrac{\text{umass}}{\text{ulength}^3}}{\mu_u \dfrac{\text{umass}}{\text{ulength}\cdot\text{utime}}} = \frac{V_u D_u \rho_u}{\mu_u}, \qquad (1.19)$$

where all the units have cancelled—no surprise, since we know the Reynolds number is dimensionless. There must exist pure numbers a, b, c, such that

$$\text{ulength} = a \cdot \text{Ulength},$$

$$\text{utime} = b \cdot \text{Utime},$$

$$\text{umass} = c \cdot \text{Umass}.$$

It follows that

$$V_u \frac{\text{ulength}}{\text{utime}} = V_u \frac{a \cdot \text{Ulength}}{b \cdot \text{Utime}} = \left(\frac{a V_u}{b}\right) \frac{\text{Ulength}}{\text{Utime}},$$

$$D_u \text{ulength} = (a D_u) \text{Ulength},$$

$$\rho_u \frac{\text{umass}}{\text{ulength}^3} = \left(\rho_u \frac{c}{a^3}\right) \frac{\text{Umass}}{\text{Ulength}^3},$$

$$\mu_u \frac{\text{umass}}{\text{ulength}\cdot\text{utime}} = \left(\mu_u \frac{c}{ab}\right) \frac{\text{Umass}}{\text{Ulength}\cdot\text{Utime}};$$

so that, consistent with Rule I, we get

$$V_U = \frac{aV_u}{b}, \quad D_U = aD_u, \quad \rho_U = \rho_u\left(\frac{c}{a^3}\right), \text{ and } \mu_U = \left(\frac{c}{ab}\right)\mu_u. \quad (1.20)$$

Having already calculated R_u, the Reynolds number in the u triplet of units, we now use the definition of the Reynolds number [Eq. (1.17)] to calculate R_U, the Reynolds number in the U triplet of units:

$$R_U = \frac{V_U\dfrac{\text{Ulength}}{\text{Utime}} \cdot D_U\text{Ulength} \cdot \rho_U\dfrac{\text{Umass}}{\text{Ulength}^3}}{\mu_U\dfrac{\text{Umass}}{\text{Ulength} \cdot \text{Utime}}} = \frac{V_U D_U \rho_U}{\mu_U}.$$

Substitution of Eq. (1.20) into this expression gives

$$R_U = \frac{\left(V_u\dfrac{a}{b}\right)(D_u a)\left(\rho_u\dfrac{c}{a^3}\right)}{\left(\mu_u\dfrac{c}{ab}\right)} = \frac{V_u D_u \rho_u}{\mu_u}. \quad (1.21)$$

Comparison of Eq. (1.21) to Eq. (1.19) shows that

$$R_U = R_u, \quad (1.22)$$

demonstrating that the Reynolds number is the same pure number in all systems of units. A similar analysis of the dimensionless number ξ in Eq. (1.16) shows that

$$\xi_U = \xi_u. \quad (1.23)$$

In fact, all dimensionless numbers are invariant with respect to the fundamental units used to measure their components. This is convenient for scientists and engineers, who can use dimensionless parameters to analyze the behavior of physical systems without specifying the units of the physical quantities involved.

1.5 THE CGS AND MKS MECHANICAL SYSTEMS OF UNITS

Having settled on a set of fundamental dimensions, we can give each dimension a fundamental unit and from those construct a set of derived units. Choosing 1 centimeter (cm) for the unit of length, 1 gram (gm) for the unit of mass, and 1 second (sec) for the unit of time, we create the centimeter-gram-second or cgs system of units. Choosing 1 meter (m) for the unit of length, 1 kilogram (kg) for the unit of mass, and 1 sec for the unit of time, we create the meter-kilogram-second or mks system of units. In each system, some combinations of units occur

Table 1.1 Mechanical units of the cgs system.

gm (gram)
cm (centimeter)
sec (second)
dyne $= $ gm $\cdot$ cm/sec^2
erg $=$ dyne $\cdot$ cm $=$ gm $\cdot$ cm^2/sec^2
poise $=$ dyne $\cdot$ sec/cm$^2 =$ gm/(cm $\cdot$ sec)
barye $=$ dyne/cm$^2 =$ gm/(cm $\cdot$ sec^2)
stokes $=$ cm^2/sec
rhe $=$ poise$^{-1} =$ (cm $\cdot$ sec/gm)

Table 1.2 Mechanical units of the mks system.

kg (kilogram)
m (meter)
sec (second)
newton (newton), newton $=$ kg $\cdot$ m/sec^2
joule (joule), joule $=$ newton $\cdot$ m $=$ kg $\cdot$ m^2/sec^2
watt (watt), watt $=$ joule/sec $=$ kg $\cdot$ m^2/sec^3
poiseulle (poiseulle),
poiseulle $=$ newton $\cdot$ sec/m$^2 =$ kg/(m $\cdot$ sec)
pascal (pascal), pascal $=$ newton/m$^2 =$ kg/(m $\cdot$ sec^2)
stere (stere), stere $=$ m^3

Table 1.3 Conversion between the cgs and mks mechanical systems of units. (The unit of time, sec, is the same for both systems.)

kg $= 10^3 \cdot$ gm
m $= 10^2 \cdot$ cm
newton $= 10^5 \cdot$ dyne
joule $= 10^7 \cdot$ erg
poiseulle $= 10 \cdot$ poise
pascal $= 10 \cdot$ barye

so frequently that they are given their own names. These units are called derived units. The derived units of force in the cgs and mks systems are called dyne and newton, respectively, with

$$1 \text{ dyne} = \frac{\text{gm} \cdot \text{cm}}{\text{sec}^2} \quad \text{and} \quad 1 \text{ newton} = \frac{\text{kg} \cdot \text{m}}{\text{sec}^2}.$$

The derived units of energy in the cgs and mks systems are called erg and joule, respectively, with

$$1\,\text{erg} = 1\,\text{dyne} \cdot \text{cm} = \frac{\text{gm} \cdot \text{cm}^2}{\text{sec}^2} \quad \text{and} \quad 1\,\text{joule} = 1\,\text{newton} \cdot \text{m} = \frac{\text{kg} \cdot \text{m}^2}{\text{sec}^2}.$$

The unit of power in the mks system is

$$1\,\text{watt} = \frac{\text{joule}}{\text{sec}} = \frac{\text{kg} \cdot \text{m}^2}{\text{sec}^3}.$$

With derived units there is no need to reduce all physical quantities to their fundamental units. In the discussion after Eq. (1.11) the spring constant k is found to have dimensions mass $\cdot$ time^{-2}, so in the cgs system k has units of gm $\cdot$ sec^{-2}. These units can also be written as dyne/cm or erg/cm^2:

$$1\,\frac{\text{dyne}}{\text{cm}} = 1\,\frac{\text{erg}}{\text{cm}^2} = 1\,\frac{\text{gm}}{\text{sec}^2}.$$

The numeric part of k is the same in all three cases:

$$k = k_{[\text{dyne}\cdot\text{cm}^{-1}]}\frac{\text{dyne}}{\text{cm}} = k_{[\text{erg}\cdot\text{cm}^{-2}]}\frac{\text{erg}}{\text{cm}^2} = k_{[\text{gm}\cdot\text{sec}^{-2}]}\frac{\text{gm}}{\text{sec}^2},$$

so that

$$k_{[\text{dyne}\cdot\text{cm}^{-1}]} = k_{[\text{erg}\cdot\text{cm}^{-2}]} = k_{[\text{gm}\cdot\text{sec}^{-2}]}.$$

For this reason we could, if desired, just write k_{cgs} for the numeric part of k in the cgs system of units:

$$k_{\text{cgs}} = k_{[\text{dyne}\cdot\text{cm}^{-1}]} = k_{[\text{erg}\cdot\text{cm}^{-2}]} = k_{[\text{gm}\cdot\text{sec}^{-2}]}.$$

From now on we will adopt the convention that a lowercase subscript attached to the name of a physical quantity indicates the numeric part of that quantity in the system of units specified by the subscript. As an example of how this works, for a distance r, a force F, and an energy E, we can write

$$r_{\text{cgs}} = r_{[\text{cm}]}, \ F_{\text{cgs}} = F_{[\text{dyne}]}, \ \text{and} \ E_{\text{cgs}} = E_{[\text{erg}]}.$$

In the mks system of units we can write

$$r_{\text{mks}} = r_{[\text{m}]}, \ F_{\text{mks}} = F_{[\text{newton}]}, \ \text{and} \ E_{\text{mks}} = E_{[\text{joule}]}.$$

The derived units of the cgs and mks systems provide a useful flexibility. As an example, consider the physical quantity torque used to describe the twist produced by a force on a physical system.

$$\vec{\tau} = \vec{r} \times \vec{F} \text{ or } |\vec{\tau}| = |\vec{r}| \cdot |\vec{F}| \cdot \sin\theta, \tag{1.24}$$

where vector $\vec{F}$ is the force applied to the system a distance $\vec{r}$ from the point about which the torque is produced, and θ is the smaller angle between $\vec{F}$ and $\vec{r}$. The dimensions of torque are clearly length $\cdot$ (dimensions of force) or mass $\cdot$ length^2 $\cdot$ time^{-2}. In the mks system the fundamental units of torque are kg $\cdot$ m^2 $\cdot$ sec^{-2}, the same as the fundamental units of energy. When working with torques an engineer can write the torque in units of newton $\cdot$ m, to remind readers of Eq. (1.24) and emphasize that it is a torque rather than some amount of energy. If, later on, energy is to be specified, the engineer can use joules, which are the mks units of energy. Although it is technically correct to measure torque in joules, because

$$1 \text{ joule} = 1 \text{ newton} \cdot \text{m} = 1 \frac{\text{kg} \cdot \text{m}^2}{\text{sec}^2},$$

it would almost always be confusing and very bad style. The derived units also let us adjust numeric parts of physical quantities to a convenient size. Returning to the example of torque, since

$$1 \text{ newton} \cdot \text{m} = 10^7 \text{ dyne} \cdot \text{cm}$$

it is quite possible to have physical situations where both τ_{cgs} and τ_{mks} are respectively too large and too small for everyday use. By mixing together mks and cgs derived units,

$$\tau_{\text{cgs}} \text{dyne} \cdot \text{cm} = (\tau_{\text{cgs}} 10^{-5}) \text{newton} \cdot \text{cm} = (\tau_{\text{cgs}} 10^{-2}) \text{dyne} \cdot \text{m},$$

we have more options for putting the decimal point in a convenient place. We shall see in Chapter 2 that exactly this consideration—having conveniently sized units—ended up driving the development of electromagnetic units during the second half of the nineteenth century.

1.6 The U and N Operators

When working with fundamental systems of units it is helpful to define both a U and an N operator. When applied to a physical quantity, the U operator returns its units in some specified system of fundamental units, and the N operator applied to a physical quantity returns the numeric part in some system of fundamental units. We say that $\underset{\text{cgs}}{\text{U}}$, $\underset{\text{cgs}}{\text{N}}$ are the U and N operators for the cgs system of units; and $\underset{\text{mks}}{\text{U}}$,

$\underset{\text{mks}}{\text{N}}$ are the U and N operators for the mks system of units. To use U and N we need to know the definitions of the physical quantities on which they operate. Because we know from Eq. (1.1) that L is a length,

$$\underset{\text{cgs}}{\text{U}}(L) = \text{cm},$$

$$\underset{\text{cgs}}{\text{N}}(L) = 8.5,$$

$$\underset{\text{mks}}{\text{U}}(L) = \text{m},$$

and

$$\underset{\text{mks}}{\text{N}}(L) = 0.085.$$

If the numeric part of a length y is not known then we have

$$\underset{\text{cgs}}{\text{U}}(y) = \text{cm},$$

$$\underset{\text{cgs}}{\text{N}}(y) = y_{\text{cgs}} = y_{[\text{cm}]},$$

$$\underset{\text{mks}}{\text{U}}(y) = \text{m},$$

and

$$\underset{\text{mks}}{\text{N}}(y) = y_{\text{mks}} = y_{[\text{m}]}.$$

When using this notation we know that the N operators always produce pure numbers, so y_{cgs} and y_{mks} are numerics; and, because y is a length, y_{cgs} and y_{mks} must be the numeric parts of y when y is measured in centimeters and meters, respectively.

The U and N operators can be applied to complicated as well as simple physical quantities. Working with γ and k from Eq. (1.11), we get

$$\underset{\text{cgs}}{\text{U}}(\gamma) = \frac{\text{gm}}{\text{sec}},$$

$$\underset{\text{cgs}}{\text{N}}(y) = \gamma_{\text{cgs}} = \gamma_{[\text{gm·sec}^{-1}]},$$

$$\underset{\text{mks}}{\text{U}}(\gamma) = \frac{\text{kg}}{\text{sec}},$$

$$\underset{\text{mks}}{\text{N}}(\gamma) = \gamma_{\text{mks}} = \gamma_{[\text{kg·sec}^{-1}]};$$

and

$$\underset{\text{cgs}}{U}(k) = \frac{\text{gm}}{\text{sec}^2} = \frac{\text{dyne}}{\text{cm}} = \frac{\text{erg}}{\text{cm}^2},$$

$$\underset{\text{mks}}{U}(k) = \frac{\text{kg}}{\text{sec}^2} = \frac{\text{newton}}{\text{m}} = \frac{\text{joule}}{\text{m}^2},$$

$$\underset{\text{cgs}}{N}(k) = k_{\text{cgs}} = k_{[\text{gm}\cdot\text{sec}^{-2}]} = k_{[\text{dyne}\cdot\text{cm}^{-1}]} = k_{[\text{erg}\cdot\text{cm}^{-2}]},$$

$$\underset{\text{mks}}{N}(k) = k_{\text{mks}} = k_{[\text{kg}\cdot\text{sec}^{-2}]} = k_{[\text{newton}\cdot\text{m}^{-1}]} = k_{[\text{joule}\cdot\text{m}^{-2}]}.$$

The U operators applied to k show that the answer can be given in terms of the derived units belonging to some fundamental system of units instead of the fundamental units themselves. The N operators can also have their numeric answer written with a more specific subscript than "cgs" or "mks"—the subscripts can use square brackets to specify the actual units, either fundamental or derived, in which the physical quantity is measured.

The U and N operators turn out to be some of the most flexible operators known to mathematics—virtually any algebraic manipulation is allowed. When the U and N operators are applied to the product of two physical quantities a and b,

$$\underset{\text{cgs}}{U}(a \cdot b) = \underset{\text{cgs}}{U}(a) \cdot \underset{\text{cgs}}{U}(b),$$

$$\underset{\text{mks}}{U}(a \cdot b) = \underset{\text{mks}}{U}(a) \cdot \underset{\text{mks}}{U}(b);$$

and

$$\underset{\text{cgs}}{N}(a \cdot b) = \underset{\text{cgs}}{N}(a) \cdot \underset{\text{cgs}}{N}(b),$$

$$\underset{\text{mks}}{N}(a \cdot b) = \underset{\text{mks}}{N}(a) \cdot \underset{\text{mks}}{N}(b).$$

To show that these properties apply to the U and N operators for any system of units, not just the cgs and mks systems, we write the operator equalities without any subscripts:

$$U(a \cdot b) = U(a) \cdot U(b), \tag{1.25a}$$

$$N(a \cdot b) = N(a) \cdot N(b). \tag{1.25b}$$

Equations (1.25a, b) also hold true when a is a scalar and $\vec{b}$ is a vector:

$$U(a\vec{b}) = U(a) \cdot U(\vec{b}) \text{ and } N(a\vec{b}) = N(a) \cdot N(\vec{b}).$$

The U operator treats the vector cross product and vector dot product in the same way:

$$U(\vec{a} \cdot \vec{b}) = U(\vec{a} \times \vec{b}) = U(|\vec{a}|)U(|\vec{b}|); \tag{1.25c}$$

and the N operator preserves the form of the vector dot and cross products:

$$N(\vec{a} \cdot \vec{b}) = N(\vec{a}) \cdot N(\vec{b}), \tag{1.25d}$$

$$N(\vec{a} \times \vec{b}) = N(\vec{a}) \times N(\vec{b}). \tag{1.25e}$$

When U and N are applied to the sum of two physical quantities a and b we get

$$U(a + b) = U(a) = U(b) \tag{1.26a}$$

and

$$N(a + b) = N(a) + N(b). \tag{1.26b}$$

The U and N operators behave in a formally different manner in Eqs. (1.26a, b) because we want the sum of like combinations of units to reduce to the unit combination itself, whereas the N operator must preserve the form of the original physical equation (see Rule IV). When U and N are applied to a physical quantity b raised to a pure number x, we get

$$U(b^x) = [U(b)]^x, \tag{1.27a}$$

$$N(b^x) = [N(b)]^x. \tag{1.27b}$$

The last two rules to be specified show what happens when the U and N operators are applied to a pure number p:

$$U(p) = 1, \tag{1.28a}$$

$$N(p) = p. \tag{1.28b}$$

We can deduce one last general equality for the N operator from Eq. (1.28b). For any formal function f, such as the sin, cos, $\tan^{-1}$, ln, etc.,

$$N[f(x, y, \ldots)] = f(N(x), N(x), \ldots). \tag{1.28c}$$

In a way, Eq. (1.28b) and Rule V make Eq. (1.28c) trivially true, since Rule V states that the value of any function f is always a pure number, as are its arguments; therefore $N(f) = f$, $N(x) = x$, $N(y) = y$, and so on, from which it follows that Eq. (1.28c) is exactly the same as saying $f(x, y, \ldots) = f(x, y, \ldots)$. Nevertheless, Eq. (1.28c) is still worth writing down explicitly because it shows that N operators jump inside functions and then act on the individual physical quantities which combine to form the function's dimensionless arguments.

As an example of how to use these operators, we apply them to both sides of Eq. (1.15). By Rule V and Eqs. (1.25a) and (1.28a), a U operator applied to both sides of Eq. (1.15) gives

$$U(y) = U\left(\frac{V}{\sqrt{\dfrac{k}{m} + \dfrac{\gamma^2}{4m^2}}}\right) \cdot U\left[e^{-t(\gamma/2m)}\right] \cdot U\left[\sin\left(t\sqrt{\frac{k}{m} - \frac{\gamma^2}{4m^2}}\right)\right]$$

$$= U\left(\frac{V}{\sqrt{\dfrac{k}{m} + \dfrac{\gamma^2}{4m^2}}}\right) \cdot$$

By Eqs. (1.25a) and (1.27a) this becomes

$$U(y) = U(V) \cdot \left[U\left(\frac{k}{m} + \frac{\gamma^2}{4m^2}\right)\right]^{-1/2}$$

and by Eqs. (1.26a), (1.25a), and (1.27a),

$$U(y) = U(V)\sqrt{\frac{U(m)}{U(k)}}.$$

To go any further with this example we need to specify the fundamental system of units, either cgs or mks, of operator U. Since the dimensions of k are mass $\cdot$ time^{-2}, and V is a velocity and m is a mass by Eq. (1.14), we have

$$\underset{\text{cgs}}{U}(y) = \underset{\text{cgs}}{U}(V)\sqrt{\frac{\underset{\text{cgs}}{U}(m)}{\underset{\text{cgs}}{U}(k)}} = \frac{\text{cm}}{\text{sec}} \cdot \left[\frac{\text{gm}}{\text{gm} \cdot \text{sec}^{-2}}\right]^{1/2} = \text{cm} \qquad (1.29a)$$

and

$$\underset{\text{mks}}{U}(y) = \underset{\text{mks}}{U}(V)\sqrt{\frac{\underset{\text{mks}}{U}(m)}{\underset{\text{mks}}{U}(k)}} = \frac{\text{m}}{\text{sec}} \cdot \left[\frac{\text{kg}}{\text{kg} \cdot \text{sec}^{-2}}\right]^{1/2} = \text{m}. \qquad (1.29b)$$

The dimensions mass, length, and time behave exactly like units in algebraic manipulations of this sort, so we can define for any physical quantity b,

$$\underset{\text{mlt}}{U}(b) = \text{dimensional formula for } b \qquad (1.29c)$$

to create an operator obeying all the rules for U operators in Eqs. (1.25a) through (1.28a). The $\underset{\text{mlt}}{U}$ operator can be used to analyze dimensions the way other U operators are used to analyzed units. Because the $\underset{\text{mlt}}{U}$ operator obeys the same rules as other U operators, we get the same formula when it is applied to y:

$$\underset{\text{mlt}}{U}(y) = \underset{\text{mlt}}{U}(V)\sqrt{\frac{\underset{\text{mlt}}{U}(m)}{\underset{\text{mlt}}{U}(k)}} = \frac{\text{length}}{\text{time}} \cdot \left[\frac{\text{mass}}{\text{mass} \cdot \text{time}^{-2}}\right]^{1/2} = \text{length}.$$

By Eqs. (1.25b) and (1.28c), an N operator applied to both sides of Eq. (1.15) gives

$$N(y) = N\left(\frac{V}{\sqrt{\dfrac{k}{m} - \dfrac{\gamma^2}{4m^2}}}\right) e^{N[-t(\gamma/2m)]} \sin\left[N\left(t\sqrt{\frac{k}{m} - \frac{\gamma^2}{4m^2}}\right)\right].$$

From Eqs. (1.25b) and (1.27b),

$$N(y) = \frac{N(V)}{\sqrt{N\left(\dfrac{k}{m} - \dfrac{\gamma^2}{4m^2}\right)}} e^{N(-1/2)N(\gamma)N(t)/(N(m))} \sin\left[N(t) \cdot \sqrt{N\left(\frac{k}{m} - \frac{\gamma^2}{4m^2}\right)}\right],$$

and from Eqs. (1.26b), (1.25b), and (1.28b)

$$N(y) = \frac{N(V)}{\sqrt{N\left(\dfrac{k}{m}\right) - \dfrac{1}{4}N\left(\dfrac{\gamma^2}{m^2}\right)}} e^{(-1/2)N(\gamma)N(t)/(N(m))} \sin\left[N(t) \cdot \sqrt{N\left(\frac{k}{m}\right) - \frac{1}{4}N\left(\frac{\gamma^2}{m^2}\right)}\right].$$

Using Eqs. (1.25b) and (1.27b), and the fundamental definition of operator N in the cgs and mks systems of units, we get

$$\underset{\text{cgs}}{N}(y) = \frac{V_{\text{cgs}}}{\sqrt{\dfrac{k_{\text{cgs}}}{m_{\text{cgs}}} - \dfrac{\gamma^2_{\text{cgs}}}{4m^2_{\text{cgs}}}}} e^{-(\gamma_{\text{cgs}} t_{\text{cgs}}/2m_{\text{cgs}})} \sin\left(t_{\text{cgs}}\sqrt{\frac{k_{\text{cgs}}}{m_{\text{cgs}}} - \frac{\gamma^2_{\text{cgs}}}{4m^2_{\text{cgs}}}}\right)$$

and

$$\underset{\text{mks}}{N}(y) = \frac{V_{\text{mks}}}{\sqrt{\dfrac{k_{\text{mks}}}{m_{\text{mks}}} - \dfrac{\gamma^2_{\text{mks}}}{4m^2_{\text{mks}}}}} e^{-(\gamma_{\text{mks}} t_{\text{mks}}/2m_{\text{mks}})} \sin\left(t_{\text{mks}}\sqrt{\frac{k_{\text{mks}}}{m_{\text{mks}}} - \frac{\gamma^2_{\text{mks}}}{4m^2_{\text{mks}}}}\right).$$

It may seem that all the N operator does is add a cgs or mks subscript to the variables; but it will be useful later on, especially when we discuss different systems of electromagnetic units, because it makes very clear what is and what is not a pure number. Students who have spent significant amounts of time analyzing dimensional problems will find Eqs. (1.25a) through (1.28c) superfluous, because for them the rules have become second nature. Novices, however, may find them helpful—and the U, N operators themselves make it easier to specify how units based on different sets of fundamental dimensions relate to each other.

1.7 TEMPERATURE UNITS

There are physical quantities which do not fit comfortably into any system based on the three fundamental dimensions of mass, length, and time. One example of this is temperature, which in equations and formulas containing other physical quantities is almost always treated as a fourth fundamental dimension.

Temperature is typically given units of either degrees Kelvin (degK) or degrees Rankine (degR).* Historically these two temperature scales are modifications of, respectively, the Celsius temperature scale (which used to be called the centigrade temperature scale) and the Fahrenheit temperature scale. When units based on mass, length, and time were created, there was no question what a zero amount of mass, a zero amount of length, zero amount of time, etc., was; but the Celsius and Fahrenheit scales were created before it was known that there was such a thing as an absolute zero of temperature—the coldest and lowest temperature that can exist.[†] Consequently, no one knew how to avoid expressing very cold temperatures in the Celsius and Fahrenheit scales as physical quantities with negative numeric components, and both temperature scales ended up specifying a different physical temperature as zero. Since zero degrees Celsius is not the same temperature as zero degrees Fahrenheit, we have to specify which temperature scale is being used when specifying absolute zero temperature. Therefore the Kelvin temperature scale is defined to be the Celsius scale with its zero point shifted to absolute zero, 273.15 degrees Celsius below zero degrees in the Celsius scale; and the Rankine temperature scale is defined to be the Fahrenheit scale with its zero point shifted to absolute zero, 459.67 degrees Fahrenheit below zero degrees in the Fahrenheit scale. Consequently, temperatures measured in either degK or degR are always positive (that is, always have a positive numeric component). The physical temperature T can be written in either degK or degR as

$$T = T_{[\mathrm{degK}]} \cdot \mathrm{degK} = T_{[\mathrm{degR}]} \cdot \mathrm{degR}.$$

* By international convention, the abbreviation for degrees Kelvin is K. In this book we use "degK" rather than "K" to avoid confusion with "k", the traditional letter used for Boltzmann's constant, and also to make degK stand out in equations as the name of a unit rather than a one-letter variable representing a physical quantity.

[†] Even the name "absolute zero" hints at the surprise attending its discovery; from the viewpoint of dimensional analysis it is just zero temperature, the same sort of thing as zero area, zero mass, zero volume, etc.

The conversion from one temperature scale to the other is given by

$$1 \, \mathrm{degK} = \frac{9}{5} \mathrm{degR},$$

so that, by Rule I

$$T_{[\mathrm{degK}]} = \frac{5}{9} T_{[\mathrm{degR}]},$$

where $T_{[\mathrm{degK}]}$ and $T_{[\mathrm{degR}]}$ are the numeric parts of temperature T when T is measured in units of degK and degR, respectively. We note that

$$1 \, \mathrm{degK} = 1 \, \text{degree Celsius}$$

and

$$1 \, \mathrm{degR} = 1 \, \text{degree Fahrenheit}$$

because the only difference between the Kelvin and Celsius—or between the Rankine and Fahrenheit scales—is the physical temperature labeled as zero.

1.8 DIMENSIONLESS UNITS

The ideal gas law is

$$PV = nRT, \tag{1.30}$$

where P is the pressure—the force per unit area—exerted by a gas on the walls of a container of volume V, n is the number of moles of gas molecules, R is the universal gas constant, and T is the gas temperature (in degK or degR). The P and V on the left-hand side of Eq. (1.30) fit comfortably into the cgs and mks system of units, with P measured in $\mathrm{dyne/cm}^2$ in the cgs system or

$$\text{pascal} = \mathrm{newton/m}^2$$

in the mks system, and V measured in cm^3 in the cgs system or m^3 in the mks system. The right-hand side of Eq. (1.30), however, has two interesting variables, n and R, which deserve further discussion.

Although it is the official position of the United States government that the unit mole has dimensions,[3] students of chemistry know that a mole can be treated like a pure number:[4]

$$1 \, \text{mole} = 6.023 \times 10^{23}.$$

When thought of this way, we see that 0.385 moles of gas molecules are the same thing as

$$0.385 \cdot 1 \text{ mole} = 0.385 \cdot 6.023 \times 10^{23} \cong 2.319 \times 10^{23}$$

gas molecules. In this sense, the mole is an example of a dimensionless unit. Moles are used to put the decimal point in a convenient place; it is, after all, easier to say that a container holds approximately 2 moles of O_2 than to say it holds approximately 12×10^{23} oxygen molecules.

The universal gas constant R is a dimensional constant which, unlike the dimensional quantities previously encountered, cannot be changed—it is a law of nature that R has the value it does. Chemistry textbooks often quote its value as

$$R = 0.0821 \frac{\text{liter} \cdot \text{atm}}{\text{mole} \cdot \text{degK}}.$$

The numerator contains volume and pressure units that are popular in chemistry,

$$1 \text{ liter} = 10^{-3} \cdot \text{m}^3 = 10^3 \cdot \text{cm}^3 \qquad (1.31a)$$

and

$$1 \text{ atm} = 1.01325 \times 10^5 \frac{\text{newton}}{\text{m}^2}. \qquad (1.31b)$$

The liter is a convenient unit of volume intermediate in size between m^3 and cm^3; and the abbreviation atm stands for atmosphere, with 1 atm being the standard pressure of the earth's atmosphere at sea level. The units of R have mole in the denominator. This can be used as a mnemonic, alerting those using R that getting to a final result usually involves multiplication by some number of moles, cancelling the moles in the denominator. The mole in the denominator plays the same role as the degK in the denominator, which as an ordinary dimensional unit suggests that R is often multiplied by a temperature. In the right-hand side of the ideal gas law,

$$nRT = \left(n_{[\text{mole}]} \cdot \cancel{\text{mole}} \right) \cdot \left(0.0821 \frac{\text{liter} \cdot \text{atm}}{\cancel{\text{mole}} \cdot \cancel{\text{degK}}} \right) \cdot \left(T_{[\text{degK}]} \cdot \cancel{\text{degK}} \right)$$
$$= \left(n_{[\text{mole}]} T_{[\text{degK}]} \right) \cdot (0.0821 \text{ liter} \cdot \text{atm}).$$

The mole cancellation can be regarded as shorthand for

$$nRT = \left(n_{[\text{mole}]} \cdot \cancel{6.023 \times 10^{23}} \right) \cdot \left(0.0821 \frac{\text{liter} \cdot \text{atm}}{\cancel{6.023 \times 10^{23}} \cdot \cancel{\text{degK}}} \right) \cdot \left(T_{[\text{degK}]} \cdot \cancel{\text{degK}} \right)$$
$$= \left(n_{[\text{mole}]} T_{[\text{degK}]} \right) \cdot (0.0821 \text{ liter} \cdot \text{atm}).$$

The final units of nRT, liter · atm, are easily converted to either the mks or cgs system of units. We divide both sides of Eq. (1.31a) by liter to get

$$1 = \frac{10^{-3}\mathrm{m}^3}{\mathrm{liter}},$$

and both sides of (1.31b) by atm to get

$$1 = \frac{1.01325 \times 10^5 \, \mathrm{newton}}{\mathrm{atm} \cdot \mathrm{m}^2}.$$

The quantity nRT has the same value if it is multiplied by one twice, so

$$nRT \cdot 1 \cdot 1 = \left(n_{[\mathrm{mole}]}T_{[\mathrm{degK}]}\right) \cdot (0.0821 \, \mathrm{liter} \cdot \mathrm{atm}) \cdot \frac{10^{-3}\mathrm{m}^3}{\mathrm{liter}} \cdot \frac{1.01325 \cdot 10^5 \, \mathrm{newton}}{\mathrm{atm} \cdot \mathrm{m}^2}$$

$$\cong 8.32\left(n_{[\mathrm{mole}]}T_{[\mathrm{degK}]}\right) \mathrm{joules},$$

where in the last step we have cancelled the units liter, atm, and m^2, recognizing that newton · m = joule. The method used here to convert nRT to the mks system is perhaps the most popular way of converting physical quantities from one set of units to another. It clearly gives the same result as the method used in Eq. (1.3), because substitution of Eqs. (1.31a, b) into the expression for nRT also gives

$$nRT = \left(n_{[\mathrm{mole}]}T_{[\mathrm{degK}]}\right) \cdot \left[0.0821\left(10^{-3}\mathrm{m}^3\right) \cdot \left(\frac{1.01325 \times 10^5 \, \mathrm{newton}}{\mathrm{m}^2}\right)\right]$$

$$\cong 8.32\left(n_{[\mathrm{mole}]}T_{[\mathrm{degK}]}\right)\mathrm{joules}. \tag{1.31c}$$

No matter how we convert to the mks system of units, nRT ends up in units of energy. Clearly the liter · atm unit has dimensions of energy and the gas constant R has dimensions of energy per unit temperature interval when the moles unit is regarded as being dimensionless.

The radian (rad) is another unit regarded as dimensionless. To get an angle's measure in radians we draw a circle centered on the angle's vertex which, as shown in Fig. 1.1, specifies an arc length s of the circle's perimeter lying inside the angle. The angle in radians, θ, is the ratio s/r, where r is the circle's radius and s, r are measured with the same units of length. This ratio is the same pure number no matter what size radius the circle has:

$$\theta = \frac{s}{r}. \tag{1.32}$$

The angle θ, being the ratio of two lengths, is clearly dimensionless; and θ is said to be measured in radians, making radians a truly dimensionless unit. If $\gamma = 0$ in

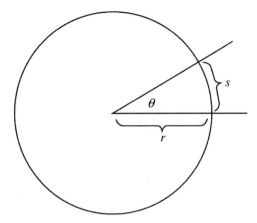

Figure 1.1 The angle θ in radians is the arc length s divided by the circle's radius r.

Eq. (1.11), the solution given in Eq. (1.15) becomes

$$y = \frac{V}{\omega} \sin(\omega t), \tag{1.33a}$$

where

$$\omega = \sqrt{\frac{k}{m}}. \tag{1.33b}$$

We have already found the dimension of ω to be time^{-1}. Often an oscillatory function, such as $\sin(\omega t)$ in Eq. (1.33a), is written as $\sin(2\pi f t)$, where f is a pure frequency :

$$f = \frac{\omega}{2\pi} = \frac{1}{2\pi}\sqrt{\frac{k}{m}}.$$

Since 2π is dimensionless, the dimension of f is also time^{-1}. To distinguish ω and f, the units of ω are often given in the cgs and mks systems as rad/sec instead of just sec^{-1}. A pure frequency f is then given in hertz (Hz),* defined as sec^{-1} in the sense of cycles per sec, where cycle is another dimensionless unit. Of course, we could just as easily refer to ω by its proper name, the angular frequency, as opposed to f, which is just a plain (or pure) frequency; but it reduces the chance of confusion when we reiterate what ω and f are by giving ω units of rad/sec and f units of Hz or cycles/sec.

The geometrical idea of an angle probably came from ancient astronomers, who used it to measure the "distance" in the sky between objects such as stars or planets. Here, distance refers not to the length separating the objects, but rather to the change in direction when looking from one object to another. An observer,

---

* The hertz was named in honor of Heinrich Hertz (1857–1894).

for example, might see a night-flying airplane cross in front of a star, so that the "distance" in the sky between them is zero, whereas the star is so far away that the physical length separating the airplane and the star is almost as large as the physical length separating the observer and the star.

Just as an angle can be used to define a "distance" in the sky, so can a solid angle be used to define an "area" in the sky. Solid angles are measured in terms of another dimensionless unit, the steradian (sr). To get a solid angle's measure in steradians, we imagine a sphere centered on the vertex of the solid angle, as shown in Fig. 1.2, and note that an area A of the sphere lies inside the solid angle. The solid angle in steradians, Ω, is then the ratio A/r^2, where r is the sphere's radius and A and r are measured in the same system of units. This ratio is the same pure number no matter what the radius of the sphere:

$$\Omega = \frac{A}{r^2}. \tag{1.34}$$

The steradian must be a dimensionless unit, because Ω is the ratio of two areas. The area of a sphere of radius r is $4\pi r^2$, so the solid-angle Ω representing the entire sky is $(4\pi r^2)/r^2 = 4\pi$ steradians.

In optics and radiometry, steradians are included in the units of physical quantities to show that the quantity is measured per unit solid angle. When dealing with the radiant energy emitted by a surface, the radiant exitance of the surface is given in units of joules/sec/m^2 in the mks system. This shows it is the total radiant energy emitted each second per unit area of the luminous surface. The radiance of that same surface, however, is given in units of joules/sec/m^2/sr in the mks system, showing that it is the radiant energy emitted each second into a unit solid angle per unit area of the luminous surface. The difference between the quantities is important because the radiance is a good measure of what we subjectively experience as "brightness" when looking at a luminous surface; whereas the exitance, because it measures the optical energy per second leaving a surface in all directions, helps to specify the rate at which the surface cools. The concept of solid angle is so basic to

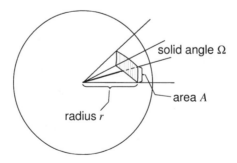

Figure 1.2 The solid angle Ω in steradians is given by the area A divided by the square of the radius r.

optics and radiometry that when checking whether the units balance in radiometric equations it is a good idea to treat the steradian like a dimensional unit—an unbalanced unit of steradians in a radiometric formula is almost always an indication that a mistake has been made.

1.9 REMOVAL OF THE UNIVERSAL GAS CONSTANT FROM THE IDEAL GAS LAW

Dimensional constants can be removed from physical equations by using a system of units in which they become dimensionless and take on the numerical value 1. This maneuver has become quite common over the last 50 years in relativistic and quantum physics, but is not usually encountered in other fields of mathematical sciences and engineering. We go into it in some detail in the next several sections, because it must be mastered in order to convert electromagnetic equations from one system of units to another.

We choose as our first example the universal gas constant R, which, by the way, is almost never made dimensionless and removed from the equations of chemistry and statistical mechanics. From the ideal gas law, Eq. (1.30), we know that R has the dimensions of energy per unit temperature interval [see discussion after Eq. (1.31c)]. Using Eqs. (1.31a, b) to convert R to units of joule/mole/degK gives

$$R = 0.0821 \frac{\text{liter} \cdot \text{atm}}{\text{mole} \cdot \text{degK}} = 0.0821 \frac{10^{-3}\,\text{m}^3 \cdot (1.01325 \times 10^5\,\text{newton/m}^2)}{\text{mole} \cdot \text{degK}}$$

$$\cong 8.32 \frac{\text{joules}}{\text{mole} \cdot \text{degK}},$$

since 1 joule = 1 newton $\cdot$ m. We define the mksK system of units to be the mks system of units with all temperatures given a separate dimensionality and all temperatures measured in units of degK. As long as physical equations and formulas do not involve temperature, the mks and mksK systems are identical. Only when physical quantities such as the ideal gas constant R enter the picture is there a difference between the mks and mksK systems of units; the expressions $\underset{\text{mks}}{U}(R)$ and $\underset{\text{mks}}{N}(R)$ are undefined, whereas, because we are regarding moles as being dimensionless,

$$\underset{\text{mksK}}{U}(R) = \frac{\text{joule}}{\text{degK}} \tag{1.35a}$$

and

$$\underset{\text{mksK}}{N}(R) = R_{\text{mksK}} \cong 8.32. \tag{1.35b}$$

The first step in turning R into 1 is to create a unit of temperature, which we call degR1, in which the numeric part of R is 1. If we define

$$\text{degR1} = (R_{\text{mksK}})^{-1}\text{degK}, \tag{1.36}$$

then Rule I can be used to convert the temperature units of R to degR1:

$$R = R_{\text{mksK}}\frac{\text{joule}}{\text{degK}} = R_{\text{mksK}}\frac{\text{joule}}{R_{\text{mksK}}\text{degR1}} = 1\frac{\text{joule}}{\text{degR1}}. \tag{1.37}$$

In Eq. (1.37) we have dropped the mole because we are treating it as dimensionless. We define the mksR1 system of units to be the same as the mksK system of units, but with temperature measured in degR1 instead of degK. The $\underset{\text{mksR1}}{U}$ and $\underset{\text{mksR1}}{N}$ operators applied to R give

$$\underset{\text{mksR1}}{N}(R) = 1 \tag{1.38a}$$

and

$$\underset{\text{mksR1}}{U}(R) = \frac{\text{joule}}{\text{degR1}}. \tag{1.38b}$$

Applying the $\underset{\text{mksR1}}{N}$ operator to both sides of Eq. (1.30), the ideal gas law, we get

$$\underset{\text{mksR1}}{N}(P) = \underset{\text{mks}}{N}(P) = P_{\text{mksR1}} = P_{\text{mks}}, \tag{1.39a}$$

$$\underset{\text{mksR1}}{N}(V) = \underset{\text{mks}}{N}(V) = V_{\text{mksR1}} = V_{\text{mks}}, \tag{1.39b}$$

$$\underset{\text{mksR1}}{N}(n) = n, \tag{1.39c}$$

$$\underset{\text{mksR1}}{N}(R) = 1, \tag{1.39d}$$

$$\underset{\text{mksR1}}{N}(T) = T_{\text{mksR1}}, \tag{1.39e}$$

so that

$$P_{\text{mks}}V_{\text{mks}} = nT_{\text{mksR1}}. \tag{1.39f}$$

In Eqs. (1.39a, b) the numeric parts of the mechanical quantities P and V, which do not involve temperature, are the same in the mksR1 and mks systems of units; and Eq. (1.39d) uses Eq. (1.37). Note that Eq. (1.38b) requires $\underset{\text{mksR1}}{U}(R) = \text{joule/degR1}$, so the gas constant R has not yet become the dimensionless numeric 1.

The second step in turning R into 1 is to make R dimensionless by giving temperature the dimensions of energy. Can we get away with this? The number of independent dimensions assigned to a given system of units can be arbitrarily chosen to be anything we want, as long as the self-consistency of the equations and formulas is not violated.[5,6] It turns out, even though it is almost never done, that it makes sense to measure temperature in units of energy. In chemistry and statistical physics, temperature is intimately associated with a quantity called entropy. In classical thermodynamics, the infinitesimal change in the entropy dS of an isolated system is defined to be

$$dS = \frac{dQ}{T}, \tag{1.40a}$$

where dQ is the amount of heat energy added to the system when it is at temperature T. Conventionally, S is assigned dimensions of energy over temperature (remember that the derivative operator "d" is dimensionless), so its units might be, for example, joule/degK. If T has dimensions of energy, then S becomes a dimensionless quantity. In quantum statistical mechanics, the entropy from classical thermodynamics is shown to be

$$S = k\ln(\Gamma), \tag{1.40b}$$

where $k \cong 1.38 \times 10^{-23}$ joules/degK is Boltzmann's constant and Γ is a dimensionless quantity representing the number of different quantum states that the closed system might be occupying. We can use the U and N operators to write

$$\underset{\text{mksK}}{\text{U}}(k) = \frac{\text{joule}}{\text{degK}} \tag{1.41a}$$

and

$$\underset{\text{mksK}}{\text{N}}(k) = k_{\text{mksK}} \cong 1.38 \times 10^{-23}. \tag{1.41b}$$

By Rule V the quantity $\ln(\Gamma)$ is dimensionless, so S takes on the units of k. From Eq. (1.41a) we see that when the temperature is measured in units of energy, k becomes dimensionless, making the entropy dimensionless. This is consistent with our conclusion based on Eq. (1.40a). As a final check on the validity of giving temperature the dimension of energy, we consider one of the formal definitions of temperature from classical statistical mechanics,

$$T = \left(\frac{\partial S}{\partial E}\right)^{-1}\Bigg|_{\text{constant volume}}, \tag{1.42}$$

where $\partial S/\partial E$ is the change in a constant-volume system's entropy with respect to its energy.[7] Equation (1.42) can be thought of as a disguised form of Eq. (1.40a).

We already know that entropy becomes dimensionless when temperature is given dimensions of energy, so $\partial S/\partial E$ then has dimensions of energy^{-1} and Eq. (1.42) is dimensionally balanced.

Now that temperature has lost its separate dimensionality, the degR1 unit of temperature is just another unit of energy. To avoid confused thinking we call this unit of energy EdegR1 and reserve degR1 for that temperature unit equal to $(R_{\mathrm{mksK}})^{-1}$degK and having the dimension of temperature. We say that

$$\mathrm{degR1} \neq \mathrm{EdegR1} \tag{1.43a}$$

because degR1 has dimensions of temperature and EdegR1 has dimensions of energy (otherwise we have to discard Rule II above). We can, however, define an mksER1 system of units, which is the standard mks system of units with all the temperatures measured in the units of energy EdegR1. Because the only difference between the mksR1 and the mksER1 systems of units is that the temperature T has lost its separate dimensionality, we expect

$$\mathop{\mathrm{N}}_{\mathrm{mksER1}}(T) = \mathop{\mathrm{N}}_{\mathrm{mksR1}}(T), \tag{1.43b}$$

so that

$$T_{\mathrm{mksER1}} = T_{\mathrm{mksR1}}. \tag{1.43c}$$

In fact, it makes sense to define $\mathop{\mathrm{N}}_{\mathrm{mksER1}}$ by saying that

$$\mathop{\mathrm{N}}_{\mathrm{mksER1}}(b) = \mathop{\mathrm{N}}_{\mathrm{mksR1}}(b), \tag{1.43d}$$

for any physical quantity b. To enforce our decision to replace degR1 by EdegR1 in the units for all physical quantities b, we define

$$\mathop{\mathrm{U}}_{\mathrm{mksER1}}(b) = \left(\frac{\mathrm{EdegR1}}{\mathrm{degR1}}\right)^{v} \mathop{\mathrm{U}}_{\mathrm{mksR1}}(b), \tag{1.43e}$$

where

$$\mathop{\mathrm{U}}_{\mathrm{mksR1}}(b) = (\mathrm{degR1})^{v}(\mathrm{kg})^{x}(\mathrm{m})^{y}(\mathrm{sec})^{z} \tag{1.43f}$$

for real exponents x, y, z, and v. Definition (1.43e) forces degR1 to be replaced by EdegR1 in any unit expression, since $(\mathrm{EdegR1})^{v}$ is multiplied into $\mathop{\mathrm{U}}_{\mathrm{mksR1}}(b)$ after $(\mathrm{degR1})^{v}$ has been cancelled out. Because EdegR1 has units of energy, we know that R is dimensionless; so from Eq. (1.28a)

$$\mathop{\mathrm{U}}_{\mathrm{mksER1}}(R) = 1. \tag{1.43g}$$

From Eq. (1.43g)—and Eqs. (1.39d) and (1.43d), which make the numeric part of R equal to 1—we conclude that in the mksER1 system of units

$$R = 1. \qquad (1.43h)$$

To find the units of EdegR1, we apply $\underset{\text{mksER1}}{U}$ to both sides of the ideal gas law in Eq. (1.30) and use (1.43g) and (1.28a) to get

$$\underset{\text{mksER1}}{U}(P) \cdot \underset{\text{mksER1}}{U}(V) = 1 \cdot 1 \cdot \underset{\text{mksER1}}{U}(T)$$

or

$$\underset{\text{mksER1}}{U}(T) = \frac{\text{newton}}{\text{m}^2} \cdot \text{m}^3 = \text{newton} \cdot \text{m} = \text{joule}. \qquad (1.44a)$$

The units of T must be joules for this equation to have balanced units in the mksER1 system, so EdegR1 must be the same as joule:

$$1 \, \text{EdegR1} = 1 \, \text{joule}. \qquad (1.44b)$$

Multiplying both sides of Eq. (1.39f) by joule or EdegR1—they are, after all, the same unit of energy—we get, using $T_{\text{mksER1}} = T_{\text{mksR1}}$ from Eq. (1.43c), that

$$P_{\text{mks}} V_{\text{mks}} \, \text{joule} = n T_{\text{mksER1}} \, \text{joule},$$

or, using that $1 \, \text{joule} = \text{newton} \cdot \text{m}$,

$$\left(P_{\text{mks}} \frac{\text{newton}}{\text{m}^2} \right) (V_{\text{mks}} \text{m}^3) = n (T_{\text{mksER1}} \, \text{EdegR1}).$$

The expressions inside the parentheses are all dimensional physical quantities, so the ideal gas law in the mksER1 system of units must be

$$PV = nT. \qquad (1.45)$$

This is exactly the result we would expect from substituting $R = 1$ into $PV = nRT$, the ideal gas law written in units where $R \neq 1$.

Inequality [Eq. (1.43a)] prevents us from creating an equation relating the EdegR1, or joule, to the degK; but nothing stops us from setting up an equation between their numeric parts (because they are pure numbers). Using Rule I and Eq. (1.36) we get

$$\underset{\text{mksK}}{N}(T) = (R_{\text{mksK}})^{-1} \underset{\text{mksR1}}{N}(T).$$

This becomes, using Eqs. (1.35b) and (1.43b),

$$\underset{\text{mksK}}{\text{N}}(T) = (R_{\text{mksK}})^{-1} \underset{\text{mksER1}}{\text{N}}(T) \cong \frac{1}{8.32} \underset{\text{mksER1}}{\text{N}}(T) \qquad (1.46a)$$

or

$$T_{[\text{degK}]} = T_{\text{mksK}} = \frac{1}{R_{\text{mksK}} T_{\text{mksER1}}} \cong \frac{1}{8.32} T_{\text{mksER1}}. \qquad (1.46b)$$

Equations (1.46a, b) show why using a system of units where a prominent dimensional constant has the dimensionless value of 1 simplifies calculations; the chain of unit equalities breaks where the dimensionality changes, as in Eq. (1.43a), but the chain of numeric equalities extends indefinitely. We could build all thermometers scaled to measure temperature in degR1, call the unit EdegR1, and never have to worry about the value of R again.

Equation (1.45), the ideal gas law with $R = 1$, is written using the physical quantities P, V, and T. For this to be meaningful all three variables must behave like physical quantities, making Eq. (1.45) true in any set of units where the temperature is given dimensions of energy in such a way that $R = 1$. Only the temperature unit has been modified, so the only physical quantity whose behavior needs to be examined is T. Figure 1.3 shows that up to now we have followed the path A-B-C to get a system of units where $R = 1$. We can repeat the process using the equally good liter $\cdot$ atm energy unit, following path A-B$'$-C$'$ in Fig. 1.3 to get a system of units where $R = 1$. We now show that the C, C$'$ systems of units, in both of which $R = 1$, behave like true systems of units because we can use Rule I to go from the temperature in the C system (which is, of course, the mksER1 system of units) to the temperature in the C$'$ system.

If we had created the temperature unit for which $R = 1$ starting with the value of R in (liter $\cdot$ atm)/degK/mole,

$$R = 0.0821 \frac{\text{liter} \cdot \text{atm}}{\text{mole} \cdot \text{degK}},$$

then instead of Eq. (1.36) the first step would have been to create a new temperature unit degR1LA, making the numeric part of R equal to 1 when R is measured in units of (liter $\cdot$ atm)/degR1LA/mole,

$$\text{degR1LA} = (0.0821)^{-1} \text{degK}.$$

The second step, deciding to measure temperature in energy units, would give instead of Eq. (1.44b)

$$\text{EdegR1LA} = \text{liter} \cdot \text{atm}.$$

In place of Eq. (1.46b) we would have gotten that

R has dimensions of energy divided by temperature when the mole is regarded as a dimensionless unit (see Section 1.8 of Chapter 1).

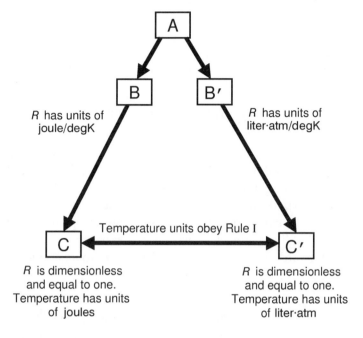

Figure 1.3 The temperature units obey Rule I no matter which path is used to make the ideal gas constant R dimensionless and equal to 1.

$$T_{[\text{degK}]} = \frac{1}{0.0821} T_{[\text{EdegR1LA}]}. \tag{1.47}$$

This completes path A-B$'$-C$'$ in Fig. 1.3. We now check that this is consistent with the results obtained using path A-B-C by using Rule I to go directly from C to C$'$ in Fig. 1.3. Equations (1.31a, b) and Rule I require the temperature in EdegR1 = joule = newton $\cdot$ m to be, in liter $\cdot$ atm,

$$T_{\text{mksER1}} \text{newton} \cdot \text{m} = T_{\text{mksER1}} \left(\frac{\text{newton}}{\text{m}^2} \right) \cdot (\text{m}^3)$$

$$= T_{\text{mksER1}} \cdot \frac{10^{-2}}{1.01325} \text{liter} \cdot \text{atm}$$

$$= T_{[\text{liter}\cdot\text{atm}]} \text{liter} \cdot \text{atm}$$

with

$$T_{[\text{liter}\cdot\text{atm}]} = T_{\text{mksER1}} \cdot \frac{10^{-2}}{1.01325}. \tag{1.48a}$$

If $T_{[\text{liter}\cdot\text{atm}]}$ is truly the numeric part of the same temperature in the C' system, then we must have

$$T_{[\text{liter}\cdot\text{atm}]} \overset{?}{=} T_{[\text{EdegR1LA}]}.$$

Substitution of Eq. (1.46b) into Eq. (1.48a) gives

$$T_{[\text{liter}\cdot\text{atm}]} = \frac{10^{-2}}{1.01325} \cdot \left(8.32 T_{[\text{degK}]}\right) = 0.0821 T_{[\text{degK}]}. \tag{1.48b}$$

Comparison of Eqs. (1.47) and (1.48b) shows that $T_{[\text{liter}\cdot\text{atm}]}$ is indeed the same number as $T_{[\text{EdegR1LA}]}$, so the temperature unit EdegR1LA is indeed the same as the temperature unit we get using Rule I to go directly from C to C' in Fig. 1.3. We conclude that when the temperature unit is modified to make the universal gas constant equal to 1, the temperature behaves like a physical quantity no matter what energy units—joules, liter $\cdot$ atm, etc.—we use to measure it.

It is straightforward to return Eq. (1.45), $PV = nT$, to its original form. Separated into units and numeric parts, Eq. (1.45) is

$$P_{\text{mks}} \frac{\text{newton}}{\text{m}^2} \cdot V_{\text{mks}} \, \text{m}^3 = n T_{\text{mksER1}} \, \text{joule} = n T_{\text{mksER1}} \text{EdegR1}, \tag{1.49}$$

which is clearly balanced in its dimensions, since 1 joule $= 1$ newton $\cdot$ m. Returning to the temperature its separate dimensionality, we write

$$T = T_{\text{mksR1}} \text{degR1}. \tag{1.50}$$

From Eq. (1.43c) we know that $T_{\text{mksER1}} = T_{\text{mksR1}}$, so we multiply and divide the right-hand side of Eq. (1.49) by degR1 to get, using 1 joule $= 1$ EdegR1,

$$P_{\text{mks}} \frac{\text{newton}}{\text{m}^2} \cdot V_{\text{mks}} \, \text{m}^3 = n T_{\text{mksER1}} \text{degR1} \cdot \frac{\text{EdegR1}}{\text{degR1}}$$
$$= n T_{\text{mksR1}} \text{degR1} \cdot \frac{\text{joule}}{\text{degR1}}. \tag{1.51}$$

Rule I and Eq. (1.36) give

$$\left(P_{\text{mks}} \frac{\text{newton}}{\text{m}^2}\right) \cdot \left(V_{\text{mks}} \, \text{m}^3\right) = n T_{\text{mksR1}} (R_{\text{mksK}})^{-1} \text{degK} \cdot \frac{\text{EdegR1}}{(R_{\text{mksK}})^{-1} \text{degK}}$$
$$= n \left[T_{\text{mksR1}} \cdot (R_{\text{mksK}})^{-1}\right] \text{degK} \cdot \left(R_{\text{mksK}} \frac{\text{joule}}{\text{degK}}\right)$$
$$= n (T_{\text{mksK}} \text{degK}) \cdot \left(R_{\text{mksK}} \frac{\text{joule}}{\text{degK}}\right). \tag{1.52}$$

Since

$$R = R_{\mathrm{mksK}} \frac{\mathrm{joule}}{\mathrm{degK}}$$

and the expressions inside the parentheses are dimensional physical quantities, we get

$$PV = nRT. \qquad (1.53)$$

This return to the original form of the ideal gas law follows naturally from giving back the temperature its separate dimension. Indeed, we can go directly from Eq. (1.51) to Eq. (1.53) by pointing out that

$$P = P_{\mathrm{mks}} \frac{\mathrm{newton}}{\mathrm{m}^2}, \quad V = V_{\mathrm{mks}}\,\mathrm{m}^3,$$

and

$$T = T_{\mathrm{mksR1}}\mathrm{degR1}, \ \text{ and } 1\frac{\mathrm{joule}}{\mathrm{degR1}} = R$$

are all physical quantities, so Eq. (1.51) is the same as $PV = nRT$ in the mksR1 system of units. This result is all we need to prove Eq. (1.53) true, because physical quantities that are equal in mksR1 units must be equal in all systems of units where temperature has a separate dimension.

1.10 REMOVAL OF THE SPEED OF LIGHT FROM RELATIVISTIC EQUATIONS

The next example we look at is setting the speed of light equal to 1. According to relativity theory, all observers, no matter how fast they are moving, measure the same value for c, the speed of light; it is a universal constant appearing in a wide variety of formulas. The most famous equation of twentieth-century physics,

$$E = mc^2, \qquad (1.54a)$$

states that an object of mass m at rest with respect to an observer is equivalent to an amount of energy mc^2. What is perhaps less well known is that this is a special case of the formula

$$E = \frac{mc^2}{\sqrt{1 - \dfrac{v^2}{c^2}}}, \qquad (1.54b)$$

which gives the total energy—kinetic energy plus rest energy—of that same object moving at velocity v with respect to an observer. [Equation (1.54b) reduces to (1.54a) when $v = 0$.] The momentum of an object of mass m moving at velocity v is

$$p = \frac{mv}{\sqrt{1 - \dfrac{v^2}{c^2}}}, \tag{1.54c}$$

and if a luminous object is emitting light at frequency f_0 (in Hz = cycles/sec) then an observer moving toward the object at velocity v measures the frequency of that light as

$$f = f_0 \cdot \left(\frac{\sqrt{1 - \dfrac{v^2}{c^2}}}{1 - \dfrac{v}{c}} \right). \tag{1.54d}$$

The expressions inside the square roots of Eqs. (1.54b–d) are never negative because one consequence of elementary relativity theory is that material objects never travel faster than the speed of light.* Equations (1.54a–d) are typical formulas of relativistic physics, and it is clear they would be greatly simplified if the speed of light were made equal to 1. In fact, the same procedure used to remove the universal gas constant from the ideal gas law can be used to remove c from the formulas of relativistic physics. So popular has this maneuver become among physicists that the presence of the phrase "working in units where the velocity of light is equal to 1" has become a sure sign that the student has begun an advanced, rather than introductory, treatment of relativistic physics or cosmology.

The first step in removing the speed of light is to choose units of length and time such that the numeric part of c is 1. In cgs units

$$c = c_{cgs} \cdot \frac{cm}{sec}, \tag{1.55a}$$

where

$$\underset{cgs}{N}(c) = c_{cgs} \cong 2.99792 \times 10^{10}. \tag{1.55b}$$

---

* Although it would be a digression to get into the details of special relativity, we can note that the momentum in Eq. (1.54c) becomes infinitely large as the velocity gets ever closer to the speed of light. One basic property of a force in physics is that it increases the momentum of moveable objects. Hence (1.54c) suggests we can apply an arbitrarily large force an arbitrarily long time—increasing the momentum to an arbitrarily large value—without ever accelerating an object up to the speed of light.

Choosing a new unit of time, which we call cmtime$_3$, such that

$$1\,\text{cmtime}_3 = (c_{\text{cgs}})^{-1}\,\text{sec}, \tag{1.56a}$$

we note that in units of cm and cmtime$_3$ the speed of light becomes

$$c = c_{\text{cgs}}\frac{\text{cm}}{\text{sec}} = \frac{\text{cm}}{\text{cmtime}_3} = 1\frac{\text{cm}}{\text{cmtime}_3}. \tag{1.56b}$$

The subscript 3 of cmtime$_3$ indicates that it belongs to a system of units where mass, length, and time are three separate dimensions. We call the system of units where mass is measured in gm, length is measured in cm, and time is measured in cmtime$_3$, the cgc system of units.

The second step, just as when we made $R = 1$ in the ideal gas law, is to remove the separate dimensionality that prevents c from becoming a dimensionless constant. We give time the dimension of length, so that now cmtime is also a unit of length. We call this new unit of length cmtime$_2$ to show that it belongs to a system of units where mass and length are the only two separate dimensions. We note that

$$\text{cmtime}_2 \neq \text{cmtime}_3 \tag{1.57a}$$

and

$$\text{cmtime}_2 = \text{cm}, \tag{1.57b}$$

where Eq. (1.57a) follows from Rule II, which does not allow physical quantities of different dimensionality to be equal, and Eq. (1.57b) comes from the desire to have the units cancel in Eq. (1.56b) when cmtime$_3 \to$ cmtime$_2$. This creates a new system of units, which we call the centimeter-gram or cg system of units. The numeric parts of all physical quantities are identical in the cg and cgc systems of units, so

$$\mathop{\text{N}}_{\text{cg}}(b) = \mathop{\text{N}}_{\text{cgc}}(b) \tag{1.58a}$$

for any physical quantity b. The units of b, however, are different whenever b contains some non-zero power of time as part of its dimensionality.

$$\mathop{\text{U}}_{\text{cg}}(b) \neq \mathop{\text{U}}_{\text{cgc}}(b) \tag{1.58b}$$

when

$$\mathop{\text{U}}_{\text{mlt}}(b) = \text{mass}^x\,\text{length}^y\,\text{time}^z$$

for real exponents x, y, z, with $z \neq 0$. Since cmtime$_3$ is replaced by cmtime$_2$ = cm in the cg system of units, we define $\underset{cg}{U}$ by the rule

$$\underset{cg}{U}(b) = \left(\frac{cm}{cmtime_3} \right)^z \underset{cgc}{U}(b).$$ (1.58c)

A little thought shows that an equivalent definition of $\underset{cg}{U}$ is

$$\underset{cg}{U}(b) = \left(\frac{cm}{sec} \right)^z \underset{cgs}{U}(b),$$ (1.58d)

for any physical quantity b. Now we can formally show that from Eqs. (1.56b) and (1.58a) we have $\underset{cg}{N}(c) = 1$, and from Eq. (1.58c) or (1.58d) we get $\underset{cg}{U}(c) = 1$; so in the cg system of units

$$c = 1.$$ (1.59)

Rule VI summarizes the two-step procedure used to make dimensional constants such as c or R equal to 1.

RULE VI

The first step in setting a dimensional constant equal to 1 is to rescale, using Rule I, the units of one of its dimensions so that the numeric part of the constant becomes equal to 1. We call this unit that has been rescaled unitA. The second step is to treat unitA as formally equivalent to some other fundamental or derived unit, which we call unitB. UnitB may be multidimensional and is chosen to make the dimensional constant dimensionless. The unit that is treated as formally equivalent to unitB should be given a different name, say unitA$'$, because Rule II forbids two physical quantities, for example unitA and unitB, from being equal when they have different dimensions.

The method used to make the universal gas constant R equal to 1 and the method used to make the speed of light equal to 1 both follow the procedure given in Rule VI. The first step in making $R = 1$ is to convert from degK to degR1, making its numeric part equal to 1. This conversion takes us from the mksK system of units to the mksR1 system of units. We then treat the temperature unit degR1 as formally equivalent to 1 joule, a unit of energy, giving it the new name EdegR1. This takes us to the mksER1 system of units. The numeric parts of all physical quantities in the mksER1 system of units are the same as in the mksR1 system of units, but the temperature unit EdegR1 is now equal to 1 joule so temperature has the dimensions of energy. Note that we are careful to say that degR1 $\neq$ EdegR1 in Eq. (1.43a). The first step in making the velocity of light equal to 1 is to convert from sec to cmtime$_3$, making its numeric part equal to 1. This conversion takes us from the cgs to the cgc system of units; for the second step we make the cmtime unit formally equivalent to the cm by replacing the cmtime$_3$ with the cmtime$_2$, going from the cgc to the cg system of units and making the velocity of light dimensionless. Again, we are careful to say that cmtime$_3 \neq$ cmtime$_2$ [see Eq. (1.57a)].

Giving length and time the same dimensions has consequences beyond allowing us to set $c = 1$ in all the equations of relativistic physics. Breaking Eq. (1.54d) up into numeric parts and units in the cgs system of units gives

$$f_{\mathrm{cgs}}\,\mathrm{sec}^{-1} = (f_0)_{\mathrm{cgs}}\mathrm{sec}^{-1}\left[\frac{\sqrt{1-\dfrac{(v_{\mathrm{cgs}})^2\,\mathrm{cm}^2/\mathrm{sec}^2}{(c_{\mathrm{cgs}})^2\,\mathrm{cm}^2/\mathrm{sec}^2}}}{1-\dfrac{v_{\mathrm{cgs}}\,\mathrm{cm}/\mathrm{sec}}{c_{\mathrm{cgs}}\,\mathrm{cm}/\mathrm{sec}}}\right]$$

$$\tag{1.60a}$$

$$= (f_0)_{\mathrm{cgs}}\mathrm{sec}^{-1}\left[\frac{\sqrt{1-\dfrac{(v_{\mathrm{cgs}})^2}{(c_{\mathrm{cgs}})^2}}}{1-\dfrac{v_{\mathrm{cgs}}}{c_{\mathrm{cgs}}}}\right],$$

where

$$\underset{\mathrm{cgs}}{\mathrm{N}}(f) = f_{[\mathrm{Hz}]} = f_{\mathrm{cgs}}, \quad \underset{\mathrm{cgs}}{\mathrm{U}}(f) = \mathrm{sec}^{-1},$$

$$\underset{\mathrm{cgs}}{\mathrm{N}}(f_0) = (f_0)_{\mathrm{cgs}}, \quad \underset{\mathrm{cgs}}{\mathrm{U}}(f_0) = \mathrm{sec}^{-1},$$

$$\underset{\mathrm{cgs}}{\mathrm{N}}(v) = v_{\mathrm{cgs}}, \quad \underset{\mathrm{cgs}}{\mathrm{U}}(c) = \underset{\mathrm{cgs}}{\mathrm{U}}(v) = \mathrm{cm}/\mathrm{sec}.$$

Converting from cgs to cgc units, Eq. (1.60a) becomes

$$f_{\mathrm{cgs}}\frac{1}{c_{\mathrm{cgs}}}\mathrm{cmtime}_3^{-1} = (f_0)_{\mathrm{cgs}}\cdot\left(\frac{1}{c_{\mathrm{cgs}}}\mathrm{cmtime}_3^{-1}\right)\cdot\left[\frac{\sqrt{1-\dfrac{(v_{\mathrm{cgs}})^2}{(c_{\mathrm{cgs}})^2}}}{1-\dfrac{v_{\mathrm{cgs}}}{c_{\mathrm{cgs}}}}\right],$$

and converting from cgc to cg units gives

$$\left(\frac{f_{\mathrm{cgs}}}{c_{\mathrm{cgs}}}\right) \mathrm{cm}^{-1} = \frac{(f_0)_{\mathrm{cgs}}}{c_{\mathrm{cgs}}} \cdot \mathrm{cm}^{-1} \cdot \left[\frac{\sqrt{1 - \dfrac{(v_{\mathrm{cgs}})^2}{(c_{\mathrm{cgs}})^2}}}{1 - \dfrac{v_{\mathrm{cgs}}}{c_{\mathrm{cgs}}}}\right]. \tag{1.60b}$$

In cg units all frequencies f have the dimension of length^{-1}, so f and f_0 have units of cm^{-1}. We see that

$$f = f_{\mathrm{cg}} \, \mathrm{cm}^{-1} \tag{1.61a}$$

and

$$f_0 = (f_0)_{\mathrm{cg}} \, \mathrm{cm}^{-1}, \tag{1.61b}$$

because

$$f_{\mathrm{cg}} = f_{\mathrm{cgs}}/c_{\mathrm{cgs}}$$

and

$$(f_0)_{\mathrm{cg}} = (f_0)_{\mathrm{cgs}}/c_{\mathrm{cgs}}. \tag{1.61c}$$

In cg units, all velocities, not just the velocity of light, are dimensionless. Starting with cgs units and converting to cg units,

$$v = v_{\mathrm{cgs}} \frac{\mathrm{cm}}{\mathrm{sec}} = \frac{v_{\mathrm{cgs}}}{c_{\mathrm{cgs}}} \frac{\mathrm{cm}}{\mathrm{cmtime}_3};$$

we next give time the dimension of length to get, using Eq. (1.57b),

$$v = \frac{v_{\mathrm{cgs}}}{c_{\mathrm{cgs}}} \frac{\mathrm{cm}}{\mathrm{cmtime}_2} = \frac{v_{\mathrm{cgs}}}{c_{\mathrm{cgs}}}. \tag{1.61d}$$

Substituting Eqs. (1.61a–d) into Eq. (1.60b) gives

$$f = f_0 \cdot \left(\frac{\sqrt{1 - v^2}}{1 - v}\right). \tag{1.62}$$

Although (1.62) is formally the same as Eq. (1.54d) with $c = 1$, it is important to remember that now f and f_0 have dimensions of length^{-1} and that all velocities are dimensionless. In fact, if a physical quantity is defined to be a quantity with dimensions, then the velocity cannot be a physical quantity in any system of units where $c = 1$. This, of course, explains why in Eq. (1.62) the quantities v and v^2 can be subtracted from 1, a pure number.

Because velocity is dimensionless, the dimensions of mass, energy, and momentum are all the same in any system of units where $c = 1$. We first examine the consequences of mass and energy having the same units and finish by looking at what happens to momentum.

We use cgs units to break up Eq. (1.54a) into numeric and dimensional parts:

$$E_{[\text{erg}_3]}\text{erg}_3 = m_{[\text{gm}]}\text{gm} \cdot c_{\text{cgs}}^2 \frac{\text{cm}^2}{\text{sec}^2}, \tag{1.63a}$$

where

$$\underset{\text{cgs}}{\text{N}}(m) = m_{\text{cgs}} = m_{[\text{gm}]}, \qquad \underset{\text{cgs}}{\text{U}}(m) = \text{gm},$$
$$\underset{\text{cgs}}{\text{N}}(E) = E_{\text{cgs}} = E_{[\text{erg}_3]}, \qquad \underset{\text{cgs}}{\text{U}}(E) = \text{erg}_3.$$

The subscript "3" on ergs reminds us that erg_3 is a unit of energy from a system of units where mass, length, and time have separate dimensions. Converting Eq. (1.63a) to the cgc system of units, we get, since $1\,\text{erg}_3 = 1\,\text{gm} \cdot \text{cm}^2/\text{sec}^2$,

$$\frac{E_{[\text{erg}_3]}}{c_{\text{cgs}}^2}\frac{\text{gm} \cdot \text{cm}^2}{\text{cmtime}_3^2} = m_{[\text{gm}]}\text{gm} \cdot c_{\text{cgs}}^2 \frac{\text{cm}^2}{c_{\text{cgs}}^2\text{cmtime}_3^2}$$

or

$$E_{\text{cgc}}\frac{\text{gm} \cdot \text{cm}^2}{\text{cmtime}_3^2} = \left(m_{[\text{gm}]}\text{gm}\right) \cdot \left(1\frac{\text{cm}^2}{\text{cmtime}_3^2}\right), \tag{1.63b}$$

where

$$\underset{\text{cgc}}{\text{N}}(E) = E_{\text{cgc}} = \frac{E_{[\text{erg}_3]}}{c_{\text{cgs}}^2}. \tag{1.63c}$$

Using Eqs. (1.57b) and (1.58a) to convert to the cg system of units, we get

$$E_{\text{cgc}}\text{gm} = E_{\text{cg}}\text{gm} = m_{[\text{gm}]}\text{gm}, \tag{1.63d}$$

or, written as physical quantities,

$$E = m. \tag{1.63e}$$

Equation (1.63e) is formally the same as Eq. (1.54a) with $c = 1$, which is no surprise; and Eq. (1.63d) shows that in the cg system both energy and mass are measured in units of gm. This is a perfectly good result, but in relativistic physics—and particularly in relativistic particle physics—mass is often measured in units of energy instead of energy in units of mass. To represent both mass and energy in units of energy we need to find the conversion equation between the units of mass and energy when using units where $c = 1$.

We note that in the cgs system of units

$$1 \, \text{erg}_3 = 1 \, \text{gm} \cdot \frac{\text{cm}^2}{\text{sec}^2},$$

which becomes in cgc units

$$1 \, \text{erg}_3 = 1 \, \text{gm} \cdot c_{\text{cgs}}^{-2} \frac{\text{cm}^2}{\text{cmtime}_3^2} = c_{\text{cgs}}^{-2} \left(\text{gm} \cdot \frac{\text{cm}^2}{\text{cmtime}_3^2} \right). \tag{1.64a}$$

The cgc unit of energy is clearly $\underset{\text{cgc}}{\text{U}}(E) = \text{gm} \cdot \text{cm}^2 \cdot \text{cmtime}_3^{-2}$ and, since this involves a nonzero power of time, by Eq. (1.58b) the cgc unit of energy cannot equal the cg unit of energy. From Eq. (1.64a) the cgs unit of energy, erg_3, is proportional to the cgc unit of energy $1 \text{gm} \cdot \text{cm}^2 \cdot \text{cmtime}_3^{-2}$, so when taking the left-hand side of Eq. (1.64a) from the cgc to the cg system of units the subscript of the right-hand side has to change as well:

$$1 \, \text{erg}_2 = 1 \, \text{gm} \cdot c_{\text{cgs}}^{-2} \frac{\text{cm}^2}{\text{cmtime}_2^2} = c_{\text{cgs}}^{-2} \, \text{gm}, \tag{1.64b}$$

where

$$1 \, \text{erg}_3 \neq 1 \, \text{erg}_2. \tag{1.64c}$$

It is easy to show that the erg subscript has to change. If we do not change the subscript, we can write from Eqs. (1.64a) and (1.64b)

$$1 \, \text{gm} \cdot c_{\text{cgs}}^{-2} \cdot \frac{\text{cm}^2}{\text{cmtime}_3^2} = 1 \, \text{erg}_3 \overset{?}{=} 1 \, \text{erg}_2 = c_{\text{cgs}}^{-2} \, \text{gm} \cdot \frac{\text{cm}^2}{\text{cmtime}_2^2}$$

or

$$\text{cmtime}_3 \overset{?}{=} \text{cmtime}_2, \tag{1.65}$$

which contradicts Eq. (1.57a). Clearly, any unit of energy changes its fundamental nature when going from a system where mass, length, and time have separate dimensions to a system where only mass and length have separate dimensions. Equation (1.58b) above states that the units of any physical quantity in the cgc system which have a nonzero power of time cannot equal the units of the same quantity in the cg system. This statement, although true, is now seen to be too specific because it only applies to the transition from the cgc to the cg systems of units. We need a more general statement.

RULE VII

When the *abc* system of units lacks one of the dimensions of the *ABC* system of units, and *b* is any physical quantity which in the *ABC* system of units has a nonzero power of this dimension, then $\underset{abc}{U}(b) \neq \underset{ABC}{U}(b)$ no matter what the relationship of $\underset{abc}{N}(b)$ to $\underset{ABC}{N}(b)$.

The demonstration that erg_2 cannot equal erg_3 is an example of how Rule VII works. Rule II forbids physical quantities with different dimensions from being equal, so in the strictest sense Rule VII is just a special case of Rule II. There is no harm, however, in using Rule VII to emphasize this particular consequence of Rule II when we change the dimensionality of a set of units.

Equation (1.64b) is what we have been looking for—the conversion equation between erg_2 and gm, an energy unit and a mass unit in units where $c = 1$. Hence this is all we need to find the numeric parts of Eq. (1.63e) when both mass and energy are measured in units of erg_2. Starting with (1.63d), which is just (1.63e) broken down into numeric parts and units when both mass and energy are measured in units of gm in the cg system, we have

$$E_{\text{cgc}}\text{gm} = m_{[\text{gm}]}\text{gm}.$$

Using $E_{\text{cgc}} = E_{[\text{erg}_3]}/c_{\text{cgs}}^2$ from Eq. (1.63c) and $1\,\text{gm} = c_{\text{cgs}}^2\text{erg}_2$ from Eq. (1.64b) gives

$$\frac{E_{[\text{erg}_3]}}{c_{\text{cgs}}^2} \cdot c_{\text{cgs}}^2\text{erg}_2 = m_{[\text{gm}]}c_{\text{cgs}}^2\text{erg}_2$$

or

$$m_{[\text{erg}_2]} \cdot \text{erg}_2 = E_{[\text{erg}_3]} \cdot \text{erg}_2 \tag{1.66a}$$

with

$$m_{[\text{erg}_2]} = m_{[\text{gm}]} \cdot c_{\text{cgs}}^2. \tag{1.66b}$$

Equation (1.66b) is a specific instance of the general rule of thumb for relativistic physics: to find the mass in energy units when $c = 1$, multiply by the square of the speed of light. Equation (1.66a) is just $E = m$ with both energy and mass written in energy units rather than mass units. For this interpretation to make sense we must have

$$E_{[\mathrm{erg}_3]} = E_{[\mathrm{erg}_2]}. \tag{1.66c}$$

A little thought shows that the numeric part of a quantity of energy measured in erg_2 must indeed equal the numeric part of that same quantity of energy measured in erg_3. When comparing an unknown quantity of energy in the laboratory to another quantity of energy that we know to be 1 erg, it does not matter whether that 1 erg is 1 erg_2 or 1 erg_3—the laboratory procedure will be the same, the numerical result of the measurement will be the same, and so the numeric part assigned to the quantity of energy will be the same. The subscript given to erg is relevant only to the type of algebraic manipulations we have decided to permit in the unit equations. If erg belongs to the cgs system of units, so that

$$1\,\mathrm{erg} \neq c_{\mathrm{cgs}}^{-2}\,\mathrm{gm},$$

then $\mathrm{erg} = \mathrm{erg}_3$; and if we are working in units where $c = 1$ so that

$$1\,\mathrm{erg} = c_{\mathrm{cgs}}^{-2}\,\mathrm{gm},$$

then $\mathrm{erg} = \mathrm{erg}_2$. This is why physicists customarily neglect the difference between erg_3 and erg_2, calling both units 1 erg of energy and relying on context to distinguish between the two types of unit.

In particle physics, the mass and energy of elementary particles are often measured in eV standing for electron-volt, MeV standing for mega (or 10^6) electron-volt, and GeV standing for giga (or 10^9) electron-volt. The electron-volt is a unit of energy, and in a system of units where time has a separate dimension so that $c \neq 1$,

$$1\,\mathrm{eV}_3 = e_{[\mathrm{coul}]}\mathrm{joule}_3 \cong 1.60218 \times 10^{-19}\,\mathrm{joule}_3. \tag{1.67a}$$

The subscript 3 again shows that mass, length, and time are recognized as separate dimensions, and $e_{[\mathrm{coul}]}$ is the numeric part of an electron or proton electric charge measured in coulombs (coul).* This is, of course, just a pure number

$$e_{[\mathrm{coul}]} = 1.60218 \times 10^{-19}; \tag{1.67b}$$

and, since $1\,\mathrm{erg}_3 = 10^{-7}\,\mathrm{joule}_3$,

$$1\,\mathrm{eV}_3 = 10^7 \cdot e_{[\mathrm{coul}]} \cdot \mathrm{erg}_3. \tag{1.67c}$$

* The coulomb unit of charge is explained in Chapter 2.

Using $1 \, \text{cmtime}_3 = c_{\text{cgs}}^{-1} \sec$ from Eq. (1.56a), we can write, going from the cgs to the cgc system,

$$1 \, \text{eV}_3 = 10^7 \cdot e_{[\text{coul}]} \cdot \frac{\text{gm} \cdot \text{cm}^2}{\sec^2}$$
$$= 10^7 \cdot e_{[\text{coul}]} \cdot c_{\text{cgs}}^{-2} \cdot \frac{\text{gm} \cdot \text{cm}^2}{\text{cmtime}_3^2}. \tag{1.68a}$$

Changing over to the cg system gives

$$1 \, \text{eV}_2 = \frac{10^7 e_{[\text{coul}]}}{c_{\text{cgs}}^2} \cdot \text{gm}. \tag{1.68b}$$

Thus a mass m in gm,

$$m = m_{[\text{gm}]} \, \text{gm},$$

becomes, working in the cg system of units and using Eq. (1.68b),

$$m = m_{[\text{gm}]} \cdot \frac{c_{\text{cgs}}^2}{10^7 e_{[\text{coul}]}} \text{eV}_2 \tag{1.69a}$$
$$= m_{[\text{eV}_2]} \text{eV}_2,$$

where

$$m_{[\text{eV}_2]} = \frac{c_{\text{cgs}}^2 m_{[\text{gm}]}}{10^7 e_{[\text{coul}]}}. \tag{1.69b}$$

The mass of the electron in grams, 9.11×10^{-28} gm, transforms using Eq. (1.69b) to

$$\left[\frac{(9.11 \times 10^{-28})(2.99792 \times 10^{10})^2}{1.60218 \times 10^{-12}} \right] \text{eV}_2 \cong 5.11 \times 10^5 \, \text{eV}_2 = 0.511 \, \text{MeV}_2;$$

and the mass of the proton, 1.6727×10^{-24} gm, transforms to a mass in eV_2 of

$$\left[\frac{(1.6727 \times 10^{-24})(2.99792 \times 10^{10})^2}{1.60218 \times 10^{-12}} \right] \text{eV}_2 \cong 9.383 \times 10^8 \, \text{eV}_2 = 938.3 \, \text{MeV}_2.$$

As with ergs, it is customary to neglect the difference between eV_3 and eV_2, relying on context to show which type of energy units are being used.

In classical—that is, nonrelativistic—physics, the momentum p of an object is the product of its mass and velocity:

$$p = mv.$$

The relativistic formula for the momentum given in Eq. (1.54c) is

$$p = \frac{mv}{\sqrt{1 - \frac{v^2}{c^2}}}.$$

In the mks system, the classical formula assigns the momentum units of

$$\underset{\text{mks}}{\text{U}}(p) = \text{kg} \cdot \text{m/sec}. \tag{1.70}$$

We can break the relativistic formula for momentum into numeric parts and units to get

$$p_{\text{mks}}\left(\frac{\text{kg} \cdot \text{m}}{\text{sec}}\right) = \frac{m_{\text{mks}}\text{kg}\left(v_{\text{mks}}\frac{m}{\text{sec}}\right)}{\sqrt{1 - \frac{v^2_{\text{mks}}}{c^2_{\text{mks}}}}}, \tag{1.71a}$$

where

$$\underset{\text{mks}}{\text{N}}(p) = p_{[\text{kg}\cdot\text{m}\cdot\text{sec}^{-1}]} = p_{\text{mks}},$$

$$\underset{\text{mks}}{\text{N}}(p) = m_{[\text{kg}]} = m_{\text{mks}}, \tag{1.71b}$$

$$\underset{\text{mks}}{\text{N}}(v) = v_{[\text{m}\cdot\text{sec}^{-1}]} = v_{\text{mks}},$$

and

$$\underset{\text{mks}}{\text{N}}(c) = c_{\text{mks}} = 10^{-2} \cdot c_{\text{cgs}}. \tag{1.71c}$$

We see that in mks units the relativistic momentum is still measured in $\text{kg} \cdot \text{m/sec}$. Using $1\,\text{sec} = c_{\text{cgs}}\text{cmtime}_3$ from Eq. (1.56a) to convert sec to cmtime$_3$ gives

$$1\,\text{kg} \cdot \text{m/sec} = 1\,\frac{\text{kg} \cdot \text{m}}{c_{\text{cgs}}\text{cmtime}_3} = \frac{10^5\text{gm} \cdot \text{cm}}{c_{\text{cgs}}\text{cmtime}_3}, \tag{1.72}$$

where the last step uses $1\,\mathrm{kg} = 10^3\mathrm{gm}$ and $1\,\mathrm{m} = 10^2\,\mathrm{cm}$. Equation (1.72) equates the mks and cgc units of momentum, so by Rule I

$$\underset{\mathrm{cgc}}{\mathrm{N}}(p) = 10^5 c_{\mathrm{cgs}}^{-1} \underset{\mathrm{mks}}{\mathrm{N}}(p), \tag{1.73a}$$

or, using Eq. (1.58a),

$$\underset{\mathrm{cg}}{\mathrm{N}}(p) = 10^5 c_{\mathrm{cgs}}^{-1} \underset{\mathrm{mks}}{\mathrm{N}}(p). \tag{1.73b}$$

We know that in the cg system of units $c = 1$ and v is dimensionless,

$$\underset{\mathrm{cg}}{\mathrm{U}}(v) = 1, \tag{1.74}$$

so Eq. (1.54c) becomes

$$p = \frac{mv}{\sqrt{1 - v^2}} \tag{1.75}$$

to give, applying $\underset{\mathrm{cg}}{\mathrm{U}}$ to both sides,

$$\underset{\mathrm{cg}}{\mathrm{U}}(p) = \underset{\mathrm{cg}}{\mathrm{U}}(m) = \mathrm{gm}. \tag{1.76a}$$

This reminds us of Eq. (1.63d), where both energy and mass end up being measured in gm in the cg system of units; since energy, mass, and momentum have the same dimensions in any system of units with $c = 1$, it is no surprise to now find that momentum is also measured in gm. This is again a perfectly good result—but particle physicists often prefer to use energy-based units rather than gm to measure momentum. From Eq. (1.68b) we get

$$\underset{\mathrm{cg}}{\mathrm{U}}(p) = \frac{10^{-7} c_{\mathrm{cgs}}^2}{e_{[\mathrm{coul}]}} \mathrm{eV}_2 \tag{1.76b}$$

so that, using Rule I, the numeric part of a momentum p measured in eV_2 units is

$$p_{[\mathrm{eV}_2]} = \frac{10^{-7} c_{\mathrm{cgs}}^2}{e_{[\mathrm{coul}]}} \underset{\mathrm{cg}}{\mathrm{N}}(p)$$

or

$$p_{[\mathrm{eV}_2]} = \frac{10^{-2} c_{\mathrm{cgs}}}{e_{[\mathrm{coul}]}} \underset{\mathrm{mks}}{\mathrm{N}}(p) = \frac{10^{-2} c_{\mathrm{cgs}}}{e_{[\mathrm{coul}]}} p_{\mathrm{mks}}, \tag{1.76c}$$

where Eqs. (1.73b) and (1.71b) are used to write the right-hand side in terms of the numeric part of the momentum in mks units. Because it is important to distinguish momentum from mass and energy in relativistic particle physics, the eV$_2$ unit of momentum is often written as eV/c. The c in the denominator is a dimensionally meaningless label to show that eV energy units are being used to measure momentum. When eV/c is too small to put the decimal point in a convenient place, units of MeV/$c = 10^6$ eV/c or GeV/$c = 10^9$ eV/c are used. With this convention, Eq. (1.76c) becomes

$$p_{[\text{Mev}/c]} = \frac{10^{-8} c_{\text{cgs}}}{e_{[\text{coul}]}} p_{\text{mks}} \qquad (1.76\text{d})$$

or

$$p_{[\text{Gev}/c]} = \frac{10^{-11} c_{\text{cgs}}}{e_{[\text{coul}]}} p_{\text{mks}}. \qquad (1.76\text{e})$$

1.11 INVARIANT UNITS, CONNECTING UNITS, AND ADDITION OF EXTRA DIMENSIONS

Based on the work done so far, it makes sense to divide units into connecting units and invariant units when adding or removing dimensions from a physical equation. We say that a connecting unit is any fundamental or derived unit that gains or loses a dimension in this transition, and an invariant unit is a fundamental or derived unit that does not. In the discussion on how to make the universal gas constant R equal to 1, the units of temperature (degree Kelvin, degree Rankine, etc.), and any other derived units containing some nonzero power of temperature in their dimensional formulas, are all connecting units. The other fundamental and derived cgs and mks units—such as gm, m, newton, sec, joule, erg,   etc.—are invariant units because their dimensional formulas do not contain nonzero powers of temperature. When making the velocity of light c equal to 1, any fundamental or derived unit containing a nonzero power of time in its dimensional formula is a connecting unit, and any fundamental or derived unit that does not is an invariant unit. So erg, cmtime, newton, eV, etc., are all connecting units, and gm, kg, cm, etc., are invariant units. We note that this definition makes the sec a connecting unit, and indeed we could give an ordinary sec the label sec$_3$ and the reduced-dimension sec the label sec$_2$, with

$$1 \, \text{sec}_2 = c_{\text{cgs}} \text{cmtime}_2 = c_{\text{cgs}} \text{cm} \neq \text{sec}_3,$$

although, as we have seen, the cmtime unit is a more useful concept. That the sec$_3$ does not equal sec$_2$ points out another aspect of connecting units which distinguishes them from invariant units. A connecting unit is always really a pair of units whereas an invariant unit is not. When going from the cgc to the cg system of units, the invariant unit gm is always a gm and the invariant unit cm is always a cm, but the connecting unit cmtime is really the pair of units (cmtime$_3$, cmtime$_2$);

and we see that the other connecting units (erg_3, erg_2), (eV_3, eV_2), and so on, are also really pairs of units. Similarly, when transforming the ideal gas law to make R equal to 1, the connecting-unit pair $(degR1, EdegR1)$ takes us from the mksR1 to the mksER1 system of units, while the invariant units kg, m, newton, sec, joule, etc., do not change.

The concepts of connecting and invariant units reflect how we have decided to add and remove dimensions rather than the fundamental nature of physical reality. For example, when setting $c = 1$ we stop recognizing the separate dimension of time and instead give it the dimension of length, making the sec a connecting unit and the cm an invariant unit. We could, however, re-do this conversion by no longer recognizing the separate dimension of space, instead giving it the same dimension as time. The conversion formulas would end up the same, as would the physical equations, but now the sec (and other units of time) would be invariant units, and the cm (and other units of length) would be connecting units.

A relativistic physicist might object to calling either the cm or the sec an invariant unit, saying that what is really going on is that time and space are treated as the same physical quantity. From this physicist's perspective it is better to say that all units of time are given the combined dimension spacetime, and all units of length are also given the combined dimension spacetime. This is a perfectly valid point of view, and by pointing out that both space and time have been given a new combined dimension it reminds us that calling time a length or length a time does not make lengths and times identical—there is a difference between the two.* However, that same physicist, when shown the procedure used to make $R = 1$ in the ideal gas law, might say that giving temperature a separate dimension is a historical accident and that the true dimension of temperature really is energy. To point out the similarity between setting $c = 1$ and setting $R = 1$, we will always talk about adding and removing dimensions from the connecting units rather than combining dimensions or discovering a physical quantity's true dimension.

Adding back discarded dimensions is not difficult as long as Rule VIII is taken into account.

RULE VIII

The first step in adding a dimension to a physical equation or formula is to use the meaning of the extra dimension to specify how the connecting units attach to already existing variables. The second step is to rearrange the equation or formula so that it obeys Rules II, IV, and V in both the invariant and connecting units. Each fundamental unit for which an extra dimension is recognized must separately obey Rules II, IV, and V. The final step is to recognize the extra dimensions.

---

* This same point has been attributed to Abraham Lincoln when he said that calling a dog's tail a leg doesn't make it a leg.

To see how this works, we start with a very simple example,

$$E = m,$$

in units where $c = 1$. If both E and m are measured in the energy units erg_2, as in Eq. (1.66a), we have

$$E_{[erg_2]} \cdot erg_2 = m_{[erg_2]} \cdot erg_2, \qquad (1.77a)$$

where $E_{[erg_3]} = E_{[erg_2]}$ from Eq. (1.66c) has been used to replace $E_{[erg_3]}$ by $E_{[erg_2]}$. We note (step 1 of Rule VIII) that recognizing the separate dimension of time requires that energy and mass be measured in separate units. Since recognizing the separate dimension of time means that energy has dimensions mass $\cdot$ length^2 $\cdot$ time^{-2}, a mass unit should equal energy $\cdot$ length^{-2} $\cdot$ time^2, and the right-hand side of Eq. (1.77a) should be written as

$$E_{[erg_2]} erg_2 = m_{[erg_2]} erg_2 \left(\frac{cmtime_2}{cm} \right)^2. \qquad (1.77b)$$

We can do this to Eq. (1.77a) because cm = cmtime$_2$ in the cg system of units, so the right-hand side of the equation has just been multiplied by 1 squared. We note (step 2 of Rule VIII) that Eq. (1.77b) obeys Rule II in the cg system of units, where length and time have the same dimension; but it does not obey Rule II when the cm and cmtime$_2$ are analyzed separately, because its right-hand side has the ratio cmtime$_2$/cm, which the left-hand side does not. Another way of looking at this is to realize that if the time dimension were recognized now (the final step of Rule VIII) by replacing all the subscript 2s with subscript 3s, Eq. (1.77b) would not obey Rule II. Therefore, we multiply the right-hand side of Eq. (1.77b) by

$$\left(\frac{cm}{cmtime_2} \right)^2 = 1^2 = 1,$$

still working in units where cm = cmtime$_2$, to get

$$E_{[erg_2]} erg_2 = m_{[erg_2]} erg_2 \left(\frac{cmtime_2}{cm} \right)^2 \left(\frac{cm}{cmtime_2} \right)^2. \qquad (1.77c)$$

The units which make up erg_2 are balanced, since erg_2 appears on both sides of the equation; so Eq. (1.77c) obeys Rule II in all the fundamental units that acquire a separate dimension. Therefore, we can complete the process outlined in Rule VIII by recognizing the separate dimension of time:

$$E_{[erg_2]} erg_3 = m_{[erg_2]} erg_3 \left(\frac{cmtime_3}{cm} \right)^2 \left(\frac{cm}{cmtime_3} \right)^2. \qquad (1.77d)$$

Substitution of Eqs. (1.56a) and (1.66b) gives

$$
\begin{aligned}
E_{[\mathrm{erg}_2]}\mathrm{erg}_3 &= \left(m_{[\mathrm{gm}]}c_{\mathrm{cgs}}^2\right)\cdot\mathrm{erg}_3\cdot\left(\frac{c_{\mathrm{cgs}}^{-2}\mathrm{sec}^2}{\mathrm{cm}^2}\right)\cdot\left(\frac{\mathrm{cm}^2}{c_{\mathrm{cgs}}^{-2}\mathrm{sec}^2}\right) \\
&= m_{[\mathrm{gm}]}\left(\frac{\mathrm{erg}_3\cdot\mathrm{sec}^2}{\mathrm{cm}^2}\right)\cdot\left(c_{\mathrm{cgs}}^2\frac{\mathrm{cm}^2}{\mathrm{sec}^2}\right) \\
&= m_{[\mathrm{gm}]}\mathrm{gm}\cdot\left(c_{\mathrm{cgs}}\frac{\mathrm{cm}}{\mathrm{sec}}\right)^2,
\end{aligned}
\tag{1.77e}
$$

where the last step uses $1\,\mathrm{gm}=\mathrm{erg}_3\cdot\mathrm{sec}^2\cdot\mathrm{cm}^{-2}$. The physical quantities are now properly set up to give

$$
E = mc^2,
$$

which is the same as Eq. (1.54a), the original relativistic equality between mass and energy. This shows that the extra dimension of time has been correctly added to the formula.

We can get the same result starting with Eq. (1.63d). Because we are heading back to recognizing time as a separate dimension, we write Eq. (1.63d) as

$$
E_{\mathrm{cgc}}\mathrm{gm} = m_{[\mathrm{gm}]}\mathrm{gm}.
$$

The first step of Rule VIII leads us to multiply the left-hand side by

$$
1 = \left(\frac{\mathrm{cm}}{\mathrm{cmtime}_2}\right)^2,
$$

so that when the separate dimension of time is recognized energy has the correct mass $\cdot$ length^2 $\cdot$ time^{-2} units:

$$
E_{\mathrm{cgc}}\mathrm{gm}\left(\frac{\mathrm{cm}}{\mathrm{cmtime}_2}\right)^2 = m_{[\mathrm{gm}]}\mathrm{gm}.
$$

The second step of Rule VIII now requires us to multiply the left-hand side by

$$
1 = \left(\frac{\mathrm{cmtime}_2}{\mathrm{cm}}\right)^2,
$$

so that Rule II is obeyed when the fundamental units that acquire a new dimension are analyzed separately:

$$
E_{\mathrm{cgc}}\mathrm{gm}\left(\frac{\mathrm{cm}}{\mathrm{cmtime}_2}\right)^2\left(\frac{\mathrm{cmtime}_2}{\mathrm{cm}}\right)^2 = m_{[\mathrm{gm}]}\mathrm{gm}.
$$

Now we perform the final step of Rule VIII, recognizing the separate dimension of time to get

$$E_{\text{cgc}}\text{gm}\left(\frac{\text{cm}}{\text{cmtime}_3}\right)^2\left(\frac{\text{cmtime}_3}{\text{cm}}\right)^2 = m_{[\text{gm}]}\text{gm}.$$

Equation (1.56a) states that $\text{cmtime}_3 = c_{\text{cgs}}^{-1}\text{sec}$, allowing us to write

$$E_{\text{cgc}}\text{gm}\left(\frac{\text{cm}}{c_{\text{cgs}}^{-1}\text{sec}}\right)^2\left(\frac{c_{\text{cgs}}^{-1}\text{sec}}{\text{cm}}\right)^2 = m_{[\text{gm}]}\text{gm}.$$

Substitution of (1.63c) gives

$$E_{[\text{erg}_3]}c_{\text{cgs}}^{-2}\left(\text{gm}\frac{\text{cm}^2}{\text{sec}^2}\right)\cdot\left(\frac{\text{sec}}{\text{cm}}\right)^2 = m_{[\text{gm}]}\text{gm},$$

or, using $1\,\text{erg}_3 = \text{gm}\cdot\text{cm}^2\cdot\text{sec}^{-2}$,

$$E_{[\text{erg}_3]}\text{erg}_3\cdot\left(\frac{1}{c_{\text{cgs}}\dfrac{\text{cm}}{\text{sec}}}\right)^2 = m_{[\text{gm}]}\text{gm}.$$

Written with physical quantities, this becomes

$$\frac{E}{c^2} = m \quad\text{or}\quad E = mc^2,$$

which is again the desired result.

More complicated formulas require us to watch Rule IV when using Rule VIII. Equation (1.62) has both a dimensionless velocity v as well as frequencies f and f_0 in units of cm^{-1}:

$$f = f_0\cdot\left(\frac{\sqrt{1-v^2}}{1-v}\right).$$

When time is a separate dimension, the frequencies have the dimension time^{-1} and the velocities have dimensions of $\text{length}\cdot\text{time}^{-1}$. The first step of Rule VIII suggests we write this equation as

$$f_{\text{cg}}\text{cmtime}_2^{-1} = (f_0)_{\text{cg}}\text{cmtime}_2^{-1}\cdot\left[\frac{\sqrt{1-\left(v_{\text{cg}}\dfrac{\text{cm}}{\text{cmtime}_2}\right)^2}}{1-v_{\text{cg}}\dfrac{\text{cm}}{\text{cmtime}_2}}\right]. \qquad (1.78a)$$

In Eq. (1.78a) we are working with the cg system of units, so the dimensionless velocities can be written as $v_{cg}\text{cm}/\text{cmtime}_2$ because

$$1 = \frac{\text{cm}}{\text{cmtime}_2}.$$

Applying the second step of Rule VIII, we see that Rule IV is not obeyed separately by the fundamental units that acquire a new dimension. Therefore, we multiply the velocity v by

$$1 = \frac{\text{cmtime}_2}{\text{cm}}$$

to get

$$f_{cg}\text{cmtime}_2^{-1} = (f_0)_{cg}\text{cmtime}_2^{-1} \cdot \left[\frac{\sqrt{1 - \left(v_{cg}\dfrac{\text{cm}}{\text{cmtime}_2} \cdot \dfrac{\text{cmtime}_2}{\text{cm}} \right)^2}}{\left(1 - v_{cg}\dfrac{\text{cm}}{\text{cmtime}_2} \cdot \dfrac{\text{cmtime}_2}{\text{cm}} \right)} \right],$$

Having satisfied the second step of Rule VIII, we perform the final step of Rule VIII by recognizing the separate dimension of time to get

$$f_{cgc}\text{cmtime}_3^{-1} = (f_0)_{cgc}\text{cmtime}_3^{-1} \cdot \left[\frac{\sqrt{1 - \left(v_{cgc}\dfrac{\text{cm}}{\text{cmtime}_3} \cdot \dfrac{\text{cmtime}_3}{\text{cm}} \right)^2}}{\left(1 - v_{cgc}\dfrac{\text{cm}}{\text{cmtime}_3} \cdot \dfrac{\text{cmtime}_3}{\text{cm}} \right)} \right],$$

(1.78b)

where we have used Eq. (1.58a) to replace the numeric parts of physical quantities in the cg system of units by the numeric parts of physical quantities in the cgc system of units. The cmtime_3 unit time equals $c_{cgs}^{-1} \cdot \text{sec}$ [see Eq. (1.56a)] so Eq. (1.78b) can be written as

$$[f_{cgc}c_{cgs}]\text{sec}^{-1} = \left[(f_0)_{cgc}c_{cgs} \right]\text{sec}^{-1} \cdot \left\{ \frac{\sqrt{1 - \left(v_{cgc}c_{cgs}\dfrac{\text{cm}}{\text{sec}} \right)^2 \left(\dfrac{1}{c_{cgs}\dfrac{\text{cm}}{\text{sec}}} \right)^2}}{\left[1 - \left(v_{cgc}c_{cgs}\dfrac{\text{cm}}{\text{sec}} \right) \cdot \left(\dfrac{1}{c_{cgs}\dfrac{\text{cm}}{\text{sec}}} \right) \right]} \right\}.$$

(1.78c)

We know by Rule I that

$$f_{\text{cgs}} = f_{\text{cgc}} c_{\text{cgs}}, \quad (f_0)_{\text{cgs}} = (f_0)_{\text{cgc}} c_{\text{cgs}},$$

and

$$v_{\text{cgs}} = v_{\text{cgc}} c_{\text{cgs}}.$$

Therefore, Eq. (1.78c) can be written as

$$f = f_0 \cdot \left(\frac{\sqrt{1 - \dfrac{v^2}{c^2}}}{1 - \dfrac{v}{c}} \right), \tag{1.78d}$$

which is the same as what we started out with, Eq. (1.54d). Note that we could have gone directly from Eq. (1.78b) to (1.78d) by recognizing both that $1\,\text{cm/cmtime}_3$ is a perfectly good representation of the velocity of light c as a physical quantity in the cgc system of units, and that the velocity v as well as the frequencies f and f_0 are also correctly represented as physical quantities in the cgc system of units. Since physical equations and formulas are valid in any system of units with the same number and type of dimensions, Eq. (1.78d) follows at once.

1.12 SIMULTANEOUS REMOVAL OF $\hbar$, c, AND k

Advanced textbooks in relativistic quantum mechanics do not stop with making $c = 1$; they customarily choose units in which another dimensional constant $\hbar$, which is Planck's constant h divided by 2π, is also equal to 1. A good equation for showing how this can be useful is Planck's law for black-body radiation,

$$n_{\omega T} = \left(\frac{1}{2\pi} \right)^2 \left(\frac{\omega^2}{c^2} \right) \cdot \left[\exp\left(\frac{\hbar\omega}{kT} \right) - 1 \right]^{-1}, \tag{1.79a}$$

where $n_{\omega T} \cdot d\omega$ is the number of photons having an angular frequency (in rad/sec) between ω and $\omega + d\omega$ emitted per unit time and per unit area from a black-body surface at temperature T into a vacuum. The physical quantity $n_{\omega T} \cdot d\omega$ has units of number of photons, which is dimensionless, per unit area per unit time. Since $d\omega$ has units of radians, which is also dimensionless, per unit time, it follows that $n_{\omega T}$ has the dimension of inverse area or length^{-2}. Variable k is Boltzmann's constant, which has already been mentioned in Eqs. (1.41a, b) above; c is again the velocity of light, and

$$\hbar = \frac{h}{2\pi} \cong 1.054 \times 10^{-27} \text{erg} \cdot \text{sec}. \tag{1.79b}$$

Equation (1.79a) obeys Rules II, IV, and V, as expected.

We now go one step further than most textbooks of relativistic quantum mechanics by constructing a system of units where $\hbar = c = k = 1$. We start by setting up the cgsK system of units, which is just the cgs system of units with temperature measured in degK and the other three dimensions of length, mass, and time measured in cm, gm, and sec, respectively. It is the same as the mksK system of units used earlier in Section 1.8, only now the basic units for length and mass are cm and gm instead of m and kg. Rule I can be used to write Eqs. (1.41a, b) in cgsK units (remember that $1\,\text{erg} = 10^{-7}\text{joule}$):

$$k = k_{\text{mksK}}\frac{\text{joule}}{\text{degK}} = k_{\text{cgsK}}\frac{\text{erg}}{\text{degK}} = k_{\text{cgsK}}\frac{\text{gm} \cdot \text{cm}^2}{\text{sec}^2 \cdot \text{degK}}, \qquad (1.80a)$$

where

$$k_{\text{cgsK}} = \mathop{N}_{\text{cgsK}}(k) = 10^7 \cdot k_{\text{mksK}} \cong 1.380 \times 10^{-16}. \qquad (1.80b)$$

Equations (1.55a) and (1.79b) have the same form in cgs and cgsK units because their units do not contain a nonzero power of degK, so

$$c = c_{\text{cgs}}\frac{\text{cm}}{\text{sec}} \qquad (1.81a)$$

and

$$\hbar = \hbar_{\text{cgs}}\text{erg} \cdot \text{sec} = \hbar_{\text{cgs}}\frac{\text{gm} \cdot \text{cm}^2}{\text{sec}}, \qquad (1.81b)$$

where

$$\hbar_{\text{cgs}} = \mathop{N}_{\text{cgs}}(\hbar) = \mathop{N}_{\text{cgsK}}(\hbar) \cong 1.054 \times 10^{-27}. \qquad (1.81c)$$

The three constants $\hbar, c$, and k are eliminated from our equations by reducing the cgsK system of units to the single dimension length measured in cm. The first step is to rescale the units of mass, time, and temperature using the unspecified numeric constants α, β, and γ:

$$1\,\text{gm} = \alpha \cdot \text{umass}_4, \qquad (1.82a)$$

$$1\,\text{sec} = \beta \cdot \text{utime}_4, \qquad (1.82b)$$

$$1\,\text{degK} = \gamma \cdot \text{utemp}_4. \qquad (1.82c)$$

The subscript "4" shows that the four separate dimensions of length, mass, time, and temperature are still recognized when using the umass_4, utime_4, and utemp_4

units. We call the system of units based on the cm, $umass_4$, $utime_4$, and $utemp_4$ units for length, mass, time, and temperature the $cm\alpha\beta\gamma$ system of units. In the $cm\alpha\beta\gamma$ system of units c, $\hbar$, and k become

$$c = \left(\frac{c_{cgs}}{\beta}\right)\frac{cm}{utime_4}, \tag{1.83a}$$

$$\hbar = \left(\hbar_{cgs}\frac{\alpha}{\beta}\right)\cdot\frac{umass_4 \cdot cm^2}{utime_4}, \tag{1.83b}$$

$$k = \left(k_{cgsK}\frac{\alpha}{\beta^2\gamma}\right)\cdot\frac{umass_4 \cdot cm^2}{utime_4^2 \cdot utemp_4}. \tag{1.83c}$$

To complete the first step of Rule VI, we choose α, β, γ such that

$$\frac{c_{cgs}}{\beta} = 1, \tag{1.84a}$$

$$\frac{\alpha}{\beta}\hbar_{cgs} = 1, \tag{1.84b}$$

$$\frac{\alpha\, k_{cgsK}}{\gamma\,\beta^2} = 1. \tag{1.84c}$$

From Eq. (1.84a) we get

$$\beta = c_{cgs}, \tag{1.85}$$

which makes Eq. (1.82b)

$$1\,\sec = c_{cgs}\cdot utime_4. \tag{1.86a}$$

This is the same as Eq. (1.56a) and shows that we can regard

$$utime_4 = cmtime_3 \tag{1.86b}$$

as the same unit of time. Substitution of Eq. (1.85) into Eq. (1.84b) gives

$$\alpha = \frac{\beta}{\hbar_{cgs}} = \frac{c_{cgs}}{\hbar_{cgs}}, \tag{1.87a}$$

so that, from Eq. (1.82a)

$$1\,gm = \frac{c_{cgs}}{\hbar_{cgs}}\cdot umass_4. \tag{1.87b}$$

Substitution of Eqs. (1.85) and (1.87a) into Eq. (1.84c) gives

$$\gamma = \frac{\alpha k_{cgsK}}{\beta^2} = \frac{c_{cgs}}{\hbar_{cgs}}\cdot\frac{1}{c_{cgs}^2}\cdot k_{cgsK} = \frac{k_{cgsK}}{c_{cgs}\hbar_{cgs}}, \tag{1.88a}$$

which makes Eq. (1.82c)

$$1\,\mathrm{degK} = \frac{k_{\mathrm{cgsK}}}{c_{\mathrm{cgs}}\hbar_{\mathrm{cgs}}} \cdot \mathrm{utemp}_4.\qquad(1.88b)$$

Now we use Rule I and Eqs. (1.86a) and (1.88b) to write Eq. (1.79a) in terms of the $\mathrm{cm}\alpha\beta\gamma$ system of units:

$$\underset{\mathrm{cm}\alpha\beta\gamma}{\mathrm{N}}\,(n_\omega\,T)\cdot\mathrm{cm}^{-2} = \left(\frac{1}{2\pi}\right)^2 \left[\frac{(\omega_{\mathrm{cm}\alpha\beta\gamma})^2\mathrm{utime}_4^{-2}}{\left(1\cdot\dfrac{\mathrm{cm}}{\mathrm{utime}_4}\right)^2}\right]$$

$$\times\left\{\exp\left[\frac{\left(1\cdot\dfrac{\mathrm{umass}_4\cdot\mathrm{cm}^2}{\mathrm{utime}_4}\right)\cdot\left(\omega_{\mathrm{cm}\alpha\beta\gamma}\mathrm{utime}_4^{-1}\right)}{\left(1\cdot\dfrac{\mathrm{umass}_4\cdot\mathrm{cm}^2}{\mathrm{utime}_4^2\cdot\mathrm{utemp}_4}\right)\cdot\left(T_{\mathrm{cm}\alpha\beta\gamma}\mathrm{utemp}_4\right)}\right]-1\right\}^{-1},$$

$$(1.89a)$$

where

$$\omega_{\mathrm{cm}\alpha\beta\gamma} = \underset{\mathrm{cm}\alpha\beta\gamma}{\mathrm{N}}\,(\omega) = c_{\mathrm{cgs}}^{-1}\,\underset{\mathrm{cgsK}}{\mathrm{N}}\,(\omega) = c_{\mathrm{cgs}}^{-1}\underset{\mathrm{cgs}}{\mathrm{N}}(\omega),\qquad(1.89b)$$

$$T_{\mathrm{cm}\alpha\beta\gamma} = \underset{\mathrm{cm}\alpha\beta\gamma}{\mathrm{N}}\,(T) = \frac{k_{\mathrm{cgsK}}}{c_{\mathrm{cgs}}\hbar_{\mathrm{cgs}}}\,\underset{\mathrm{cgsK}}{\mathrm{N}}\,(T) = \frac{k_{\mathrm{cgsK}}}{c_{\mathrm{cgs}}\hbar_{\mathrm{cgs}}}\cdot T_{[\mathrm{degK}]},\qquad(1.89c)$$

$$1\cdot\frac{\mathrm{cm}}{\mathrm{utime}_4} = c,\qquad(1.89d)$$

$$1\cdot\frac{\mathrm{umass}_4\cdot\mathrm{cm}^2}{\mathrm{utime}_4} = \hbar,\qquad(1.89e)$$

and

$$1\cdot\frac{\mathrm{umass}_4\cdot\mathrm{cm}^2}{\mathrm{utime}_4^2\cdot\mathrm{utemp}_4} = k.\qquad(1.89f)$$

The argument of exp is so complicated that it is a relief to perform the second step of Rule VI and stop recognizing the separate dimensions of mass, time, and temperature. We replace the subscript 4s by subscript 1s to show that only length has a dimension:

$$1\,\mathrm{umass}_1 = 1\,\mathrm{cm}^{-1},\qquad(1.90a)$$

$$1 \, \text{utime}_1 = 1 \, \text{cm}, \qquad (1.90\text{b})$$

$$1 \, \text{utemp}_1 = 1 \, \text{cm}^{-1}. \qquad (1.90\text{c})$$

Equations (1.86a), (1.87b), and (1.88b) can now be written as

$$1 \, \text{sec}_1 = c_{\text{cgs}} \cdot \text{cm}, \qquad (1.90\text{d})$$

$$1 \, \text{gm}_1 = \frac{c_{\text{cgs}}}{\hbar_{\text{cgs}}} \cdot \text{cm}^{-1}, \qquad (1.90\text{e})$$

and

$$1 \, \text{degK}_1 = \frac{k_{\text{cgsK}}}{c_{\text{cgs}}\hbar_{\text{cgs}}} \cdot \text{cm}^{-1}. \qquad (1.90\text{f})$$

This system of units is called the cm1 system of units, and for any physical quantity b

$$\underset{\text{cm1}}{\text{N}}(b) = \underset{\text{cm}\alpha\beta\gamma}{\text{N}}(b). \qquad (1.91\text{a})$$

This requires all physical quantities to have the same numeric parts in the cm1 and the cm$\alpha\beta\gamma$ system of units. If the dimensional formula for b is

$$\text{mass}^x \text{length}^y \text{time}^z \text{temperature}^u$$

for real exponents x, y, z, and u, then the U operator for the cm1 system of units is

$$
\begin{aligned}
\underset{\text{cm1}}{\text{U}}(b) &= \left(\frac{\text{cm}^{-1}}{\text{umass}_4}\right)^x \left(\frac{\text{cm}^{-1}}{\text{utemp}_4}\right)^u \left(\frac{\text{cm}}{\text{utime}_4}\right)^z \underset{\text{cm}\alpha\beta\gamma}{\text{U}}(b) \\
&= \left(\frac{\text{cm}^{-1}}{\text{gm}}\right)^x \left(\frac{\text{cm}^{-1}}{\text{degK}}\right)^u \left(\frac{\text{cm}}{\text{sec}}\right)^z \underset{\text{cgsK}}{\text{U}}(b).
\end{aligned}
\qquad (1.91\text{b})
$$

Equation (1.89a) becomes

$$\underset{\text{cm1}}{\text{N}}(n_\omega \, T) \cdot \text{cm}^{-2} = \left(\frac{1}{2\pi}\right)^2 \left(\omega_{\text{cm1}}\text{cm}^{-1}\right)^2 \cdot \left[\exp\left(\frac{\omega_{\text{cm1}}\text{cm}^{-1}}{T_{\text{cm1}}\text{cm}^{-1}}\right) - 1\right]^{-1}, \qquad (1.92\text{a})$$

where

$$\omega_{\text{cm1}} = \omega_{\text{cm}\alpha\beta\gamma} = \underset{\text{cm1}}{\text{N}}(\omega) = \underset{\text{cm}\alpha\beta\gamma}{\text{N}}(\omega) \qquad (1.92\text{b})$$

and

$$T_{\text{cm1}} = T_{\text{cm}\alpha\beta\gamma} = \underset{\text{cm1}}{\text{N}}(T) = \underset{\text{cm}\alpha\beta\gamma}{\text{N}}(T). \qquad (1.92\text{c})$$

Because $\omega = \omega_{cm1}cm^{-1}$ and $T = T_{cm1}cm^{-1}$ are both physical quantities, Eq. (1.92a) can be written as

$$n_{\omega}T = \left(\frac{\omega}{2\pi}\right)^2 \cdot \frac{1}{e^{\omega/T} - 1}. \tag{1.93}$$

This is, as expected, Eq. (1.79a) with $\hbar, c, k$ replaced by 1.

Using Rule VIII to return to a system of units where mass, length, time, and temperature have separate dimensions is less of a challenge than it looks. The first point worth noting is that the only invariant unit is the cm; every other non-length unit is a connecting unit. Working with the ω/T exponent of e, from the first step of Rule VIII we have in the cm1 system of units

$$\frac{\omega}{T} = \frac{\omega_{cm1}\,utime_1^{-1}}{T_{cm1}\,utemp_1}. \tag{1.94}$$

By Rule V and the second step of Rule VIII, the right-hand side of Eq. (1.94) should be dimensionless in both utime and utemp, so we multiply it by

$$\frac{utemp_1}{utime_1^{-1}} = 1$$

to get

$$\frac{\omega}{T} = \frac{\omega_{cm1}}{T_{cm1}} \cdot \frac{utime_1^{-1}}{utemp_1} \cdot \frac{utemp_1}{utime_1^{-1}}. \tag{1.95a}$$

Recognizing the separate dimensions of mass, time, and temperature transforms the right-hand side of (1.95a) to

$$\frac{\omega\,cm1}{T_{cm1}} \cdot \frac{utime_4^{-1}}{utemp_4} \cdot \frac{utemp_4}{utime_4^{-1}} = \left(\frac{\omega_{cm\alpha\beta\gamma}}{T_{cm\alpha\beta\gamma}} \cdot \frac{utime_4^{-1}}{utemp_4}\right) \cdot \left(1 \cdot \frac{utemp_4}{utime_4^{-1}}\right), \tag{1.95b}$$

where Eqs. (1.92b, c) have been used to replace ω_{cm1} and T_{cm1} by their equivalent numerics in the $cm\alpha\beta\gamma$ system of units. The dimensional constant

$$\Psi = 1 \cdot \frac{utemp_4}{utime_4^{-1}} \tag{1.96}$$

does not look familiar, but it becomes recognizable if we convert it to the cgsK system of units. Equations (1.82a–c) give

$$\Psi = 1 \cdot \frac{utemp_4}{utime_4^{-1}} = \left(\gamma^{-1}\beta^{-1}\right) \cdot degK \cdot sec.$$

From Eqs. (1.85) and (1.88a)

$$\frac{c_{cgs}\hbar_{cgs}}{k_{cgsK}} \cdot \frac{1}{c_{cgs}} = \frac{\hbar_{cgs}}{k_{cgsK}}.$$

With this clue and the definitions of $\hbar$ and k, we easily see that

$$\Psi = \left(\frac{\hbar_{cgs}}{k_{cgsK}}\right) \cdot degK \cdot sec = \left(\frac{\hbar_{cgs}\dfrac{gm \cdot cm^2}{sec}}{k_{cgsK}\dfrac{gm \cdot cm^2}{sec^2 \cdot degK}}\right) = \frac{\hbar}{k}. \tag{1.97}$$

Equation (1.97) relates the physical quantities $\Psi, \hbar$, and k in any system of units recognizing mass, length, time, and temperature as its four fundamental dimensions. Therefore, $\Psi = \hbar/k$ in the $cm\alpha\beta\gamma$ system of units, and we can use Eqs. (1.96) and (1.95b) to write

$$\frac{\omega_{cm1}}{T_{cm1}} \cdot \frac{utime_4^{-1}}{utemp_4} \cdot \frac{utemp_4}{utime_4^{-1}} = \left(\frac{\omega_{cm\alpha\beta\gamma}}{T_{cm\alpha\beta\gamma}} \cdot \frac{utime_4^{-1}}{utemp_4}\right) \cdot \Psi = \left(\frac{\omega}{T}\right) \cdot \frac{\hbar}{k}, \tag{1.98a}$$

where

$$\omega = \omega_{cm\alpha\beta\gamma}\, utime_4^{-1} \tag{1.98b}$$

and

$$T = T_{cm\alpha\beta\gamma}\, utemp_4 \tag{1.98c}$$

are recognized as the physical quantities they are, showing that Rule VIII has been successfully applied to the ω/T exponent of e.

Now we apply Rule VIII to the $n_{\omega T}$ and ω^2 of Eq. (1.93). The $n_{\omega T}$ on the left-hand side of Eq. (1.93) starts with dimensions of inverse area in the cgsK system of units; and since cm is an invariant unit when going from the cgsK to the $cm\alpha\beta\gamma$ to the cm1 system of units, we know that $n_{\omega T}$ has units of cm^{-2} in Eq. (1.93). Applying $\underset{cm1}{U}$ to both sides of Eq. (1.93) gives

$$\underset{cm1}{U}(n_{\omega T}) = \underset{cm1}{U}(\omega^2) \cdot \underset{cm1}{U}\left[(2\pi)^{-2}\right] \cdot \underset{cm1}{U}\left[(e^{\omega/T} - 1)^{-1}\right]$$
$$= \underset{cm1}{U}(\omega^2) \cdot 1 \cdot 1 = \underset{cm1}{U}(\omega^2). \tag{1.99}$$

In cm1 units both $n_{\omega T}$ and ω^2 have units of cm^{-2}, balancing this equation in the cm1 system of units; but the first step of Rule VIII requires the units of ω^2 to be written as $utime_1^{-2} = cm^{-2}$ because ω is an angular frequency:

$$\omega^2 = \left(\omega_{cm1}\, utime_1^{-1}\right)^2. \tag{1.100}$$

The second step of Rule VIII requires that $n_{\omega T}$ and ω^2 be separately balanced in both the cm and utime$_1$ units, so we multiply the right-hand side of Eq. (1.100) by utime$_1^2$/cm^2, which equals 1 in cm1 units, to get

$$\omega^2 = \left(\omega_{cm1} \text{utime}_1^{-1} \cdot \frac{\text{utime}_1}{\text{cm}} \right)^2. \tag{1.101}$$

Now we are ready to apply Rule VIII to all of Eq. (1.93). Substituting Eqs. (1.95a) and (1.101) into Eq. (1.93) gives

$$n_{\omega T} = \left(\frac{1}{2\pi} \right)^2 \left(\omega_{cm1} \text{utime}_1^{-1} \cdot \frac{\text{utime}_1}{\text{cm}} \right)^2$$
$$\times \left[\exp\left(\frac{\omega_{cm1}}{T_{cm1}} \cdot \frac{\text{utime}_1^{-1}}{\text{utemp}_1} \cdot \frac{\text{utemp}_1}{\text{utime}_1^{-1}} \right) - 1 \right]^{-1}. \tag{1.102a}$$

The $n_{\omega T}$ has units of cm^{-2}, so this formula is separately balanced in the cm, utime$_1$, and utemp$_1$ units as required by the second step of Rule VIII. Performing the final step of Rule VIII, we recognize the separate dimensions of mass, time, and temperature:

$$n_{\omega T} = \left(\frac{1}{2\pi} \right)^2 \left(\omega_{cm1} \text{utime}_4^{-1} \cdot \frac{\text{utime}_4}{\text{cm}} \right)^2 \left[\exp\left(\frac{\omega_{cm1}}{T_{cm1}} \cdot \frac{\text{utime}_4^{-1}}{\text{utemp}_4} \cdot \frac{\text{utemp}_4}{\text{utime}_4^{-1}} \right) - 1 \right]^{-1}.$$

Equations (1.92b) and (1.98a) are used to get

$$n_{\omega T} = \left(\frac{1}{2\pi} \right)^2 \left(\omega_{cm\alpha\beta\gamma} \text{utime}_4^{-1} \right)^2 \left(1 \cdot \frac{\text{utime}_4}{\text{cm}} \right)^2 \left[\exp\left(\frac{\hbar\omega}{kT} \right) - 1 \right]^{-1}. \tag{1.102b}$$

The $n_{\omega T}$ left-hand side of (1.102b) does not change because it has units of cm^{-2} and the cm is an invariant unit when going from the cm1 to the cm$\alpha\beta\gamma$ system of units. From Eq. (1.89d) we see that

$$1 \cdot \frac{\text{utime}_4}{\text{cm}} = \frac{1}{c},$$

and from Eq. (1.98b)

$$\left(\omega_{cm\alpha\beta\gamma} \text{utime}_4^{-1} \right)^2 = \omega^2$$

because these are true physical quantities in any system of units based on the four fundamental dimensions mass, length, time, and temperature. Equation (1.102b) now becomes

$$n_{\omega T} = \left(\frac{\omega}{2\pi c} \right)^2 \cdot \left[\exp\left(\frac{\hbar\omega}{kT} \right) - 1 \right]^{-1}. \tag{1.102c}$$

This last step returns us to Eq. (1.79a), the original formula for $n_{\omega T}$.

REFERENCES

1. C. H. Page, "Physical Entities and Mathematical Representation," *Journal of Research of the National Bureau of Standards*—B. Mathematics and Mathematical Physics, Vol. 65B, No. 4, Oct.–Dec. 1961, pp. 227–235.
2. C. H. Page, "The Mathematical Representation of Physical Entities," *IEEE Transactions on Education*, Vol. E10, No. 2, June 1967, pp. 70–74.
3. B. N. Taylor, *Guide for the Use of the International System of Units (SI)*, NIST Special Publication 811, 1995, p. 23.
4. B. H. Mahan, *University Chemistry*, 2nd edition, Addison-Wesley Publishing Company, Inc., Reading, Mass., 1969, pp. 18–19.
5. P. W. Bridgman, *Dimensional Analysis*, Yale University Press, New Haven, Conn., 1931, p. 24.
6. R. T. Birge, "On Electric and Magnetic Units and Dimensions," *The American Physics Teacher*, Vol. 2, No. 2, 1934, pp. 41–48.
7. K. Huang, *Statistical Mechanics*, John Wiley and Sons, Inc., New York, 1963, p. 146.

UNITS ASSOCIATED WITH NINETEENTH-CENTURY ELECTROMAGNETIC THEORY

At the beginning of the nineteenth century, the study of electricity and magnetism were two separate disciplines with separate sets of units. Both sets of units recognized only mass, length, and time as fundamental physical dimensions, because the recent eighteenth-century triumphs of Newtonian physics had predisposed scientists to assume that these three dimensions were the only ones needed to describe nature. The conceptual neatness of separate units disappeared, however, during the first half of the nineteenth century, due to the discovery of ever more profound connections between electrical and magnetic phenomena, a process culminating with Maxwell's identification of light as electromagnetic radiation. Units for the new combined discipline of electromagnetism could be created either by extending the electrical units to cover magnetism, creating a new system called electrostatic units (esu), or by extending the magnetic units to cover electricity, creating a new system called electromagnetic units (emu). We shall see that both procedures gained widespread acceptance, requiring scientists to become familiar with both systems. To make the situation even more confusing, the two systems gave units that depended in different ways on mass, length, and time to the same electromagnetic physical quantities. Consequently, esu and emu units, unlike the units of Newtonian physics, are different in kind as well as size; this inevitably became a source of discontent to those who felt that the same kind of physical quantity should always be measured by the same kind of physical unit.

During the second half of the nineteenth century, the situation began to sort itself out with the growth and adoption of "practical" units based on a rescaled system of emu units. Although these units did not long survive the nineteenth century, historically speaking they paved the way for the most widespread electromagnetic system in use today, the rationalized mks system. In this chapter we not only present the system of practical units but also discuss the growing realization that electromagnetic units could be based on four, rather than three, fundamental dimensions— mass, length, and time, and something electromagnetic (for example, charge). The easiest way to show how this all fits together is to sketch in a very old-fashioned, nineteenth-century version of electromagnetic theory, showing how nineteenth-century formalism influenced the development of the first electromagnetic units. Every attempt is made to point out where modern practice differs from what is being presented; however, none of the facts or equations given here are in any sense wrong—they are just old-fashioned, occasionally looking strange to the eye of the modern engineer or physicist.* The text assumes a good but not necessarily advanced knowledge of electromagnetism, only presenting as much information as is required to understand the system of units being discussed; and, the emphasis is always on how to transform from one set of units to another, showing how the equations and formulas change when this occurs.

2.1 ELECTRIC FIELDS, MAGNETIC FIELDS, AND COULOMB'S LAW

Modern introductions to electromagnetic theory almost always begin with Coulomb's law for the force F between two point charges Q_1 and Q_2. Written in nineteenth-century notation it is

$$F = \frac{Q_1 Q_2}{\varepsilon_0 r^2}, \tag{2.1a}$$

or in vector form

$$\vec{F}_{12} = \frac{Q_1 Q_2}{\varepsilon_0 r^2} \hat{r}_{12}, \tag{2.1b}$$

where r is the distance between the point charges, $\vec{F}_{12}$ is the vector force of charge 1 on charge 2, and ε_0 is a constant that came to be called the permittivity of free space. The symbol $\hat{r}_{12}$ represents a dimensionless vector pointing from charge 1 to charge 2, whose length is the numeric value 1; vectors of this sort are written with a caret and called dimensionless unit vectors.

Perhaps the most basic nineteenth-century construct that has disappeared from today's formulations of electromagnetic theory is the point magnetic pole; nineteenth-century introductions to electromagnetic theory give as much prominence to Coulomb's law for the force between two point magnetic poles as they do to Coulomb's law for the force between two point charges. Coulomb's law for the force F between two point magnetic poles having pole strengths $(p_H)_1$ and $(p_H)_2$ is, written in nineteenth-century notation,

$$F = \frac{(p_H)_1 (p_H)_2}{\mu_0 r^2}, \tag{2.2a}$$

or in vector form

$$\vec{F}_{12} = \frac{(p_H)_1 (p_H)_2}{\mu_0 r^2} \hat{r}_{12}, \tag{2.2b}$$

---

* Fortunately we can leave out altogether the idea of the luminiferous ether, the hypothetical nineteenth-century medium through which electromagnetic radiation propagated, because it played no significant role in the development of nineteenth-century units.

where r is the distance between the point poles, $\vec{F}_{12}$ is the vector force of pole 1 on pole 2, $\hat{r}_{12}$ is the dimensionless unit vector pointing from pole 1 to pole 2, and μ_0 is a constant that came to be called the magnetic permeability of free space. When a specified point charge interacts with other point charges, the Coulomb forces from the other charges add together like vectors to give the total force on the specified charge. The same is true for the Coulomb forces of multiple magnetic poles; when a specified pole interacts with other magnetic poles, the forces from the other poles add together like vectors to give the total force on the specified pole.

Permanent magnets have two poles that are equal in magnitude and opposite in sign. Figure 2.1 shows that when a typical bar magnet or compass needle is suspended and allowed to swing freely in a plane parallel to the earth's surface, one pole is always pulled north and the other pole south. The north-seeking pole of the magnet, or north pole for short, is assigned a positive pole strength; and the south-seeking or south pole is assigned a negative pole strength. Nineteenth-century texts are always quick to point out that, unlike electric charge, a magnetic pole cannot be isolated. If a bar magnet or compass needle is cut in half we get two smaller magnets or needles, each with equal and opposite north and south magnetic poles (see Fig. 2.2). Indeed, to this day there have been no confirmed observations of isolated magnetic poles (called magnetic monopoles), although experiments are still occasionally performed to see whether they exist. For this reason, Eqs. (2.2a, b) are interpreted as describing the attractive or repulsive force between the poles of two long, thin magnets whose opposite poles are so distant that their influence can

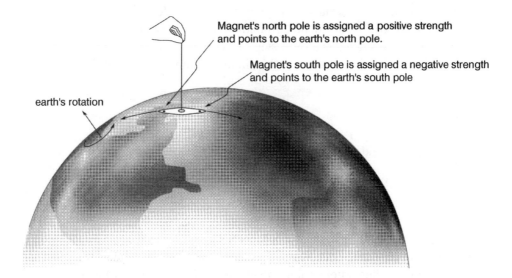

Figure 2.1 When a bar magnet or compass needle is allowed to swing freely, the positive pole of the magnet is pulled toward the earth's north pole and the negative pole of the magnet is pulled toward the earth's south pole. The positive pole is called the magnet's north (short for north-seeking) pole and the negative pole is called the magnet's south (short for south-seeking) pole.

north pole   south pole

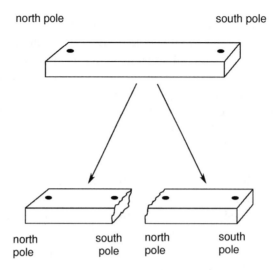

north   south   north   south
pole   pole   pole   pole

Figure 2.2 The poles of a permanent magnet cannot be isolated by breaking it into pieces.

be neglected. Figure 2.3 is a diagram of the sort of torsion balance Coulomb used to verify Eqs. (2.2a, b) for the poles of permanent magnets. Indeed, it is much easier to demonstrate the truth of Coulomb's law for magnetic poles [Eqs. (2.2a, b)] than it is to demonstrate the truth of Coulomb's law for point charges [Eqs. (2.1a, b)] because the pole of a permanent magnet, unlike the electric charge on a small object, is not affected by accidental grounding or humidity in the air.

From Coulomb's law it is a short step to electric and magnetic fields. The electric field of an isolated point charge Q is, in nineteenth-century notation,

$$\vec{E} = \frac{Q}{\varepsilon_0 r^2}\hat{r}; \tag{2.3a}$$

and the magnetic field of a magnetic pole of pole strength p_H is

$$\vec{H} = \frac{p_H}{\mu_0 r^2}\hat{r}. \tag{2.3b}$$

In Eqs. (2.3a, b) the $\vec{E}$ or $\vec{H}$ field is defined for all points in space surrounding the point charge or magnetic pole. When the electric or magnetic field is evaluated at a particular point, called a field point, the variable r in Eqs. (2.3a, b) is the distance from the position of the point charge or magnetic pole to the field point, and $\hat{r}$ is the dimensionless unit vector pointing from the point charge or magnetic pole to the field point.

The interaction of point charges can be analyzed in terms of fields. Equation (2.3a) is used to assign an electric field $\vec{E}_i$ to the i'th charge of a collection of

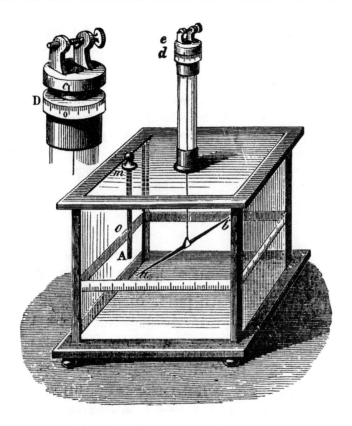

Figure 2.3 This is a diagram of the sort of torsion balance used to verify Coulomb's law for the poles of a permanent magnet.

point charges. We then add together at any desired field point the individual electric fields of all the electric charges to get the total electric field $\vec{E} = \sum_i \vec{E}_i$ for the entire collection of charges. If a test charge Q is placed at that field point, and if Q is small enough not to disturb the positions of the other charges, the force on that test charge can be written as

$$\vec{F} = Q\vec{E}. \tag{2.4}$$

Equation (2.4) follows directly from (2.1b), Coulomb's law for electric charges; these equations are just different ways of stating the same mathematical idea. Notice, however, how much more comfortable this way of looking at electrical phenomena is to the nineteenth-century experimental scientist. In his laboratory he can assemble a collection of batteries, wires, and charged metal objects—but locating all the charge is a tedious and ultimately ambiguous task. However, he can easily create a small test charge and move it around to measure the electric field; and every time the same collection of batteries, wires, etc., is assembled he finds the same electric fields at the same field points. It is quite natural for him to end up attributing as much reality to the electric field as to the charges creating it.

The interaction of magnets can also be analyzed in terms of fields. We use Eq. (2.3b) to assign a magnetic field $\vec{H}_i$ to the i'th pole of a collection of poles and then add together all the poles' individual fields to get the total magnetic field $\vec{H} = \sum_i \vec{H}_i$ for the entire collection of poles. Because isolated magnetic poles do not exist, the least-complicated magnetic field that can be constructed this way is $\vec{H}_{LEAST} = \vec{H}_1 + \vec{H}_2$, with $\vec{H}_1$ the field of a north magnetic pole and $\vec{H}_2$ the field of the corresponding south magnetic pole. Any physically realistic field $\vec{H} = \sum_i \vec{H}_i$ constructed from individual poles must also be the sum of $\vec{H}_{LEAST}$ fields corresponding to the north-south poles of individual magnets. Although the force $\vec{F}$ on an isolated pole of strength p_H would, in fact, be

$$\vec{F} = p_H \vec{H}, \tag{2.5}$$

we also know the experimenter cannot create or find such an isolated test pole. Coulomb's strategy of using long, thin magnets to lessen the influence of the un-wanted opposite pole may not work well when investigating an unknown magnetic field, because the field can influence both poles simultaneously. In practice what was done instead was to place a small magnetic needle at the field point and allow it to oscillate about its equilibrium position, with $\vec{H}$ being inversely proportional to the square of the period of oscillation (see Appendix 2.A). Although a nineteenth-century experimental scientist could easily locate the poles of all nearby permanent magnets, he too finds the field concept useful when examining the effect of the earth's magnetic poles; not having access to earth's interior, all he knows for sure is the value of the earth's magnetic field inside his laboratory. It is again natural for him to end up attributing as much reality to the magnetic field as to the magnetic poles that create it.

2.2 COMBINED SYSTEMS OF ELECTRIC AND MAGNETIC UNITS

When in 1820 it was discovered that an electric current produces a magnetic field, the need for a combined system of electric and magnetic units became clear. In nineteenth-century notation the magnetic field near a long, thin wire carrying a constant current I is

$$\vec{H} = \frac{2I}{r}\hat{e}, \tag{2.6}$$

where, as shown in Fig. 2.4, r is the distance from the wire carrying the current to the field point and the dimensionless unit vector $\hat{e}$ equals $\hat{v} \times \hat{r}$, with $\hat{r}$ the dimensionless unit vector pointing from the wire to the field point and $\hat{v}$ the di-mensionless unit vector pointing in the direction of the current. An electric current is always defined to be

$$I = \frac{dQ}{dt}, \tag{2.7}$$

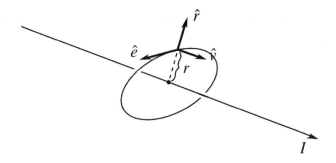

Figure 2.4 Unit-length vectors $\hat{e}, \hat{r}$, and $\hat{v}$ are used to specify the magnetic field generated by current I.

the amount of charge dQ moving through the cross-section of the wire during an amount of time dt.

Do we have enough information in Eqs. (2.1a) through (2.7) to set up a system of electromagnetic units? Suppose we keep mass, length, and time as our fundamental dimensions and formally apply the $\underset{\text{mlt}}{U}$ operator described in Chapter 1 [see Eq. (1.29c)] to Eqs. (2.1a) to (2.7). Either Eq. (2.1a) or Eq. (2.1b) gives

$$\underset{\text{mlt}}{U}(F) = \frac{\text{mass} \cdot \text{length}}{\text{time}^2} = \frac{\left[\underset{\text{mlt}}{U}(Q)\right]^2}{\underset{\text{mlt}}{U}(\varepsilon_0) \cdot \text{length}^2}$$

or

$$\underset{\text{mlt}}{U}(Q) = \left[\underset{\text{mlt}}{U}(\varepsilon_0)\right]^{1/2} \frac{\text{mass}^{1/2} \cdot \text{length}^{3/2}}{\text{time}}. \qquad (2.8a)$$

From Eq. (2.2a) or Eq. (2.2b) we similarly get

$$\underset{\text{mlt}}{U}(p_H) = \left[\underset{\text{mlt}}{U}(\mu_0)\right]^{1/2} \frac{\text{mass}^{1/2} \cdot \text{length}^{3/2}}{\text{time}}, \qquad (2.8b)$$

and from Eq. (2.3b) comes

$$\underset{\text{mlt}}{U}(H) = \frac{\underset{\text{mlt}}{U}(p_H)}{\underset{\text{mlt}}{U}(\mu_0) \cdot \text{length}^2} = \frac{\text{mass}^{1/2}}{\left[\underset{\text{mlt}}{U}(\mu_0)\right]^{1/2} \cdot \text{length}^{1/2} \cdot \text{time}}, \qquad (2.8c)$$

where in the last step we have substituted for $\underset{\text{mlt}}{U}(p_H)$ from Eq. (2.8b). From Eq. (2.7) we have

$$\underset{\text{mlt}}{U}(I) = \frac{\underset{\text{mlt}}{U}(Q)}{\text{time}} = \left[\underset{\text{mlt}}{U}(\varepsilon_0)\right]^{1/2} \frac{\text{mass}^{1/2} \cdot \text{length}^{3/2}}{\text{time}^2}, \qquad (2.8d)$$

where in the last step $\underset{\text{mlt}}{U}(Q)$ is replaced by the right-hand side of Eq. (2.8a). Applying $\underset{\text{mlt}}{U}$ to Eq. (2.6) gives

$$\underset{\text{mlt}}{U}(H) = \frac{\underset{\text{mlt}}{U}(I)}{\text{length}}, \tag{2.8e}$$

and substitution of Eqs. (2.8c, d) into Eq. (2.8e) leads to

$$\frac{\text{mass}^{1/2}}{\left[\underset{\text{mlt}}{U}(\mu_0)\right]^{1/2} \cdot \text{length}^{1/2} \cdot \text{time}} = \left[\underset{\text{mlt}}{U}(\varepsilon_0)\right]^{1/2} \frac{\text{mass}^{1/2} \cdot \text{length}^{1/2}}{\text{time}^2}$$

or

$$\left[\underset{\text{mlt}}{U}(\mu_0)\right] \cdot \left[\underset{\text{mlt}}{U}(\varepsilon_0)\right] = \left(\frac{\text{time}}{\text{length}}\right)^2. \tag{2.8f}$$

No matter how the equations of nineteenth-century electromagnetic theory are pushed around—even when we include all of Maxwell's equations—this is the result that shows up. We always find that the product $\varepsilon_0 \mu_0$ has the dimensions of a squared inverse velocity, length^{-2}· time^2, but we cannot pin down the dimensions of either ε_0 or μ_0 as separate physical quantities. Quite naturally this was often taken to mean that more research was needed; some new law of nature yet to be discovered was expected to reveal the true dimensions of both ε_0 and μ_0. Table 2.1 shows what happens when this outlook is adopted wholeheartedly and the dimensions of all the common electrical quantities are evaluated in terms of mass, length, and time and the yet-to-be-discovered true dimensions of ε_0 and μ_0. We use

$$\text{eps} = \underset{\text{mlt}}{U}(\varepsilon_0) \tag{2.9a}$$

and

$$\text{mu} = \underset{\text{mlt}}{U}(\mu_0) \tag{2.9b}$$

to label the unknown true dimensions of ε_0 and μ_0. In the second column of Table 2.1, Eq. (2.8f) is used to eliminate μ_0 from all the dimensional formulas; and in the third column of Table 2.1, Eq. (2.8f) is used to eliminate ε_0 from all the dimensional formulas. Unfortunately for this point of view, no one has (so far) been able to find that one extra law of nature revealing the "true" dimensions of ε_0 and μ_0. In a way, all physics and engineering students confused by the multiple systems of electromagnetic units in use today are still paying for this nondiscovery, because it

Table 2.1 Dimensions of electromagnetic physical quantities written in terms of mass, length, time, and the unknown dimensions of $\text{eps} = \underset{mlt}{U}(\varepsilon_0)$ and $\text{mu} = \underset{mlt}{U}(\mu_0)$.

Physical quantity	Dimensions using eps	Dimensions using mu
A (magnetic vector potential)	$\text{eps}^{-1/2}\text{mass}^{1/2}\text{length}^{-1/2}$	$\text{mu}^{1/2}\text{mass}^{1/2}\text{length}^{1/2}\text{time}^{-1}$
B (magnetic induction)	$\text{eps}^{-1/2}\text{mass}^{1/2}\text{length}^{-3/2}$	$\text{mu}^{1/2}\text{mass}^{1/2}\text{length}^{-1/2}\text{time}^{-1}$
C (capacitance)	$\text{eps}\cdot\text{length}$	$\text{mu}^{-1}\text{length}^{-1}\text{time}^2$
D (electric displacement)	$\text{eps}^{1/2}\text{mass}^{1/2}\text{length}^{-1/2}\text{time}^{-1}$	$\text{mu}^{-1/2}\text{mass}^{1/2}\text{length}^{-3/2}$
E (electric field)	$\text{eps}^{-1/2}\text{mass}^{1/2}\text{length}^{-1/2}\text{time}^{-1}$	$\text{mu}^{1/2}\text{mass}^{1/2}\text{length}^{1/2}\text{time}^{-2}$
ε (dielectric constant)	eps	$\text{mu}^{-1}\text{length}^{-2}\text{time}^2$
ε_0 (permittivity of free space)	eps	$\text{mu}^{-1}\text{length}^{-2}\text{time}^2$
$\mathcal{F}$ (magnetomotive force)	$\text{eps}^{1/2}\text{mass}^{1/2}\text{length}^{3/2}\text{time}^{-2}$	$\text{mu}^{-1/2}\text{mass}^{1/2}\text{length}^{1/2}\text{time}^{-1}$
Φ_B (magnetic flux)	$\text{eps}^{-1/2}\text{mass}^{1/2}\text{length}^{1/2}$	$\text{mu}^{1/2}\text{mass}^{1/2}\text{length}^{3/2}\text{time}^{-1}$
G (conductance)	$\text{eps}\cdot\text{length}\cdot\text{time}^{-1}$	$\text{mu}^{-1}\cdot\text{length}^{-1}\cdot\text{time}$
H (magnetic field)	$\text{eps}^{1/2}\text{mass}^{1/2}\text{length}^{1/2}\text{time}^{-2}$	$\text{mu}^{-1/2}\text{mass}^{1/2}\text{length}^{-1/2}\text{time}^{-1}$
I (current)	$\text{eps}^{1/2}\text{mass}^{1/2}\text{length}^{3/2}\text{time}^{-2}$	$\text{mu}^{-1/2}\text{mass}^{1/2}\text{length}^{1/2}\text{time}^{-1}$
J (volume current density)	$\text{eps}^{1/2}\text{mass}^{1/2}\text{length}^{-1/2}\text{time}^{-2}$	$\text{mu}^{-1/2}\text{mass}^{1/2}\text{length}^{-3/2}\text{time}^{-1}$
$\mathcal{J}_S$ (surface current density)	$\text{eps}^{1/2}\text{mass}^{1/2}\text{length}^{1/2}\text{time}^{-2}$	$\text{mu}^{-1/2}\text{mass}^{1/2}\text{length}^{-1/2}\text{time}^{-1}$
L (inductance)	$\text{eps}^{-1}\cdot\text{length}^{-1}\cdot\text{time}^2$	$\text{mu}\cdot\text{length}$
m_H (permanent-magnet dipole moment)	$\text{eps}^{-1/2}\text{mass}^{1/2}\text{length}^{3/2}$	$\text{mu}^{1/2}\text{mass}^{1/2}\text{length}^{5/2}\text{time}^{-1}$
m_I (current-loop magnetic dipole moment)	$\text{eps}^{1/2}\text{mass}^{1/2}\text{length}^{7/2}\text{time}^{-2}$	$\text{mu}^{-1/2}\text{mass}^{1/2}\text{length}^{5/2}\text{time}^{-1}$
M_H (permanent-magnet dipole density)	$\text{eps}^{-1/2}\text{mass}^{1/2}\text{length}^{-3/2}$	$\text{mu}^{1/2}\text{mass}^{1/2}\text{length}^{-1/2}\text{time}^{-1}$

Table 2.1 (Continued).

Physical quantity	Dimensions using eps	Dimensions using mu
M_I (current-loop magnetic dipole density)	$eps^{1/2}mass^{1/2}length^{1/2}time^{-2}$	$mu^{1/2}mass^{1/2}length^{-1/2}time^{-1}$
μ (magnetic permeability)	$eps^{-1}length^{-2}time^2$	mu
μ_0 (magnetic permeability of free space)	$eps^{-1}length^{-2}time^2$	mu
p_H (magnetic pole strength)	$eps^{-1/2}mass^{1/2}length^{1/2}$	$mu^{1/2}mass^{1/2}length^{3/2}time^{-1}$
p (electric dipole moment)	$eps^{1/2}mass^{1/2}length^{5/2}time^{-1}$	$mu^{-1/2}mass^{1/2}length^{3/2}$
P (electric dipole density)	$eps^{1/2}mass^{1/2}length^{-1/2}time^{-1}$	$mu^{-1/2}mass^{1/2}length^{-3/2}$
$\mathcal{P}$ (permeance)	$eps^{-1}length^{-1}time^2$	$mu \cdot length$
Q (charge)	$eps^{1/2}mass^{1/2}length^{3/2}time^{-1}$	$mu^{-1/2}mass^{1/2}length^{1/2}$
R (resistance)	$eps^{-1}length^{-1}time$	$mu \cdot length \cdot time^{-1}$
$\mathcal{R}$ (reluctance)	$eps \cdot length \cdot time^{-2}$	$mu^{-1} \cdot length^{-1}$
ρ_Q (volume charge density)	$eps^{1/2}mass^{1/2}length^{-3/2}time^{-1}$	$mu^{-1/2}mass^{1/2}length^{-5/2}$
ρ_R (resistivity)	$eps^{-1}time$	$mu \cdot length^2 \cdot time^{-1}$
S (elastance)	$eps^{-1} \cdot length^{-1}$	$mu \cdot length \cdot time^{-2}$
S_Q (surface charge density)	$eps^{1/2}mass^{1/2}length^{-1/2}time^{-1}$	$mu^{-1/2}mass^{1/2}length^{-3/2}$
σ (conductivity)	$eps \cdot time^{-1}$	$mu^{-1} \cdot length^{-2} \cdot time$
V (electric potential)	$eps^{-1/2}mass^{1/2}length^{1/2}time^{-1}$	$mu^{1/2}mass^{1/2}length^{3/2}time^{-2}$
Ω_H (magnetic scalar potential)	$eps^{1/2}mass^{1/2}length^{3/2}time^{-2}$	$mu^{-1/2}mass^{1/2}length^{1/2}time^{-1}$

is still not compellingly obvious how electromagnetic units should be related to the three mechanical dimensions of mass, length, and time. The problem is an irritating one, because not knowing the true dimensions of ε_0 and μ_0 has experimental implications. Mathematical procedures used to get Eq. (2.8f) can be repeated with careful measurements of all the physical quantities involved. Just as no equation like

$$\underset{\text{mlt}}{U}(\varepsilon_0) = \left\{ \text{mechanical dimensions not including } \underset{\text{mlt}}{U}(\mu_0) \right\}$$

or

$$\underset{\text{mlt}}{U}(\mu_0) = \left\{ \text{mechanical dimensions not including } \underset{\text{mlt}}{U}(\varepsilon_0) \right\},$$

can be found, so we can never get an experimental result of the form

$$\varepsilon_0 = \left\{ \text{mechanical quantities not including } \mu_0 \right\}$$

or

$$\mu_0 = \left\{ \text{mechanical quantities not including } \varepsilon_0 \right\}.$$

We literally cannot measure ε_0 and μ_0 as separate quantities; the best that can be done is to measure their product, which turns out to be one over the speed of light squared:

$$\varepsilon_0 \mu_0 = \frac{1}{c^2}. \tag{2.10}$$

This is no accident, of course. Maxwell's equations show that time-varying electric and magnetic fields can propagate as waves travelling at a speed

$$v = (\varepsilon_0 \mu_0)^{-1/2},$$

so if light is electromagnetic radiation—and it is—then Eq. (2.10) is exactly what we expect to find.

2.3 The ESU and EMU Systems of Units

Since nature does not assign unique values to both ε_0 and μ_0, nineteenth-century scientists had to do the job. Two choices were made, one leading to the cgs electrostatic system of units and one leading to the cgs electromagnetic system of units. Both systems of units use cm, gm, sec as their fundamental units of length, mass, and time. In the cgs electrostatic system of units, called esu units, we choose

$$\varepsilon_0 = 1, \tag{2.11a}$$

so that, to satisfy Eq. (2.10),

$$\mu_0 = \frac{1}{c^2}. \tag{2.11b}$$

In the cgs electromagnetic system of units, called emu units, we choose

$$\mu_0 = 1, \tag{2.12a}$$

so that, to satisfy Eq. (2.10),

$$\varepsilon_0 = \frac{1}{c^2}. \tag{2.12b}$$

In esu units, ε_0 is dimensionless and equal to 1, disappearing from all of the electromagnetic formulas. These are convenient units for analyzing electrostatic phenomena. In emu units, μ_0, dimensionless and equal to 1, disappears from all the electromagnetic formulas; these are convenient units for analyzing magnetostatic phenomena.

As was explained in Chapter 1, we can—for any system of units—define U and N operators for examining the units and numeric parts of physical quantities. We say that $\underset{esu}{U}$ and $\underset{esu}{N}$ are the U and N operators for the cgs esu system of units and that $\underset{emu}{U}$ and $\underset{emu}{N}$ are the U and N operators for the cgs emu system of units. We require

$$\underset{esu}{U}(b_{MECH}) = \underset{cgs}{U}(b_{MECH}), \tag{2.13a}$$

$$\underset{esu}{N}(b_{MECH}) = \underset{cgs}{N}(b_{MECH}), \tag{2.13b}$$

$$\underset{emu}{U}(b_{MECH}) = \underset{cgs}{U}(b_{MECH}), \tag{2.13c}$$

$$\underset{emu}{N}(b_{MECH}) = \underset{cgs}{N}(b_{MECH}) \tag{2.13d}$$

for any physical quantity b_{MECH}, which is purely mechanical in nature. Physical quantities such as mass m, length L, time t, force F, velocity v, acceleration a, etc., are purely mechanical physical quantities because they can be understood and measured without worrying about charge, magnetic pole strength, electric current, electric and magnetic fields, or any of the other electromagnetic physical quantities described later in this chapter. We use a lowercase esu or emu subscript to indicate the numeric component of any electromagnetic or physical quantity b in the esu or emu system of units:

$$\underset{esu}{N}(b) = b_{esu}, \tag{2.13e}$$

$$\underset{emu}{N}(b) = b_{emu}. \tag{2.13f}$$

This just continues the convention used in Chapter 1 where a lowercase subscript (such as cgs or mks) indicated the numeric component of a physical quantity in the system of units specified by the subscript. From Eqs. (2.13b) and (2.13d) we see that for mechanical physical quantities b_{MECH}

$$\underset{esu}{N}(b_{MECH}) = (b_{MECH})_{esu} = \underset{cgs}{N}(b_{MECH}) = (b_{MECH})_{cgs}$$

and

$$\underset{emu}{N}(b_{MECH}) = (b_{MECH})_{emu} = \underset{cgs}{N}(b_{MECH}) = (b_{MECH})_{cgs}.$$

All U and N operators have to obey the rules described in Section 1.6. We now apply $\underset{esu}{U}, \underset{esu}{N}$ and $\underset{emu}{U}, \underset{emu}{N}$ to Eqs. (2.1a) through (2.7) to find out what they mean when operating on electromagnetic physical quantities such as Q, p_H, E, H, etc.

Starting with Eqs. (2.1a) and (2.13a), we apply $\underset{esu}{U}$ to both sides to get

$$\underset{esu}{U}(F) = \underset{cgs}{U}(F) = dynes = \frac{gm \cdot cm}{sec^2} = \frac{\underset{esu}{U}(Q)^2}{\underset{esu}{U}(\varepsilon_0) \cdot cm^2}.$$

Equation (2.11a) shows that $\underset{esu}{U}(\varepsilon_0) = 1$, which gives

$$\underset{esu}{U}(Q) = \frac{gm^{1/2} \cdot cm^{3/2}}{sec}. \tag{2.14a}$$

This is an odd-looking result; but in principle nothing prevents us from using fractional powers of fundamental units, so we continue. From Eqs. (2.12b) and (2.13c),

$$\underset{emu}{U}(\varepsilon_0) = \underset{cgs}{U}(c^{-2}) = \frac{sec^2}{cm^2},$$

so applying $\underset{emu}{U}$ to both sides of Eq. (2.1a) gives

$$\frac{gm \cdot cm}{sec^2} = \frac{\underset{emu}{U}(Q)^2}{sec^2}$$

or

$$\underset{emu}{U}(Q) = gm^{1/2} \cdot cm^{1/2}. \tag{2.14b}$$

This takes care of finding $\underset{esu}{U}(Q)$ and $\underset{emu}{U}(Q)$ for any charge Q.

To find $\underset{\text{esu}}{\text{N}}(Q)$ and $\underset{\text{emu}}{\text{N}}(Q)$ for any charge Q, we compare two point charges Q_A and Q_B to a third unchanging point charge Q_C, adjusting Q_A and Q_B until they both experience the same Coulomb force at the same distance from Q_C. Then, we know that $Q_A = Q_B = Q'$, and Eq. (2.1a) shows the repulsive force between Q_A and Q_B to be, when separated by a distance r,

$$F = \frac{Q'^2}{\varepsilon_0 r^2}. \tag{2.14c}$$

Applying $\underset{\text{esu}}{\text{N}}$ to Eq. (2.14e) gives

$$F_{\text{esu}} = F_{\text{cgs}} = \frac{\underset{\text{esu}}{\text{N}}(Q')^2}{\underset{\text{esu}}{\text{N}}(\varepsilon_0) \cdot r_{\text{cgs}}^2},$$

or, since $\varepsilon_0 = 1$ in esu units,

$$\underset{\text{esu}}{\text{N}}(Q') = Q'_{\text{esu}} = r_{\text{cgs}}\sqrt{F_{\text{cgs}}}. \tag{2.14d}$$

This result specifies $\underset{\text{esu}}{\text{N}}(Q')$ in terms of the already known numerics r_{cgs} and F_{cgs}. The numerical component $\underset{\text{esu}}{\text{N}}(Q_D)$ of any other point charge Q_D can now be found from the Coulomb force between Q_D and Q'. We say that the Coulomb force between Q_D and Q' when they are a distance R apart is

$$f = \frac{Q_D Q'}{R^2},$$

and applying $\underset{\text{esu}}{\text{N}}$ to both sides of this gives

$$f_{\text{cgs}} = \frac{\underset{\text{esu}}{\text{N}}(Q_D) Q'_{\text{esu}}}{(R_{\text{cgs}})^2}$$

or

$$\underset{\text{esu}}{\text{N}}(Q_D) = \frac{f_{\text{cgs}} R_{\text{cgs}}^2}{Q'_{\text{esu}}}. \tag{2.14e}$$

The equations of electromagnetism are self-consistent, so applying $\underset{\text{esu}}{\text{N}}$ to any other formula containing the point charge Q_D, permittivity ε_0, and other mechanical quantities must give the same numbers for $\underset{\text{esu}}{\text{N}}(Q_D)$, the numeric component of the point charge in esu units. Because all charge can be analyzed as a collection of one or more point charges, we can now regard $\underset{\text{esu}}{N}$ as a well-defined quantity when operating on charge.

To find $\underset{\text{emu}}{N}(Q)$, we again compare two point charges Q_A and Q_B to a third unchanging point charge Q_C, adjusting Q_A and Q_B until they both experience the same force at the same distance from Q_C. Once more, $Q_A = Q_B = Q'$, and we can apply $\underset{\text{emu}}{N}$ to Eq. (2.14c) to get

$$F_{\text{emu}} = F_{\text{cgs}} = \frac{\underset{\text{emu}}{N}(Q')^2}{\underset{\text{emu}}{N}(\varepsilon_0) \cdot r_{\text{cgs}}^2},$$

or, since $\varepsilon_0 = c^{-2}$ in emu units,

$$\underset{\text{emu}}{N}(Q') = Q'_{\text{emu}} = \frac{r_{\text{cgs}}}{c_{\text{cgs}}}\sqrt{F_{\text{cgs}}}. \tag{2.14f}$$

This specifies $\underset{\text{emu}}{N}(Q')$ in terms of the already known numerics r_{cgs}, F_{cgs}, and c_{cgs}. Just as in esu units, the Coulomb force f between Q' and any other charge Q_D when Q' and Q_D are separated by a distance R is

$$f = \frac{Q_D Q'}{\varepsilon_0 R^2}.$$

Applying $\underset{\text{emu}}{N}$ to both sides of the formula gives

$$\underset{\text{emu}}{N}(Q_D) = \frac{\underset{\text{emu}}{N}(\varepsilon_0) f_{\text{cgs}} R_{\text{cgs}}^2}{Q'_{\text{emu}}},$$

or, since $\underset{\text{emu}}{N}(\varepsilon_0) = c_{\text{cgs}}^{-2}$,

$$\underset{\text{emu}}{N}(Q_D) = \frac{f_{\text{cgs}} R_{\text{cgs}}^2}{Q'_{\text{emu}} c_{\text{cgs}}^2}. \tag{2.14g}$$

This specifies $\underset{\text{emu}}{N}(Q_D)$ in terms of already-known numeric components. Again we note that any other formula for the charge in terms of ε_0 and mechanical quantities must give the same number for $\underset{\text{emu}}{N}(Q_D)$, because the electromagnetic equations are self-consistent. Since all charge can be analyzed as a collection of point charges, $\underset{\text{emu}}{N}$ is now a well-defined quantity when operating on charge.

Comparing Eqs. (2.14d) and (2.14f) shows that we can write

$$\underset{\text{esu}}{N}(Q') = c_{\text{cgs}} \cdot \underset{\text{emu}}{N}(Q'). \tag{2.14h}$$

This means that $Q'_{esu} = c_{cgs} Q'_{emu}$. Comparison of Eqs. (2.14e) and (2.14g) gives

$$\underset{emu}{N}(Q_D) = \frac{f_{cgs} R^2_{cgs}}{(c^{-1}_{cgs} Q'_{esu}) c^2_{cgs}} = \frac{1}{c_{cgs}} \underset{esu}{N}(Q_D), \qquad (2.14i)$$

which shows that Eq. (2.14h) still holds true when Q_D replaces Q'; clearly Eq. (2.14h) must be true for the numerical component of any point charge. Because all charge can be regarded as a collection of point charges, we conclude that Eq. (2.14h) holds true for the numerical components of any type of electric charge.

Applying $\underset{esu}{U}$ to Eq. (2.2a) for magnetic poles gives

$$\underset{esu}{U}(p_H)^2 = \frac{gm \cdot cm^3}{sec^2} \underset{esu}{U}(\mu_0).$$

From Eq. (2.11b) we get

$$\underset{esu}{U}(\mu_0) = \frac{sec^2}{cm^2},$$

so for magnetic poles

$$\underset{esu}{U}(p_H) = gm^{1/2} \cdot cm^{1/2}. \qquad (2.15a)$$

From Eq. (2.12a) we know that $\underset{emu}{U}(\mu_0)$ must be 1, so applying $\underset{emu}{U}$ to both sides of Eq. (2.2a) gives

$$\frac{gm \cdot cm^3}{sec^2} = \underset{emu}{U}(p_H)^2$$

or

$$\underset{emu}{U}(p_H) = \frac{gm^{1/2} \cdot cm^{3/2}}{sec}. \qquad (2.15b)$$

Creating two equal-strength magnetic poles to the accuracy required by nineteenth-century scientists is simple: just break a long, thin bar magnet in half, as shown in Fig. 2.2. This produces two long, thin bar magnets, with both magnets having the same-strength north poles and same-strength south poles. (To confirm this, compare the pole strengths of the new magnets to that of a third.) Coulomb's torsion balance in Fig. 2.3 can then be used to measure the force

$$F = \frac{(p_H)^2}{\mu_0 r^2} \qquad (2.15c)$$

between two identical poles of strength p_H a distance r apart. Applying $\underset{\text{esu}}{N}$ to Eq. (2.15c) gives

$$\underset{\text{esu}}{N}(p_H)^2 = F_{\text{cgs}} \cdot r_{\text{cgs}}^2 \cdot \underset{\text{esu}}{N}(\mu_0)$$

or, since $\underset{\text{esu}}{N}(\mu_0) = \underset{\text{esu}}{N}(c^{-2}) = c_{\text{cgs}}^{-2}$,

$$\underset{\text{esu}}{N}(p_H) = \frac{r_{\text{cgs}}}{c_{\text{cgs}}} \sqrt{F_{\text{cgs}}}. \tag{2.15d}$$

This specifies $\underset{\text{esu}}{N}(p_H) = (p_H)_{\text{esu}}$ in terms of the already-known numerics r_{cgs}, F_{cgs}, and c_{cgs}. The numeric part of the pole strength for any third magnet, $\underset{\text{esu}}{N}(p'_H)$, can be found by using the torsion balance to measure the force f between it and one of the two magnets of known pole strength $(p_H)_{\text{esu}}$. When the poles are separated by a distance R, which is short compared to the lengths of both magnets,

$$f = \frac{(p_H)(p'_H)}{\mu_0 R^2},$$

so that, applying $\underset{\text{esu}}{N}$ to this formula, we get

$$\underset{\text{esu}}{N}(p'_H) = \frac{\underset{\text{esu}}{N}(\mu_0)\, f_{\text{cgs}} R_{\text{cgs}}^2}{(p_H)_{\text{esu}}} = \frac{f_{\text{cgs}} R_{\text{cgs}}^2}{(p_H)_{\text{esu}} c_{\text{cgs}}^2}. \tag{2.15e}$$

Equations (2.15d, e) show how to find the numeric component of the pole strength in the esu system of units for any pole p'_H. Therefore, $\underset{\text{esu}}{N}$ is a well-defined quantity when operating on pole strengths. Repeating the process in the emu system of units, we apply $\underset{\text{emu}}{N}$ to both sides of Eq. (2.15c) and use $\underset{\text{emu}}{N}(\mu_0) = 1$ to get

$$F_{\text{cgs}} = \frac{\underset{\text{emu}}{N}(p_H)^2}{r_{\text{cgs}}^2}$$

or

$$\underset{\text{emu}}{N}(p_H) = r_{\text{cgs}} \sqrt{F_{\text{cgs}}}. \tag{2.15f}$$

The numeric part of the pole strength for any third magnet in emu units, $\underset{\text{emu}}{N}(p'_H)$, is specified by comparing it to the known value of

$$\underset{\text{emu}}{N}(p_H) = (p_H)_{\text{emu}}.$$

We apply $\underset{\text{emu}}{N}$ to the formula for the Coulomb force f between p_H and p'_H when separated by a distance R much smaller than the magnets' lengths to get [remember $\underset{\text{emu}}{N}(\mu_0) = 1$]

$$\underset{\text{emu}}{N}\left(p'_H\right) = \frac{f_{\text{cgs}} R^2_{\text{cgs}}}{(p_H)_{\text{emu}}}. \tag{2.15g}$$

The self-consistency of electromagnetic theory ensures that any other formula for the pole strength in terms of μ_0 and mechanical quantities must give the same numbers for its numeric component as Eqs. (2.15d–g). This makes $\underset{\text{emu}}{N}$ a well-defined quantity when operating on pole strengths. Comparison of Eq. (2.15d) to Eq. (2.15f) shows that

$$\underset{\text{emu}}{N}(p_H) = c_{\text{cgs}} \cdot \underset{\text{esu}}{N}(p_H) \quad \text{or} \quad (p_H)_{\text{emu}} = c_{\text{cgs}} \cdot (p_H)_{\text{esu}};$$

and comparison of Eq. (2.15g) to Eq. (2.15e) shows that

$$\underset{\text{emu}}{N}\left(p'_H\right) = \frac{f_{\text{cgs}} R^2_{\text{cgs}}}{(p_H)_{\text{emu}}} = \frac{f_{\text{cgs}} R^2_{\text{cgs}}}{c_{\text{cgs}}(p_H)_{\text{esu}}} = c_{\text{cgs}} \cdot \underset{\text{esu}}{N}\left(p'_H\right).$$

Since the numeric component of the pole strength in emu is c_{cgs} times the numeric component of the pole stength in esu both for the original pole p_H and for any other pole p'_H, we conclude that for any size magnetic pole p''_H

$$\underset{\text{emu}}{N}\left(p''_H\right) = c_{\text{cgs}} \cdot \underset{\text{esu}}{N}\left(p''_H\right). \tag{2.15h}$$

Moving on to electric field, Eq. (2.4) shows that

$$\underset{\text{esu}}{U}(F) = \underset{\text{esu}}{U}(Q) \cdot \underset{\text{esu}}{U}(E)$$

or, using Eq. (2.14a) and dyne $=$ gm $\cdot$ cm $\cdot$ sec^{-2},

$$\underset{\text{esu}}{U}(E) = \frac{\text{dynes} \cdot \text{sec}}{\text{gm}^{1/2} \cdot \text{cm}^{3/2}} = \frac{\text{gm}^{1/2}}{\text{cm}^{1/2} \cdot \text{sec}}. \tag{2.16a}$$

Applying $\underset{\text{esu}}{N}$ to Eq. (2.4) gives

$$\underset{\text{esu}}{N}(E) = \frac{\underset{\text{esu}}{N}(F)}{\underset{\text{esu}}{N}(Q)} = \frac{F_{\text{cgs}}}{Q_{\text{esu}}}. \tag{2.16b}$$

The $\underset{\text{emu}}{U}$ operator can be applied to Eq. (2.4) to get, using Eq. (2.14b),

$$\underset{\text{emu}}{U}(E) = \frac{\text{gm}^{1/2} \cdot \text{cm}^{1/2}}{\text{sec}^2}; \tag{2.16c}$$

and the $\underset{\text{emu}}{N}$ operator can be applied to Eq. (2.4) to get

$$\underset{\text{emu}}{N}(E) = \frac{\underset{\text{emu}}{N}(F)}{\underset{\text{emu}}{N}(Q)} = \frac{F_{\text{cgs}}}{Q_{\text{emu}}}. \tag{2.16d}$$

These equations give the esu and emu units of the E field and specify

$$\underset{\text{esu}}{N}(E) = E_{\text{esu}} \quad \text{and} \quad \underset{\text{emu}}{N}(E) = E_{\text{emu}}$$

in terms of the already-known numerics Q_{esu}, Q_{emu}, and F_{cgs}. Hence, $\underset{\text{esu}}{U}$, $\underset{\text{emu}}{U}$, $\underset{\text{esu}}{N}$ and $\underset{\text{emu}}{N}$ are well defined when applied to an electric field E.

Examining the magnetic field, we get from Eqs. (2.5), (2.15a), and (2.15b) that

$$\underset{\text{esu}}{U}(H) = \frac{\text{gm}^{1/2} \cdot \text{cm}^{1/2}}{\text{sec}^2} \tag{2.17a}$$

and

$$\underset{\text{emu}}{U}(H) = \frac{\text{gm}^{1/2}}{\text{cm}^{1/2} \cdot \text{sec}}. \tag{2.17b}$$

We cannot apply $\underset{\text{esu}}{N}$ and $\underset{\text{emu}}{N}$ to Eq. (2.5) directly because we cannot create an isolated magnetic pole. Instead, we create a small magnetic needle of length l, measure the strength p_H of its north and south poles, and find the distance L between them to get, from Eq. (2.A.5b) of Appendix 2.A,

$$\underset{\text{esu}}{N}(H) = H_{\text{esu}} = \frac{\pi^2}{3} \cdot \frac{M_{\text{cgs}} l_{\text{cgs}}^2}{(p_H)_{\text{esu}} L_{\text{cgs}} T_{\text{cgs}}^2} \tag{2.17c}$$

and

$$\underset{\text{emu}}{N}(H) = H_{\text{emu}} = \frac{\pi^2}{3} \cdot \frac{M_{\text{cgs}} l_{\text{cgs}}^2}{(p_H)_{\text{emu}} L_{\text{cgs}} T_{\text{cgs}}^2}, \tag{2.17d}$$

where M_{cgs} is the numeric part of the needle's mass in cgs units (i.e., in gm), and T_{cgs} is the numeric part of the needle's period of oscillation in cgs units (i.e., in sec), when it is suspended at the field point. Equations (2.17c, d) specify H_{esu} and

H_{emu} in terms of the already-known numerics M_{cgs}, L_{cgs}, l_{cgs}, T_{cgs}, and $(p_H)_{esu}$ or $(p_H)_{emu}$. We see that now $\underset{esu}{U}$, $\underset{emu}{U}$, $\underset{esu}{N}$, and $\underset{emu}{N}$ are well defined when applied to any magnetic field H.

When comparing Eqs. (2.14a) and (2.15b), we notice that

$$\underset{esu}{U}(Q) = \underset{emu}{U}(p_H), \tag{2.18a}$$

and from Eqs. (2.14b) and (2.15a) we see that

$$\underset{esu}{U}(p_H) = \underset{emu}{U}(Q). \tag{2.18b}$$

Similarly, Eqs. (2.16a), (2.17b), (2.16c), and (2.17a) show that

$$\underset{esu}{U}(E) = \underset{emu}{U}(H) \tag{2.18c}$$

and

$$\underset{esu}{U}(H) = \underset{emu}{U}(E). \tag{2.18d}$$

In these equations, the esu units for an electrical quantity (such as charge or electric field) are the same as the emu units for the corresponding magnetic quantity (such as magnetic pole strength or magnetic field), and the esu units for magnetic quantities are the same as the emu units for the corresponding electrical quantities. This symmetry comes from the basic similarity of the formulas used to elaborate nineteenth-century electromagnetic theory—Coulomb's laws for electric charges and magnetic poles, Eqs. (2.4) and (2.5) defining the electric and magnetic fields, and Eqs. (2.11a) through (2.12b) for ε_0 and μ_0 in the esu and emu systems. Table 2.2 lists the major electrical and magnetic physical quantities having the symmetry specified in Eqs. (2.18a–d); we see that this pattern does not end with the few basic electromagnetic physical quantities that have been defined so far.

We also see from this preliminary analysis that the numbers given by the operation of $\underset{esu}{N}$ and $\underset{emu}{N}$ on electromagnetic physical quantities can always be calculated from the values given by $\underset{esu}{N}$ and $\underset{emu}{N}$ operating on previously defined terms. In Eqs. (2.1a, b), Q is specified in terms of r, F, and ε_0, and in Eqs. (2.14f–i) this is used to define $\underset{esu}{N}(Q)$ and $\underset{emu}{N}(Q)$. In Eqs. (2.2a, b), p_H is specified in terms of r, F, and μ_0; and in Eqs. (2.15d–g) this is used to define $\underset{esu}{N}(p_H)$ and $\underset{emu}{N}(p_H)$. In Eq. (2.4), E is specified in terms of F and Q; and in Eqs. (2.16b, d) this is used to define $\underset{esu}{N}(E)$ and $\underset{emu}{N}(E)$. Even when the equation introducing a new electromagnetic quantity b_{new} cannot be used to define the operation of $\underset{esu}{N}$ and $\underset{emu}{N}$ on b_{new}, the self-consistency of electromagnetic theory eventually produces a way to calculate $\underset{esu}{N}(b_{new})$ and $\underset{emu}{N}(b_{new})$ from already known physical quantities. For example, the

Table 2.2 Symmetry in esu and emu systems for selected electric and magnetic physical quantities.

$\underset{esu}{U}(\varepsilon_0) = \underset{emu}{U}(\mu_0) = 1$	$\underset{esu}{U}(\mu_0) = \underset{emu}{U}(\varepsilon_0) = \dfrac{sec^2}{cm^2}$
$\underset{esu}{U}(E) = \underset{emu}{U}(H) = \dfrac{gm^{1/2}}{cm^{1/2} \cdot sec}$	$\underset{esu}{U}(H) = \underset{emu}{U}(E) = \dfrac{gm^{1/2} \cdot cm^{1/2}}{sec^2}$
$\underset{esu}{U}(D) = \underset{emu}{U}(B) = \dfrac{gm^{1/2}}{cm^{1/2} \cdot sec}$	$\underset{esu}{U}(B) = \underset{emu}{U}(D) = \dfrac{gm^{1/2}}{cm^{3/2}}$
$\underset{esu}{U}(Q) = \underset{emu}{U}(p_H) = \dfrac{gm^{1/2} \cdot cm^{3/2}}{sec}$	$\underset{esu}{U}(p_H) = \underset{emu}{U}(Q) = gm^{1/2} \cdot cm^{1/2}$
$\underset{esu}{U}(p) = \underset{emu}{U}(m_H) = \dfrac{gm^{1/2} \cdot cm^{5/2}}{sec}$	$\underset{esu}{U}(m_H) = \underset{emu}{U}(p) = gm^{1/2} \cdot cm^{3/2}$
$\underset{esu}{U}(P) = \underset{emu}{U}(M_H) = \dfrac{gm^{1/2}}{cm^{1/2} \cdot sec}$	$\underset{esu}{U}(M_H) = \underset{emu}{U}(P) = \dfrac{gm^{1/2}}{cm^{3/2}}$
$\underset{esu}{U}(V) = \underset{emu}{U}(\Omega_H) = \dfrac{gm^{1/2} \cdot cm^{1/2}}{sec}$	$\underset{esu}{U}(\Omega_H) = \underset{emu}{U}(V) = \dfrac{gm^{1/2} \cdot cm^{3/2}}{sec^2}$

magnetic field in Eq. (2.5) is calculated using Eqs. (2.17c, d) from Appendix 2.A. Because there must always be some way of measuring any newly introduced physical quantity b_{new}, we can always determine $\underset{esu}{N}(b_{new})$ and $\underset{emu}{N}(b_{new})$ from the operation of $\underset{esu}{N}$ and $\underset{emu}{N}$ on previously measured physical quantities. In this sense, there is no real need to examine any further the behavior of the $\underset{esu}{N}$ and $\underset{emu}{N}$ operators when they are applied to electromagnetic physical quantites, because they automatically define themselves as we elaborate electromagnetic theory using esu and emu units. The same thing, of course, could be said about the action of the $\underset{esu}{U}$ and $\underset{emu}{U}$ operators—that they automatically define themselves as we elaborate electromagnetic theory using esu or emu units. Our goal, however, is to learn how to transform an equation in esu units to the corresponding equation in emu units, or to transform an equation in emu units to the corresponding equation in esu units. Given an equation, we can always assume that all electromagnetic physical quantities have some sort of numeric components given by $\underset{esu}{N}$ and $\underset{emu}{N}$, but we need to write down the exact form of the units to transform equations from one system to the other. Consequently, it makes sense to continue analyzing the esu and emu units of electromagnetic physical quantities. From this point on, we assume the existence of well-defined $\underset{esu}{N}$ and $\underset{emu}{N}$ operators as we use nineteenth-century ideas—and the $\underset{esu}{U}$ and $\underset{emu}{U}$ operators—to find the units of standard electromagnetic physical quantities.

2.4 THE D AND B FIELDS

Nineteenth-century electromagnetic theory used two auxiliary fields, the D field and the B field, to describe the interaction of the electric and magnetic fields with bulk matter. To understand these fields we have to introduce the concepts of electric and magnetic dipole moments and dipole moment densities.

A bar magnet is defined to have a permanent-magnet dipole moment

$$\vec{m}_H = (|p_H| \cdot L)\hat{e}, \tag{2.19a}$$

where $|p_H|$ is the magnitude of the north and south poles' magnetic pole strength, L is the distance between the poles, and $\hat{e}$ is a dimensionless unit vector pointing from the negative pole to the positive pole. We insist on the somewhat awkward phrase "permanent-magnet dipole moment" because during the first half of the twentieth century another definition of magnetic dipole moment almost completely replaced the nineteenth-century one given in Eq. (2.19a). The definition of dipole moment in Eq. (2.19a) is a mathematical concept; it is defined for any pair of poles that are equal in magnitude, opposite in sign, and separated by a distance L. The next step is to shrink L while increasing $|p_H|$ in such a way that the product $|p_H| \cdot L$ remains constant. This leads to the definition of the permanent-magnet dipole moment at a point,

$$\vec{m}_H = \lim_{\substack{|p_H| \to \infty \\ L \to 0}} \left[(|p_H| \cdot L)\hat{e} \right]. \tag{2.19b}$$

The concept of a permanent-magnet dipole moment at a point can be used to define a density of permanent-magnet dipole moments. We say that the total permanent-magnet dipole moment inside an infinitesimal volume dV at any field point is

$$\vec{m}_H = dV \cdot \vec{M}_H, \tag{2.19c}$$

where $\vec{M}_H$ is the permanent-magnet dipole density field at that field point. In empty space, of course, $\vec{m}_H$ and $\vec{M}_H$ are always zero. The esu and emu units for $\vec{m}_H$ and $\vec{M}_H$ come directly from their definitions and are given in Table 2.2.

The dipole density field is a useful concept because permanent magnets can be modeled as a volume of space where $\vec{M}_H$ is not zero. Inside a magnet the positive poles of the point dipole moments lie next to the negative poles of neighboring point dipoles [see Figs. 2.5(a) and 2.5(b)], which means their individual fields tend to cancel when $\vec{M}_H$ is approximately constant. Only at the boundaries of a region of nonzero $\vec{M}_H$ are the positive or negative poles of the point dipole moments un-cancelled by the corresponding negative or positive poles of adjacent point dipoles, creating a strong positive or negative polarity. A bar magnet, for example, can be approximated as having a constant interior $\vec{M}_H$ field always pointing along the axis

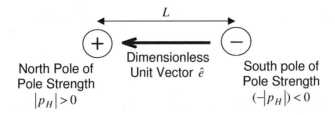

Figure 2.5a The permanent-magnet dipole moment at a point is defined to be $\vec{m}_H = [\lim_{\substack{|p_H| \to \infty \\ L \to 0}} (|p_H| \cdot L)]\hat{e}$.

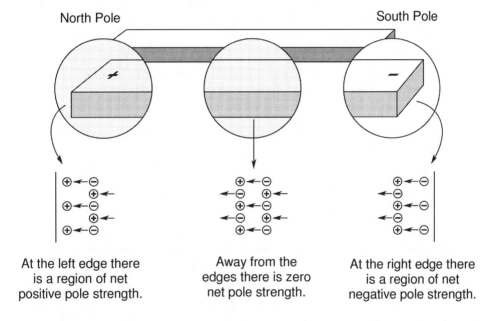

Figure 2.5b A bar magnet can be approximated as a bar-shaped volume of constant permanent-magnet dipole density.

of the bar [see Fig. 2.5(b)]. This not only accounts for the north and south poles at the ends of the bar, but it also explains why when a bar magnet is broken, both pieces always have north and south poles rather than just one isolated pole. The action of magnetic fields on unmagnetized iron or steel objects can be explained in terms of the magnetic field inducing (i.e., creating) a temporary dipole density field $\vec{M}_H$ in the unmagnetized iron or steel. When dealing with complicated systems of permanent and induced $\vec{M}_H$ fields, nineteenth-century physicists found it convenient to define a new type of magnetic field $\vec{B}$ called the magnetic induction :

$$\vec{B} = \mu_0 \vec{H} + 4\pi \vec{M}_H. \tag{2.19d}$$

Note that in emu units μ_0 is 1, so that Eq. (2.19d) specializes to

$$\vec{B} = \vec{H} + 4\pi \vec{M}_H, \tag{2.19e}$$

and using emu units in empty space gives $\vec{B} = \vec{H}$ because $\vec{M}_H = 0$.

One of the major attractions of $\vec{B}$ is that its divergence is always zero, no matter what the distribution of permanent and induced $\vec{M}_H$ fields happens to be. Readers familiar with advanced electromagnetic theory know that zero-divergence fields can be very useful when solving complicated electromagnetic problems. From a historical perspective, it is worth noting that in the nineteenth century the $\vec{H}$ field and its associated magnetic poles were regarded as the basic theoretical ideas, while the $\vec{B}$ field was treated as more of a useful auxiliary construct. As we shall see, in the twentieth century the $\vec{H}$ and $\vec{B}$ fields switched roles, with the $\vec{B}$ field taken to be fundamental and the $\vec{H}$ field treated as a useful auxiliary construct. Their mathematical definitions remained the same, however, which is why in empty space the $\vec{H}$ field is now defined, using Eq. (2.19d) with $\vec{M}_H$ equal to 0, as

$$\vec{H} = \frac{1}{\mu_0} \vec{B} = \vec{B} \text{ divided by a constant}, \tag{2.19f}$$

instead of the more straightforward definition

$$\vec{H} = \vec{B} \text{ multiplied by a constant}.$$

Both electric dipole moments and magnetic dipole moments rely on the same basic mathematical idea; hence, we can define an electric dipole moment to be

$$\vec{p} = (|Q| \cdot L)\hat{e}, \tag{2.20a}$$

where $|Q|$ is the magnitude of equal positive and negative point charges separated by a distance L, and $\hat{e}$ is a dimensionless unit vector pointing from the negative charge to the positive charge. By shrinking L while increasing $|Q|$ in such a way as to keep the $|Q| \cdot L$ product constant, we define the electric dipole moment at a point to be

$$\vec{p} = \lim_{\substack{|Q| \to \infty \\ L \to 0}} \left[(|Q| \cdot L)\hat{e} \right]. \tag{2.20b}$$

Again following the pattern of the magnetic-dipole discussion, we define a density field of point electric dipoles such that the electric dipole moment inside an infinitesimal volume dV at any field point is

$$\vec{p} = dV \cdot \vec{P}, \tag{2.20c}$$

where $\vec{P}$ is the electric dipole density field at that field point. Both $\vec{p}$ and $\vec{P}$ are zero in empty space. The esu and emu units for $\vec{p}$ and $\vec{P}$ are given in Table 2.2.

The $\vec{P}$ field, like the $\vec{M}_H$ field, is used to describe the interaction of matter with electric fields. There do exist substances called electrets whose interiors can have self-maintaining nonzero $\vec{P}$ fields analogous to the self-maintaining $\vec{M}_H$ fields of permanent magnets. The interiors of most substances, however, can only have $\vec{P}$ fields when under the influence of outside $\vec{E}$ fields. We say these $\vec{P}$ fields are induced by the $\vec{E}$ fields.

When dealing with complicated systems of $\vec{P}$ and $\vec{E}$ fields, it is often useful to divide the charge in the system into bound charge and unbound charge. The unbound charge can be moved from one object to another using either mechanical forces or $\vec{E}$ fields, but the bound charge only changes its position by microscopic amounts under the influence of mechanical forces or $\vec{E}$ fields. Nineteenth-century electrical scientists were primarily interested in the behavior of the unbound charge, so they found it convenient to define an auxiliary field called the electric displacement:

$$\vec{D} = \varepsilon_0 \vec{E} + 4\pi \vec{P}, \tag{2.20d}$$

which has the useful property that its divergence at any field point is proportional to the density of unbound charge at that field point. In esu units ε_0 is 1, so Eq. (2.20d) becomes

$$\vec{D} = \vec{E} + 4\pi \vec{P}; \tag{2.20e}$$

and in empty space where $\vec{P}$ is zero we have, using esu units, $\vec{D} = \vec{E}$. During the twentieth century, as physicists learned more about the electronic structure of matter, the distinction between bound and unbound charge became less useful, making the $\vec{D}$ field seem less real—helpful in some types of problems, no doubt, but basically just a mathematical construct. Although the $\vec{B}$ field has gained status since the nineteenth century while the $\vec{D}$ field has—if anything—lost status, Table 2.2 shows their units in the esu and emu systems still have the same nineteenth-century symmetry as the electromagnetic quantities Q, p_H and $\vec{E}$, $\vec{H}$:

$$\underset{\text{esu}}{\text{U}}(D) = \underset{\text{emu}}{\text{U}}(B),$$

$$\underset{\text{esu}}{\text{U}}(B) = \underset{\text{emu}}{\text{U}}(D).$$

2.5 THE ELECTRIC AND MAGNETIC POTENTIALS

Although electromagnetic theory deals primarily with point charges, dipoles, magnetic and electric fields, in the laboratory even the simplest experiments depend heavily on batteries, circuits, electric-current meters, etc. This was also the case in the nineteenth century, and as engineers and scientists became more familiar with electric circuits they settled on two physical quantities as being of fundamental importance for everyday work. One, the electric current I, has already been discussed in Eqs. (2.6) and (2.7), but we have not yet talked about the other—the electrical potential, or voltage, V. When charge travels around a circuit, it loses potential energy. The potential energy difference per unit charge between points a and a' in the circuit is defined to be

$$V_{aa'} = -\int_a^{a'} \vec{E} \cdot d\vec{s}. \tag{2.21a}$$

where the line integral between a and a' is taken along the circuit connecting the two points. In general we can define for any static $\vec{E}$ field—one that does not change with time—a potential field V throughout the laboratory using the differential relation

$$\vec{E} = -\vec{\nabla}V. \tag{2.21b}$$

Equation (2.21b) only defines the potential field up to an arbitrary additive constant, of course (see Appendix 2.B for a quick review of the properties of the vector differential operator $\vec{\nabla}$). Applying $\underset{\text{emu}}{U}$ and $\underset{\text{esu}}{U}$ to Eqs. (2.21a, b), we find that, using Table 2.2,

$$\underset{\text{esu}}{U}(V) = \text{cm} \cdot \underset{\text{esu}}{U}(\vec{E}) = \frac{\text{gm}^{1/2} \cdot \text{cm}^{1/2}}{\text{sec}} \tag{2.22a}$$

and

$$\underset{\text{emu}}{U}(V) = \text{cm} \cdot \underset{\text{emu}}{U}(\vec{E}) = \frac{\text{gm}^{1/2} \cdot \text{cm}^{3/2}}{\text{sec}^2}. \tag{2.22b}$$

A magnetic-potential field Ω_H can be defined so that for many types of static magnetic fields over large regions of space,

$$\vec{H} = -\vec{\nabla}\Omega_H. \tag{2.23}$$

As a general rule, Ω_H is not as useful a concept as V because there are static magnetic fields, such as the circular fields around long, straight wires carrying constant currents (see Fig. 2.4) that cannot be described at every point around the wire by the same Ω_H field. The magnetic fields of point magnetic dipoles and of permanent magnets outside the magnets' interiors can, however, be described quite adequately in terms of a magnetic potential Ω_H. Table 2.2 shows that V and Ω_H have the expected symmetry in esu and emu units, with

$$\underset{\text{esu}}{U}(V) = \underset{\text{emu}}{U}(\Omega_H)$$

and

$$\underset{\text{esu}}{\text{U}}(\Omega_H) = \underset{\text{emu}}{\text{U}}(V).$$

2.6 THE SYSTEM OF PRACTICAL UNITS

Because today's textbooks almost always begin with electrostatics, they leave the impression that it must be more natural to use electrostatic units; but, in fact, electromagnetic units were more popular during the nineteenth century. Not only were electric currents measured with instruments based on the interaction of current-carrying coils of wire with small permanent magnets, but also nineteenth-century cartography and navigation demanded an ever more exact knowledge of the earth's magnetic field. When, however, emu units for electric current and potential were used in typical nineteenth-century circuits, it quickly became clear that they have inconvenient sizes—something that is definitely more than a minor drawback when all calculations must be done by hand. Therefore, toward the end of the nineteenth century a new system of units was proposed, called the practical system of units, which was based on the emu units and designed for use with electric circuits.

The ampere was the name given to the practical unit of I, the electric current.[*] It was defined to be one tenth the size of the emu unit of current.
We have, according to Eq. (2.7)

$$1\,\text{amp} = 1/10\,(1\,\text{emu unit of current}) = (1/10)\,\underset{\text{emu}}{\text{U}}(dQ)/\underset{\text{emu}}{\text{U}}(dt)$$

$$= (1/10)\,\underset{\text{emu}}{\text{U}}(Q)/\underset{\text{emu}}{\text{U}}(t) = (1/10)(\text{gm}^{1/2}\cdot\text{cm}^{1/2})/\text{sec}. \qquad (2.24a)$$

The volt was the name given to the practical unit of electric potential.[†] It was defined to be 10^8 times larger than the emu unit of potential:

$$1\,\text{volt} = 10^8\,(1\,\text{emu unit of potential}) = 10^8\,\frac{\text{gm}^{1/2}\cdot\text{cm}^{3/2}}{\text{sec}^2}. \qquad (2.24b)$$

The terms "1 emu unit of current" or "1 emu unit of potential" are, by the way, entirely conventional—it is always acceptable to specify electromagnetic quantities by the system of units used to measure them. This turn of phrase can be applied to any electromagnetic system, not just the esu or emu systems; it is, in fact, most likely to be encountered when reading articles about uncommon systems of electromagnetic units where there are no widely accepted names for the individual units of current, potential, etc.

When talking about practical units, we descend from the complete theory of electromagnetism to the more restricted domain of circuit theory. The equations of

---

[*] This unit was named in honor of Andre Marie Ampere (1775–1836).

[†] This unit was named in honor of Alessandro Giuseppe Antonio Anastasio Volta (1745–1827).

circuit theory have the same form no matter what units are used. The equations of elementary circuit theory assume that all circuits are composed of separate circuit elements. Each circuit element has two terminals and its behavior is characterized by both the electric potential between the terminals and the current flowing through the circuit element (see Fig. 2.6 for an example of how this works). To discuss the practical system of units, we create U and N operators called $\underset{\mathrm{prac}}{\mathrm{U}}$ and $\underset{\mathrm{prac}}{\mathrm{N}}$ that can be used only with circuit-theory equations. Having already defined the practical units for current I and potential V in Eqs. (2.24a, b), we must have

$$\underset{\mathrm{prac}}{\mathrm{U}}(I) = \mathrm{amp}, \tag{2.24c}$$

$$\underset{\mathrm{prac}}{\mathrm{U}}(V) = \mathrm{volt}; \tag{2.24d}$$

and from Rule I,

$$\underset{\mathrm{prac}}{\mathrm{N}}(I) = 10 \cdot \underset{\mathrm{emu}}{\mathrm{N}}(I) = I_{[\mathrm{amp}]}, \tag{2.24e}$$

$$\underset{\mathrm{prac}}{\mathrm{N}}(V) = 10^{-8} \cdot \underset{\mathrm{emu}}{\mathrm{N}}(V) = V_{[\mathrm{volt}]}. \tag{2.24f}$$

If we also define for any time t that

$$\underset{\mathrm{prac}}{\mathrm{U}}(t) = \sec \quad \text{and} \quad \underset{\mathrm{prac}}{\mathrm{N}}(t) = t_{[\sec]}, \tag{2.24g}$$

then we have all that is needed to set up the complete practical system of electromagnetic units. Any physical quantity $b_{CIRCUIT}$ defined in circuit theory must, from the self-consistency of the theory, be measurable in some way using volt, amp, and sec. We define operator $\underset{\mathrm{prac}}{\mathrm{N}}$ by saying that the number coming from the measurement of $b_{CIRCUIT}$ must be $\underset{\mathrm{prac}}{\mathrm{N}}(b_{CIRCUIT})$. As for operator $\underset{\mathrm{prac}}{\mathrm{U}}$, the equations of circuit theory are themselves straightforward, making the units of the practical system easy to define.

Like most textbooks on elementary circuit theory, we start with Ohm's law,

$$V = I R. \tag{2.25a}$$

In this equation, V is the electric potential across a circuit element, with I the electric current flowing through the circuit element, and R its resistance. There are specialized circuit elements called resistors designed to provide a specified value of R across their terminals. The practical unit called the ohm is used to measure resistance:*

$$\underset{\mathrm{prac}}{\mathrm{U}}(R) = \mathrm{ohm}.$$

---

* This unit was named in honor of Georg Simon Ohm (1787–1854).

Circuit Elements

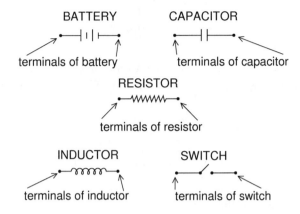

An Example of a Simple Circuit Network

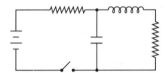

Figure 2.6 Wires are used to connect the terminals of circuit elements, forming circuit networks.

Applying $\underset{\text{prac}}{U}$ to both sides of Eq. (2.25a) gives

$$\underset{\text{prac}}{U}(V) = \underset{\text{prac}}{U}(I) \cdot \underset{\text{prac}}{U}(R)$$

so that, using Eqs. (2.24c, d),

$$\text{volt} = \text{amp} \cdot \text{ohm}$$

or

$$\text{ohm} = \frac{\text{volt}}{\text{amp}}. \qquad (2.25b)$$

Equation (2.25b) can be used as the definition of ohm. When Eq. (2.25a) is written as

$$I = \frac{V}{R} = GV,$$

the conductance $G = R^{-1}$ has units of ohm^{-1}, often called the mho:

$$1 \text{ mho} = \text{ohm}^{-1}. \tag{2.25c}$$

We have already seen in Eq. (2.7) that the current is dQ/dt, the time rate of change of the electric charge. Applying $\underset{\text{prac}}{\text{U}}$ to both sides of Eq. (2.7) gives

$$\underset{\text{prac}}{\text{U}}(I) = \frac{\underset{\text{prac}}{\text{U}}(dQ)}{\underset{\text{prac}}{\text{U}}(dt)} = \frac{\underset{\text{prac}}{\text{U}}(Q)}{\underset{\text{prac}}{\text{U}}(t)}. \tag{2.26a}$$

We have from Eq. (2.24g) that

$$\underset{\text{prac}}{\text{U}}(t) = \text{sec}, \tag{2.26b}$$

which means that the unit of charge, called the coulomb, in the practical system of units must be*

$$\underset{\text{prac}}{\text{U}}(Q) = \underset{\text{prac}}{\text{U}}(I) \cdot \text{sec} = \text{amp} \cdot \text{sec}.$$

Using coul for the abbreviation of coulomb, we define

$$1 \text{ coul} = 1 \text{ amp} \cdot \text{sec}. \tag{2.26c}$$

Charge can be stored on circuit elements called capacitors. Indeed, calculating the capacitance of metallic objects with simple geometric shapes is a standard textbook exercise, and nonideal circuit elements of all types may have measurable amounts of capacitance. A capacitor with an electric potential V across its terminals stores an amount of charge

$$Q = CV, \tag{2.27a}$$

where C is the constant capacitance of the capacitor. Applying $\underset{\text{prac}}{\text{U}}$ to Eq. (2.27a) we find that

$$\underset{\text{prac}}{\text{U}}(C) = \frac{\underset{\text{prac}}{\text{U}}(Q)}{\underset{\text{prac}}{\text{U}}(V)} = \frac{\text{coul}}{\text{volt}}.$$

---

* This unit is named in honor of Charles Augustine Coulomb (1736–1806). He also gave his name to Coulomb's law for electric charges and magnetic poles.

We now define $\underset{\text{prac}}{U}(C) = \text{farad}$ to be the practical unit of capacitance.[*]

$$1 \text{ farad} = \frac{\text{coul}}{\text{volt}}. \qquad (2.27b)$$

When Eq. (2.27a) is written as

$$V = C^{-1}Q = SQ,$$

the elastance $S = C^{-1}$ is measured in units of farad^{-1}. Another name for farad^{-1}, following the pattern of ohm^{-1} = mho, was daraf; but this usage had fallen out of favor by the second half of the twentieth century. The last type of circuit element given a separate practical unit is the inductor. When the current through an inductor has a time rate of change dI/dt, the potential across the inductor is

$$V = L\frac{dI}{dt}, \qquad (2.28a)$$

where L is the inductor's coefficient of induction, or just its induction for short. We see that

$$\underset{\text{prac}}{U}(V) = \underset{\text{prac}}{U}(L) \cdot \frac{\underset{\text{prac}}{U}(I)}{\underset{\text{prac}}{U}(t)},$$

or

$$\underset{\text{prac}}{U}(L) = \frac{\text{volt} \cdot \text{sec}}{\text{amp}}. \qquad (2.28b)$$

Equation (2.28b) defines the henry, the practical unit of induction:[†]

$$1 \text{ henry} = \frac{\text{volt} \cdot \text{sec}}{\text{amp}}. \qquad (2.28c)$$

A later addition to the system of practical units is the weber, useful in measuring magnetic quantities:[‡]

$$1 \text{ weber} = \text{volt} \cdot \text{sec}. \qquad (2.29)$$

[*] This unit is named in honor of Michael Faraday (1791–1867). The practical unit of current was sometimes called the farad before an international commission gave it the name ampere in 1881.
[†] This unit is named in honor of Joseph Henry (1797–1878).
[‡] This unit is named in honor of Wilhelm Eduard Weber (1795–1878). The practical units of current and charge were sometimes called the weber before an international commission gave them the names ampere and coulomb in 1881.

2.7 THE "AB-" AND "STAT-" PREFIXES

The practical units became so popular that by the beginning of the twentieth century their names were borrowed to describe the units of the esu and emu systems. This practice never really caught on in Europe but became nearly universal in the United States. Units in the esu system were given the same name as the corresponding practical unit preceded by the prefix "stat", for electro-*stat*-ic system of units. The unit of charge in esu units became the statcoulomb, which we abbreviate as statcoul,

$$1 \text{ statcoul} = \frac{\text{gm}^{1/2} \cdot \text{cm}^{3/2}}{\text{sec}}; \tag{2.30a}$$

the unit of current in the esu system became the statampere (the statamp for short),

$$1 \text{ statamp} = \frac{\text{gm}^{1/2} \cdot \text{cm}^{3/2}}{\text{sec}^2}; \tag{2.30b}$$

the unit of potential in the esu system became the statvolt,

$$1 \text{ statvolt} = \frac{\text{gm}^{1/2} \cdot \text{cm}^{1/2}}{\text{sec}}; \tag{2.30c}$$

and so on. The emu system of units was called the absolute system, and emu units were given the same name as the practical unit preceded by the prefix "ab" for *ab*-solute. In the absolute system of units—that is, in the emu system—the unit of charge became the abcoulomb (which we abbreviate as abcoul)

$$1 \text{ abcoul} = \text{gm}^{1/2} \cdot \text{cm}^{1/2}, \tag{2.31a}$$

the unit of current became the abampere (which we abbreviate as abamp)

$$1 \text{ abamp} = \frac{\text{gm}^{1/2} \cdot \text{cm}^{1/2}}{\text{sec}^2}; \tag{2.31b}$$

the unit of electric potential became the abvolt:

$$1 \text{ abvolt} = \frac{\text{gm}^{1/2} \cdot \text{cm}^{3/2}}{\text{sec}^2}; \tag{2.31c}$$

and so on. Table 2.3 gives the definitions of the esu and emu units corresponding to the practical units, and Tables 2.4 and 2.5 give the unit names of common electromagnetic physical quantities following this convention.* In Table 2.5 for

---

* The third columns of Tables 2.4 and 2.5 will become clear when we discuss the esuq and emuq systems of units.

Table 2.3 Names of esu and emu units using "stat" and "ab" prefixes.

esu units	emu units
$1 \text{ statcoul} = \dfrac{\text{gm}^{1/2} \cdot \text{cm}^{3/2}}{\text{sec}}$	$1 \text{ abcoul} = \text{gm}^{1/2} \cdot \text{cm}^{1/2}$
$1 \text{ statamp} = \dfrac{\text{gm}^{1/2} \cdot \text{cm}^{3/2}}{\text{sec}^2}$	$1 \text{ abamp} = \dfrac{\text{gm}^{1/2} \cdot \text{cm}^{1/2}}{\text{sec}}$
$1 \text{ statvolt} = \dfrac{\text{gm}^{1/2} \cdot \text{cm}^{1/2}}{\text{sec}}$	$1 \text{ abvolt} = \dfrac{\text{gm}^{1/2} \cdot \text{cm}^{3/2}}{\text{sec}^2}$
$1 \text{ statfarad} = \text{cm}$	$1 \text{ abfarad} = \dfrac{\text{sec}^2}{\text{cm}}$
$1 \text{ stathenry} = \dfrac{\text{sec}^2}{\text{cm}}$	$1 \text{ abhenry} = \text{cm}$
$1 \text{ statohm} = \dfrac{\text{sec}}{\text{cm}}$	$1 \text{ abohm} = \dfrac{\text{cm}}{\text{sec}}$
$1 \text{ statweber} = \text{gm}^{1/2} \cdot \text{cm}^{1/2}$	$1 \text{ abweber} = \dfrac{\text{gm}^{1/2} \cdot \text{cm}^{3/2}}{\text{sec}}$

the emu units, several of the magnetic quantities (for example B, $\mathcal{F}$, Φ_B, and H) have names that do not follow the ab-prefix convention (their unit names are gauss, gilbert, maxwell, and oersted, respectively*). These magnetic quantities were more often referred to by their non-ab names; the ab prefix versions, however, have been listed because they are helpful when converting equations to other systems of units.

Textbooks using the ab and stat notation for emu and esu units almost always have equations like

$$1 \text{ statvolt} \overset{?}{=} 3 \times 10^{10} \text{ abvolt} \tag{2.32a}$$

or

$$1 \text{ statcoul} \overset{?}{=} \frac{1}{3 \times 10^{10}} \text{ abcoul,} \tag{2.32b}$$

to describe the conversion of physical quantities from one system of units to the other. There are usually footnotes stating that all factors of 3×10^{10} are just an approximation for 2.998×10^{10}, the numeric part of the speed of light in cm/sec. A quick glance at Eqs. (2.30a) and (2.31a) or Eqs. (2.30c) and (2.31c) shows what is wrong with this convention from our point of view; if, for example, we accept

---

* These units were named in honor of Karl Friedrich Gauss (1777–1855), William Gilbert (1544–1603), James Clerk Maxwell (1831–1879), and Hans Christian Oersted (1777–1851).

Table 2.4 Electromagnetic physical quantities in esu and esuq units.

Physical quantity	esu units	esuq units
A (magnetic vector potential)	$\underset{esu}{U}(A) = \dfrac{gm^{1/2}}{cm^{1/2}} = \dfrac{statweber}{cm}$	$\underset{esuq}{U}(A) = \dfrac{statweberq}{cm}$
B (magnetic induction)	$\underset{esu}{U}(B) = \dfrac{gm^{1/2}}{cm^{3/2}} = \dfrac{statweber}{cm^2}$	$\underset{esuq}{U}(B) = \dfrac{statweberq}{cm^2}$
C (capacitance)	$\underset{esu}{U}(C) = cm = statfarad$	$\underset{esuq}{U}(C) = statfaradq$
D (electric displacement)	$\underset{esu}{U}(D) = \dfrac{gm^{1/2}}{cm^{1/2} \cdot sec}$ $= \dfrac{statcoul}{cm^2}$	$\underset{esuq}{U}(D) = \dfrac{statcoulq}{cm^2}$
E (electric field)	$\underset{esu}{U}(E) = \dfrac{gm^{1/2}}{cm^{1/2} \cdot sec}$ $= \dfrac{statvolt}{cm}$	$\underset{esuq}{U}(E) = \dfrac{statvoltq}{cm}$
ε (dielectric constant)	$\underset{esu}{U}(\varepsilon) = 1 = \dfrac{statfarad}{cm}$	$\underset{esuq}{U}(\varepsilon) = \dfrac{statfaradq}{cm}$
ε_0 (permittivity of free space)	$\underset{esu}{U}(\varepsilon_0) = 1 = \dfrac{statfarad}{cm}$	$\underset{esuq}{U}(\varepsilon_0) = \dfrac{statfaradq}{cm}$
$\mathcal{F}$ (magnetomotive force)	$\underset{esu}{U}(\mathcal{F}) = \dfrac{gm^{1/2} \cdot cm^{3/2}}{sec^2}$ $= statamp$	$\underset{esuq}{U}(\mathcal{F}) = statampq$
Φ_B (magnetic flux)	$\underset{esu}{U}(\Phi_B) = gm^{1/2} \cdot cm^{1/2}$ $= statweber$	$\underset{esuq}{U}(\Phi_B) = statweberq$
G (conductance)	$\underset{esu}{U}(G) = \dfrac{cm}{sec} = statohm^{-1}$	$\underset{esuq}{U}(G) = statohmq^{-1}$
H (magnetic field)	$\underset{esu}{U}(H) = \dfrac{gm^{1/2} \cdot cm^{1/2}}{sec^2}$ $= \dfrac{statamp}{cm}$	$\underset{esuq}{U}(H) = \dfrac{statampq}{cm}$
I (current)	$\underset{esu}{U}(I) = \dfrac{gm^{1/2} \cdot cm^{3/2}}{sec^2}$ $= statamp$	$\underset{esuq}{U}(I) = statampq$
J (volume current density)	$\underset{esu}{U}(J) = \dfrac{gm^{1/2}}{cm^{1/2} \cdot sec^2}$ $= \dfrac{statcoul}{cm^2 \cdot sec}$	$\underset{esuq}{U}(J) = \dfrac{statcoulq}{cm^2 \cdot sec}$

Table 2.4 (Continued).

Physical quantity	esu units	esuq units
$\mathcal{J}_S$ (surface current density)	$\underset{\text{esu}}{\text{U}}(\mathcal{J}_S) = \dfrac{\text{gm}^{1/2} \cdot \text{cm}^{1/2}}{\text{sec}^2}$ $= \dfrac{\text{statcoul}}{\text{cm} \cdot \text{sec}}$	$\underset{\text{esuq}}{\text{U}}(\mathcal{J}_S) = \dfrac{\text{statcoulq}}{\text{cm} \cdot \text{sec}}$
L (inductance)	$\underset{\text{esu}}{\text{U}}(L) = \dfrac{\text{sec}^2}{\text{cm}} = \text{stathenry}$	$\underset{\text{esuq}}{\text{U}}(L) = \text{stathenryq}$
m_H (permanent-magnet dipole moment)	$\underset{\text{esu}}{\text{U}}(m_H) = \text{gm}^{1/2} \cdot \text{cm}^{3/2}$ $= \text{statweber} \cdot \text{cm}$	$\underset{\text{esuq}}{\text{U}}(m_H) = \text{statweberq} \cdot \text{cm}$
m_I (current-loop magnetic dipole moment)	$\underset{\text{esu}}{\text{U}}(m_I) = \dfrac{\text{gm}^{1/2} \cdot \text{cm}^{7/2}}{\text{sec}^2}$ $= \text{statamp} \cdot \text{cm}^2$	$\underset{\text{esuq}}{\text{U}}(m_I) = \text{statampq} \cdot \text{cm}^2$
M_H (permanent-magnet dipole density)	$\underset{\text{esu}}{\text{U}}(M_H) = \dfrac{\text{gm}^{1/2}}{\text{cm}^{3/2}} = \dfrac{\text{statweber}}{\text{cm}^2}$	$\underset{\text{esuq}}{\text{U}}(M_H) = \dfrac{\text{statweberq}}{\text{cm}^2}$
M_I (current-loop magnetic dipole density)	$\underset{\text{esu}}{\text{U}}(M_I) = \dfrac{\text{gm}^{1/2} \cdot \text{cm}^{1/2}}{\text{sec}^2}$ $= \dfrac{\text{statamp}}{\text{cm}}$	$\underset{\text{esuq}}{\text{U}}(M_I) = \dfrac{\text{statampq}}{\text{cm}}$
μ (magnetic permeability)	$\underset{\text{esu}}{\text{U}}(\mu) = \dfrac{\text{sec}^2}{\text{cm}^2} = \dfrac{\text{stathenry}}{\text{cm}}$	$\underset{\text{esuq}}{\text{U}}(\mu) = \dfrac{\text{stathenryq}}{\text{cm}}$
μ_0 (magnetic permeability of free space)	$\underset{\text{esu}}{\text{U}}(\mu_0) = \dfrac{\text{sec}^2}{\text{cm}^2} = \dfrac{\text{stathenry}}{\text{cm}}$	$\underset{\text{esuq}}{\text{U}}(\mu_0) = \dfrac{\text{stathenryq}}{\text{cm}}$
p_H (magnetic pole strength)	$\underset{\text{esu}}{\text{U}}(p_H) = \text{gm}^{1/2} \cdot \text{cm}^{1/2}$ $= \text{statweber}$	$\underset{\text{esuq}}{\text{U}}(p_H) = \text{statweberq}$
p (electric dipole moment)	$\underset{\text{esu}}{\text{U}}(p) = \dfrac{\text{gm}^{1/2} \cdot \text{cm}^{5/2}}{\text{sec}}$ $= \text{statcoul} \cdot \text{cm}$	$\underset{\text{esuq}}{\text{U}}(p) = \text{statcoulq} \cdot \text{cm}$
P (electric dipole density)	$\underset{\text{esu}}{\text{U}}(P) = \dfrac{\text{gm}^{1/2}}{\text{cm}^{1/2} \cdot \text{sec}} = \dfrac{\text{statcoul}}{\text{cm}^2}$	$\underset{\text{esuq}}{\text{U}}(P) = \dfrac{\text{statcoulq}}{\text{cm}^2}$
$\mathcal{P}$ (permeance)	$\underset{\text{esu}}{\text{U}}(\mathcal{P}) = \dfrac{\text{sec}^2}{\text{cm}} = \dfrac{\text{statweber}}{\text{statamp}}$	$\underset{\text{esuq}}{\text{U}}(\mathcal{P}) = \dfrac{\text{statweberq}}{\text{statampq}}$
Q (charge)	$\underset{\text{esu}}{\text{U}}(Q) = \dfrac{\text{gm}^{1/2} \cdot \text{cm}^{3/2}}{\text{sec}}$ $= \text{statcoul}$	$\underset{\text{esuq}}{\text{U}}(Q) = \text{statcoulq}$
R (resistance)	$\underset{\text{esu}}{\text{U}}(R) = \dfrac{\text{sec}}{\text{cm}} = \text{statohm}$	$\underset{\text{esuq}}{\text{U}}(R) = \text{statohmq}$

Table 2.4 (Continued).

Physical quantity	esu units	esuq units
$\mathcal{R}$ (reluctance)	$\underset{esu}{U}(\mathcal{R}) = \dfrac{cm}{sec^2} = \dfrac{statamp}{statweber}$	$\underset{esuq}{U}(\mathcal{R}) = \dfrac{statampq}{statweberq}$
ρ_Q (volume charge density)	$\underset{esu}{U}(\rho_Q) = \dfrac{gm^{1/2}}{cm^{3/2} \cdot sec}$ $= \dfrac{statcoul}{cm^3}$	$\underset{esuq}{U}(\rho_Q) = \dfrac{statcoulq}{cm^3}$
ρ_R (resistivity)	$\underset{esu}{U}(\rho_R) = sec = statohm \cdot cm$	$\underset{esuq}{U}(\rho_R) = statohmq \cdot cm$
S (elastance)	$\underset{esu}{U}(S) = cm^{-1} = statfarad^{-1}$	$\underset{esuq}{U}(S) = statfaradq^{-1}$
S_Q (surface charge density)	$\underset{esu}{U}(S_Q) = \dfrac{gm^{1/2}}{cm^{1/2} \cdot sec}$ $= \dfrac{statcoul}{cm^2}$	$\underset{esuq}{U}(S_Q) = \dfrac{statcoulq}{cm^2}$
σ (conductivity)	$\underset{esu}{U}(\sigma) = sec^{-1} = \dfrac{statohm^{-1}}{cm}$	$\underset{esuq}{U}(\sigma) = \dfrac{statohmq^{-1}}{cm}$
V (electric potential)	$\underset{esu}{U}(V) = \dfrac{gm^{1/2} \cdot cm^{1/2}}{sec}$ $= statvolt$	$\underset{esuq}{U}(V) = statvoltq$
Ω_H (magnetic scalar potential)	$\underset{esu}{U}(\Omega_H) = \dfrac{gm^{1/2} \cdot cm^{3/2}}{sec^2}$ $= statamp$	$\underset{esuq}{U}(\Omega_H) = statampq$

Eqs. (2.30c), (2.31c), and (2.32a) as all true, it follows that

$$gm^{1/2} \cdot cm^{1/2} \overset{?}{=} (3 \times 10^{10}) \frac{gm^{1/2} \cdot cm^{3/2}}{sec}$$

or

$$1 \, sec \overset{?}{=} 3 \times 10^{10} \, cm. \tag{2.32c}$$

A statement like this is only true when time and space have the same dimension and the speed of light is equal to 1 (see Section 1.10).* It is safe to say, however, that the idea of giving time and space the same dimension was foreign to these textbook writers; they are just injudiciously applying Rule I while ignoring Rule II. What

---

* In fact, we can predict that esu and emu units will become identical when working in a system of units where the speed of light has the dimensionless value 1. Equations (2.11a, b) and (2.12a, b) show that $\varepsilon_0 = \mu_0 = 1$ automatically when $c = 1$, which means that esu and emu units will be the same.

Table 2.5 Electromagnetic physical quantities in emu and emuq units.

Physical quantity	emu units	emuq units
A (magnetic vector potential)	$\mathrm{U}_{emu}(A) = \dfrac{gm^{1/2} \cdot cm^{1/2}}{sec} = \dfrac{abweber}{cm}$	$\mathrm{U}_{emuq}(A) = \dfrac{abweberq}{cm}$
B (magnetic induction)	$\mathrm{U}_{emu}(B) = \dfrac{gm^{1/2}}{sec \cdot cm^{1/2}} = gauss = \dfrac{abweber}{cm^2}$	$\mathrm{U}_{emuq}(B) = \dfrac{abweberq}{cm^2}$
C (capacitance)	$\mathrm{U}_{emu}(C) = \dfrac{sec^2}{cm} = abfarad$	$\mathrm{U}_{emuq}(C) = abfaradq$
D (electric displacement)	$\mathrm{U}_{emu}(D) = \dfrac{gm^{1/2}}{cm^{3/2}} = \dfrac{abcoul}{cm^2}$	$\mathrm{U}_{emuq}(D) = \dfrac{abcoulq}{cm^2}$
E (electric field)	$\mathrm{U}_{emu}(E) = \dfrac{gm^{1/2} \cdot cm^{1/2}}{sec^2} = \dfrac{abvolt}{cm}$	$\mathrm{U}_{emuq}(E) = \dfrac{abvoltq}{cm}$
ε (dielectric constant)	$\mathrm{U}_{emu}(\varepsilon) = \dfrac{sec^2}{cm^2} = \dfrac{abfarad}{cm}$	$\mathrm{U}_{emuq}(\varepsilon) = \dfrac{abfaradq}{cm}$
ε_0 (permittivity of free space)	$\mathrm{U}_{emu}(\varepsilon_0) = \dfrac{sec^2}{cm^2} = \dfrac{abfarad}{cm}$	$\mathrm{U}_{emuq}(\varepsilon_0) = \dfrac{abfaradq}{cm}$
$\mathcal{F}$ (magnetomotive force)	$\mathrm{U}_{emu}(\mathcal{F}) = \dfrac{gm^{1/2} \cdot cm^{1/2}}{sec} = abamp = gilbert$	$\mathrm{U}_{emuq}(\mathcal{F}) = abampq$
Φ_B (magnetic flux)	$\mathrm{U}_{emu}(\Phi_B) = \dfrac{gm^{1/2} \cdot cm^{3/2}}{sec} = maxwell = abweber$	$\mathrm{U}_{emuq}(\Phi_B) = abweberq$
G (conductance)	$\mathrm{U}_{emu}(G) = \dfrac{sec}{cm} = abohm^{-1}$	$\mathrm{U}_{emuq}(G) = abohmq^{-1}$
H (magnetic field)	$\mathrm{U}_{emu}(H) = \dfrac{gm^{1/2}}{cm^{1/2} \cdot sec} = oersted = \dfrac{abamp}{cm}$	$\mathrm{U}_{emuq}(H) = statweberq$
I (current)	$\mathrm{U}_{emu}(I) = \dfrac{gm^{1/2} \cdot cm^{1/2}}{sec} = abamp$	$\mathrm{U}_{emuq}(I) = abweberq$
J (volume current density)	$\mathrm{U}_{emu}(J) = \dfrac{gm^{1/2}}{cm^{3/2} \cdot sec} = \dfrac{abcoul}{cm^2 \cdot sec}$	$\mathrm{U}_{emuq}(J) = \dfrac{abcoulq}{cm^2 \cdot sec}$
$\mathcal{J}_S$ (surface current density)	$\mathrm{U}_{emu}(\mathcal{J}_S) = \dfrac{gm^{1/2}}{cm^{1/2} \cdot sec} = \dfrac{abcoul}{cm \cdot sec}$	$\mathrm{U}_{emuq}(\mathcal{J}_S) = \dfrac{abcoulq}{cm \cdot sec}$

Table 2.5 (Continued).

Physical quantity	emu units	emuq units
L (inductance)	$\underset{\text{emu}}{\text{U}}\,(L) = \text{cm} = \text{abhenry}$	$\underset{\text{emuq}}{\text{U}}\,(L) = \text{abhenryq}$
m_H (permanent-magnet dipole moment)	$\underset{\text{emu}}{\text{U}}\,(m_H) = \dfrac{\text{gm}^{1/2} \cdot \text{cm}^{5/2}}{\text{sec}}$ $= \text{abweber} \cdot \text{cm}$	$\underset{\text{emuq}}{\text{U}}\,(m_H) = \text{abweberq} \cdot \text{cm}$
m_I (current-loop magnetic dipole moment)	$\underset{\text{emu}}{\text{U}}\,(m_I) = \dfrac{\text{gm}^{1/2} \cdot \text{cm}^{5/2}}{\text{sec}}$ $= \text{abamp} \cdot \text{cm}^2$	$\underset{\text{emuq}}{\text{U}}\,(m_I) = \text{abampq} \cdot \text{cm}^2$
M_H (permanent-magnet dipole density)	$\underset{\text{emu}}{\text{U}}\,(M_H) = \dfrac{\text{gm}^{1/2}}{\text{sec} \cdot \text{cm}^{1/2}}$ $= \dfrac{\text{abweber}}{\text{cm}^2}$	$\underset{\text{emuq}}{\text{U}}\,(M_H) = \dfrac{\text{abweberq}}{\text{cm}^2}$
M_I (current-loop magnetic dipole density)	$\underset{\text{emu}}{\text{U}}\,(M_I) = \dfrac{\text{gm}^{1/2}}{\text{sec} \cdot \text{cm}^{1/2}}$ $= \dfrac{\text{abamp}}{\text{cm}}$	$\underset{\text{emuq}}{\text{U}}\,(M_I) = \dfrac{\text{abampq}}{\text{cm}}$
μ (magnetic permeability)	$\underset{\text{emu}}{\text{U}}\,(\mu) = 1 = \dfrac{\text{abhenry}}{\text{cm}}$	$\underset{\text{emuq}}{\text{U}}\,(\mu) = \dfrac{\text{abhenryq}}{\text{cm}}$
μ_0 (magnetic permeability of free space)	$\underset{\text{emu}}{\text{U}}\,(\mu_0) = 1 = \dfrac{\text{abhenry}}{\text{cm}}$	$\underset{\text{emuq}}{\text{U}}\,(\mu_0) = \dfrac{\text{abhenryq}}{\text{cm}}$
p_H (magnetic pole strength)	$\underset{\text{emu}}{\text{U}}\,(p_H) = \dfrac{\text{gm}^{1/2} \cdot \text{cm}^{3/2}}{\text{sec}}$ $= \text{maxwell} = \text{abweber}$	$\underset{\text{emuq}}{\text{U}}\,(p_H) = \text{abweberq}$
p (electric dipole moment)	$\underset{\text{emu}}{\text{U}}\,(p) = \text{gm}^{1/2} \cdot \text{cm}^{3/2}$ $= \text{abcoul} \cdot \text{cm}$	$\underset{\text{emuq}}{\text{U}}\,(p) = \text{abcoulq} \cdot \text{cm}$
P (electric dipole density)	$\underset{\text{emu}}{\text{U}}\,(P) = \dfrac{\text{gm}^{1/2}}{\text{cm}^{3/2}} = \dfrac{\text{abcoul}}{\text{cm}^2}$	$\underset{\text{emuq}}{\text{U}}\,(P) = \dfrac{\text{abcoulq}}{\text{cm}^2}$
$\mathcal{P}$ (permeance)	$\underset{\text{emu}}{\text{U}}\,(\mathcal{P}) = \text{cm} = \dfrac{\text{maxwell}}{\text{gilbert}}$ $= \dfrac{\text{abweber}}{\text{abamp}}$	$\underset{\text{emuq}}{\text{U}}\,(\mathcal{P}) = \dfrac{\text{abweberq}}{\text{abampq}}$
Q (charge)	$\underset{\text{emu}}{\text{U}}\,(Q) = \text{gm}^{1/2} \cdot \text{cm}^{1/2}$ $= \text{abcoul}$	$\underset{\text{emuq}}{\text{U}}\,(Q) = \text{abcoulq}$
R (resistance)	$\underset{\text{emu}}{\text{U}}\,(R) = \dfrac{\text{cm}}{\text{sec}} = \text{abohm}$	$\underset{\text{emuq}}{\text{U}}\,(R) = \text{abohmq}$

Table 2.5 (Continued).

Physical quantity	emu units	emuq units
$\mathcal{R}$ (reluctance)	$\underset{\text{emu}}{\text{U}}(\mathcal{R}) = \text{cm}^{-1}$ $= \dfrac{\text{gilbert}}{\text{maxwell}} = \dfrac{\text{abamp}}{\text{abweber}}$	$\underset{\text{emuq}}{\text{U}}(\mathcal{R}) = \dfrac{\text{abampq}}{\text{abweberq}}$
ρ_Q (volume charge density)	$\underset{\text{emu}}{\text{U}}(\rho_Q) = \dfrac{\text{gm}^{1/2}}{\text{cm}^{5/2}} = \dfrac{\text{abcoul}}{\text{cm}^3}$	$\underset{\text{emuq}}{\text{U}}(\rho_Q) = \dfrac{\text{abcoulq}}{\text{cm}^3}$
ρ_R (resistivity)	$\underset{\text{emu}}{\text{U}}(\rho_R) = \dfrac{\text{cm}^2}{\text{sec}} = \text{abohm} \cdot \text{cm}$	$\underset{\text{emuq}}{\text{U}}(\rho_R) = \text{abohmq} \cdot \text{cm}$
S (elastance)	$\underset{\text{emu}}{\text{U}}(S) = \dfrac{\text{cm}}{\text{sec}^2} = \text{abfarad}^{-1}$	$\underset{\text{emuq}}{\text{U}}(S) = \text{abfaradq}^{-1}$
S_Q (surface charge density)	$\underset{\text{emu}}{\text{U}}(S_Q) = \dfrac{\text{gm}^{1/2}}{\text{cm}^{3/2}} = \dfrac{\text{abcoul}}{\text{cm}^2}$	$\underset{\text{emuq}}{\text{U}}(S_Q) = \dfrac{\text{abcoulq}}{\text{cm}^2}$
σ (conductivity)	$\underset{\text{emu}}{\text{U}}(\sigma) = \dfrac{\text{sec}}{\text{cm}^2} = \dfrac{\text{abohm}^{-1}}{\text{cm}}$	$\underset{\text{emuq}}{\text{U}}(\sigma) = \dfrac{\text{abohmq}^{-1}}{\text{cm}}$
V (electric potential)	$\underset{\text{emu}}{\text{U}}(V) = \dfrac{\text{gm}^{1/2} \cdot \text{cm}^{3/2}}{\text{sec}^2} = \text{abvolt}$	$\underset{\text{emuq}}{\text{U}}(V) = \text{abvoltq}$
Ω_H (magnetic scalar potential)	$\underset{\text{emu}}{\text{U}}(\Omega_H) = \dfrac{\text{gm}^{1/2} \cdot \text{cm}^{1/2}}{\text{sec}}$ $= \text{oersted} \cdot \text{cm} = \text{abamp}$	$\underset{\text{esuq}}{\text{U}}(\Omega_H) = \text{abampq}$

they mean to say is that for any potential V,

$$\underset{\text{esu}}{\text{N}}(V) = c_{\text{cgs}}^{-1} \underset{\text{emu}}{\text{N}}(V) \cong \frac{\underset{\text{emu}}{\text{N}}(V)}{3 \times 10^{10}}; \qquad (2.33a)$$

for any charge Q,

$$\underset{\text{esu}}{\text{N}}(Q) = c_{\text{cgs}} \underset{\text{emu}}{\text{N}}(Q) \cong 3 \times 10^{10} \underset{\text{emu}}{\text{N}}(Q); \qquad (2.33b)$$

and so on, for the conversion of any other electromagnetic physical quantity. Equations (2.33a, b) cannot violate any of the rules of Chapter (1) because they only involve numeric quantities such as $\underset{\text{esu}}{\text{N}}(Q)$, $\underset{\text{emu}}{\text{N}}(Q)$, $\underset{\text{esu}}{\text{N}}(V)$, $\underset{\text{emu}}{\text{N}}(V)$, and $c_{\text{cgs}} = c_{[\text{cm/sec}]}$, the numeric part of the speed of light in cgs units.

There is another sense, however, in which these textbooks anticipate an idea that became widespread—even predominant—during the twentieth century: attributing a new dimension to electromagnetic physical quantities. From this point of view, Eq. (2.32a) just recognizes that an electric potential is always an electric potential whether measured in volts, statvolts, or abvolts; and these three physical quantities must therefore have the same dimensions. This implies that, contrary to Table 2.3, one statvolt is not the same thing as $1 \text{ gm}^{1/2} \cdot \text{cm}^{1/2}/\text{sec}$, one abvolt is not the same thing as $\text{gm}^{1/2} \cdot \text{cm}^{3/2}/\text{sec}^2$, and one volt is not the same thing as $10^8 \text{gm}^{1/2} \cdot \text{cm}^{3/2}/\text{sec}^2$. Similarly, in Eq. (2.32b) electric charge is always electric charge, whether measured in coulombs, statcoulombs, or abcoulombs; and so on for all the other esu, emu, and practical electromagnetic units. This attitude matches the way electromagnetic quantities are used in the laboratory, where there is never really any need to break units down to fractional powers of grams and centimeters. There are many ways to give electromagnetic physical quantities an extra dimension, constructing a set of units based on four fundamental dimensions—mass, length, time, and something electrical. For teaching purposes it makes sense to choose electric charge, Q, as our fourth fundamental dimension since it appears in Coulomb's law for point charges, the basic rule of electricity.*

2.8 THE ESUQ AND EMUQ SYSTEMS OF UNITS

We now introduce the esuq and emuq systems of units. The N operators for the esuq and emuq systems are defined by

$$\underset{\text{esuq}}{\text{N}}(b) = \underset{\text{esu}}{\text{N}}(b) \tag{2.34a}$$

and

$$\underset{\text{emuq}}{\text{N}}(b) = \underset{\text{emu}}{\text{N}}(b), \tag{2.34b}$$

for any mechanical or electromagnetic physical quantity b. Equations (2.34a) and (2.34b) make the esuq and emuq systems identical to the esu and emu systems except that, unlike the esu and emu systems, they recognize the existence of a new fundamental dimension—electric charge. For any physical quantity b_{MECH}, which is purely mechanical, we have

$$\underset{\text{esuq}}{\text{U}}(b_{MECH}) = \underset{\text{esu}}{\text{U}}(b_{MECH}) \tag{2.34c}$$

and

$$\underset{\text{emuq}}{\text{U}}(b_{MECH}) = \underset{\text{emu}}{\text{U}}(b_{MECH}); \tag{2.34d}$$

but for any electromagnetic physical quantity b_{ELMAG}, we expect

$$\underset{\text{esuq}}{\text{U}}(b_{ELMAG}) \neq \underset{\text{esu}}{\text{U}}(b_{ELMAG}) \tag{2.34e}$$

---

* From the viewpoint of experimental scientists, electric charge is a relatively poor choice of fundamental dimension, because any fundamental unit of charge will be difficult to measure and reproduce accurately.

and

$$\underset{\text{emuq}}{U}(b_{ELMAG}) \neq \underset{\text{emu}}{U}(b_{ELMAG}). \tag{2.34f}$$

Because the cgs system is identical in all respects to the esu and emu systems for mechanical physical quantities b_{MECH}, Eqs. (2.34a–d) can be rewritten as

$$\underset{\text{esuq}}{N}(b_{MECH}) = \underset{\text{esu}}{N}(b_{MECH}) = \underset{\text{cgs}}{N}(b_{MECH}), \tag{2.34g}$$

$$\underset{\text{emuq}}{N}(b_{MECH}) = \underset{\text{emu}}{N}(b_{MECH}) = \underset{\text{cgs}}{N}(b_{MECH}), \tag{2.34h}$$

$$\underset{\text{esuq}}{U}(b_{MECH}) = \underset{\text{esu}}{U}(b_{MECH}) = \underset{\text{cgs}}{U}(b_{MECH}), \tag{2.34i}$$

$$\underset{\text{emuq}}{U}(b_{MECH}) = \underset{\text{emu}}{U}(b_{MECH}) = \underset{\text{cgs}}{U}(b_{MECH}). \tag{2.34j}$$

We define the fundamental unit of charge in the esuq system to be statcoulq; i.e.,

$$\underset{\text{esuq}}{U}(Q) = \text{statcoulq} \tag{2.35a}$$

and examine what happens to Coulomb's law, Eqs. (2.1a, b), when $\underset{\text{esuq}}{U}$ is applied to both sides:

$$\underset{\text{esuq}}{U}(F) = \frac{\underset{\text{esuq}}{U}(Q_1) \cdot \underset{\text{esuq}}{U}(Q_2)}{\underset{\text{esuq}}{U}(\varepsilon_0) \cdot \underset{\text{esuq}}{U}(r)^2} = \frac{\text{statcoulq}^2}{\underset{\text{esuq}}{U}(\varepsilon_0) \cdot \underset{\text{esuq}}{U}(r)^2}. \tag{2.35b}$$

From Eq. (2.34i), the units of the two mechanical quantities must be

$$\underset{\text{esuq}}{U}(F) = \text{dynes} \tag{2.36a}$$

and

$$\underset{\text{esuq}}{U}(r) = \text{cm} \tag{2.36b}$$

Therefore, Eq. (2.35b) becomes

$$\underset{\text{esuq}}{U}(\varepsilon_0) = \frac{\text{statcoulq}^2}{\text{dyne} \cdot \text{cm}^2}, \tag{2.37a}$$

so that ε_0 now has dimensions, making it an unarguable electromagnetic physical quantity. Equations (2.34a) and (2.11a) then require that

$$\underset{\text{esuq}}{N}(\varepsilon_0) = \underset{\text{esu}}{N}(\varepsilon_0) = 1 \tag{2.37b}$$

so that in the esuq system

$$\varepsilon_0 = 1\frac{\text{statcoulq}^2}{\text{dyne} \cdot \text{cm}^2} = 1\frac{\text{statcoulq}^2 \cdot \text{sec}^2}{\text{gm} \cdot \text{cm}^3}. \tag{2.38}$$

In the last step of Eq. (2.38) we use $1 \text{ dyne} = \text{gm} \cdot \text{cm} \cdot \text{sec}^{-2}$.
    Applying $\underset{\text{esuq}}{\text{U}}$ to both sides of Eq. (2.3a) gives

$$\underset{\text{esuq}}{\text{U}}(E) = \frac{\underset{\text{esuq}}{\text{U}}(Q)}{\underset{\text{esuq}}{\text{U}}(\varepsilon_0) \cdot \underset{\text{esuq}}{\text{U}}(r)^2} = \frac{\text{statcoulq}}{\left(\dfrac{\text{statcoulq}^2}{\text{dyne} \cdot \text{cm}^2}\right) \cdot \text{cm}^2}$$

$$= \frac{\text{dyne} \cdot \text{cm}}{\text{statcoulq} \cdot \text{cm}} = \left(\frac{\text{erg}}{\text{statcoulq}}\right) \cdot \frac{1}{\text{cm}}, \tag{2.39a}$$

where Eqs. (2.38), (2.35a), (2.36b), and $\text{erg} = \text{dyne} \cdot \text{cm}$ are used to get the final result. We now define

$$1 \text{ statvoltq} = \frac{\text{erg}}{\text{statcoulq}} \tag{2.39b}$$

to get

$$\underset{\text{esuq}}{\text{U}}(E) = \frac{\text{statvoltq}}{\text{cm}}. \tag{2.39c}$$

Applying $\underset{\text{esuq}}{\text{U}}$ to both sides of Eq. (2.21a) gives [the discussion after Eq. (1.12) shows how to analyze the integral in Eq. (2.21a)]

$$\underset{\text{esuq}}{\text{U}}(V) = \underset{\text{esuq}}{\text{U}}(E) \cdot \underset{\text{esuq}}{\text{U}}(ds) = \frac{\text{statvoltq}}{\text{cm}} \cdot \text{cm} = \text{statvoltq}. \tag{2.39d}$$

This shows that statvoltq is the unit of electric potential in the esuq system of units. Moving on to the units of the electric dipole in Eqs. (2.20a, b), we apply $\underset{\text{esuq}}{\text{U}}$ to get

$$\underset{\text{esuq}}{\text{U}}(p) = \underset{\text{esuq}}{\text{U}}(Q) \cdot \text{cm} = \text{statcoulq} \cdot \text{cm} \tag{2.40a}$$

so that the electric dipole density P in Eq. (2.20c) has units

$$\underset{\text{esuq}}{\text{U}}(P) = \frac{\underset{\text{esuq}}{\text{U}}(p)}{\underset{\text{esuq}}{\text{U}}(dV)} = \frac{\text{statcoulq} \cdot \text{cm}}{\text{cm}^3} = \frac{\text{statcoulq}}{\text{cm}^2}. \tag{2.40b}$$

Now, applying $\underset{\text{esuq}}{U}$ to Eq. (2.20d) gives

$$\underset{\text{esuq}}{U}(D) = \frac{\text{statcoulq}}{\text{cm}^2}, \tag{2.40c}$$

which we can get either from analyzing the "$4\pi P$" term using Eq. (2.40b) or from analyzing the "$\varepsilon_0 E$" term using Eqs. (2.37a), (2.39a), and erg = dyne · cm.

Suppose that back at Eq. (2.35a), we had decided to work with the emuq system of units, defining the fundamental unit of charge to be the abcoulq instead of the statcoulq:

$$\underset{\text{emuq}}{U}(Q) = \text{abcoulq}. \tag{2.41a}$$

Working through Coulomb's law using $\underset{\text{emuq}}{U}$ instead of $\underset{\text{esuq}}{U}$ gives

$$\underset{\text{emuq}}{U}(\varepsilon_0) = \frac{\text{abcoulq}^2}{\text{dyne} \cdot \text{cm}^2}. \tag{2.41b}$$

This is the same as Eq. (2.37a) with statcoulq replaced by abcoulq. The next step is different, because from Eqs. (2.12b) and (2.34b)

$$\underset{\text{emuq}}{N}(\varepsilon_0) = \underset{\text{emu}}{N}(\varepsilon_0) = c_{\text{cgs}}^{-2}. \tag{2.41c}$$

which leads to

$$\varepsilon_0 = c_{\text{cgs}}^{-2} \frac{\text{abcoulq}^2}{\text{dyne} \cdot \text{cm}^2} \tag{2.41d}$$

in the emuq system of units. After this deviation, however, the development follows the same path as before, with abcoulq replacing statcoulq everywhere. Applying $\underset{\text{emuq}}{U}$ to both sides of Eq. (2.3a) gives

$$\underset{\text{emuq}}{U}(E) = \left(\frac{\text{erg}}{\text{abcoulq}} \right) \cdot \frac{1}{\text{cm}} = \frac{\text{abvoltq}}{\text{cm}}, \tag{2.42a}$$

where we define

$$1 \text{ abvoltq} = 1 \frac{\text{erg}}{\text{abcoulq}}. \tag{2.42b}$$

This corresponds exactly to Eqs. (2.39a–c). Applying $\underset{\text{emuq}}{U}$ to both sides of Eq. (2.21a) gives

$$\underset{\text{emuq}}{U}(V) = \text{abvoltq}, \tag{2.42c}$$

which corresponds to Eq. (2.39d) and shows that abvoltq is the emuq unit of electric potential. Going through the same development used to get Eqs. (2.40a–c) with $\underset{\text{emuq}}{U}$ instead of $\underset{\text{esuq}}{U}$ gives

$$\underset{\text{emuq}}{U}(p) = \text{abcoulq} \cdot \text{cm}, \tag{2.43a}$$

$$\underset{\text{emuq}}{U}(P) = \frac{\text{abcoulq}}{\text{cm}^2}, \tag{2.43b}$$

and

$$\underset{\text{emuq}}{U}(D) = \frac{\text{abcoulq}}{\text{cm}^2}. \tag{2.43c}$$

Readers familiar with the rationalized mks system (used by almost all of today's introductory textbooks for electricity and magnetism) may by now be feeling a strong sense of *déjà vu*, because there is an obvious correspondence between the units used in the esuq, emuq, and mks systems. The mks potential has units of volts, the charge has units of coul, and the definition of a volt is joule/coul. Similarly, the esuq and emuq potential has units of statvoltq and abvoltq, the esuq and emuq charge has units of statcoulq and abcoulq, and the esuq and emuq definitions of statvoltq and abvoltq are erg/statcoulq and erg/abcoulq. The mks unit for the electric field E is volt/m and the unit for the displacement field D is coul/m^2. The esuq and emuq units for E are statvoltq/cm and abvoltq/cm, and the esuq and emuq units for D are statcoulq/cm^2 and abcoulq/cm^2. It looks as if we can get the esuq units for any electric quantity simply by taking the mks units, adding a "stat" prefix and a "q" suffix, and changing all mechanical units to their cgs counterparts (m becomes cm, joule becomes erg, etc.). The emuq units seem to follow the same recipe, except that we add an "ab" instead of a "stat" prefix. This rule of thumb is, in fact, true—and very helpful when it is not immediately clear what the correct esuq or emuq units should be. Historically speaking, this is no accident, because the mks system was first proposed at the very beginning of the twentieth century, soon after the introduction of the practical units discussed above. The logic behind these mks units—that of introducing a fourth fundamental electromagnetic dimension—was the same as that behind the esuq and emuq systems of units, and we use the stat and ab conventions to take advantage of this correspondence. Unfortunately, the mks system of units became popular in its rationalized form, and we are not yet ready to talk about the U and N operators of rationalized systems—a topic that will be postponed until Chapter 3. For now, it is enough to realize that the similarity of the esuq, emuq, and mks units exists and that it follows directly from the recognition of a fourth electromagnetic dimension. It is also worth noting that the mks units, because they do presume the existence of a fourth electromagnetic dimension, are definitely not the same as the old practical units, even though they were given the same names, and physical quantities measured with them have the same numeric components. The old practical units volt, coul, etc., discussed in Section 2.6 above are all multiples of the corresponding emu units; whereas, the mks volt, coul, etc. cannot be set equal to any of the corresponding emu units because they have a fourth electromagnetic dimension (see Section 3.6 for more discussion of this point).

Equation (2.10) is the starting point for constructing the magnetic units of the esuq and emuq systems. Applying $\underset{\text{esuq}}{U}$ to Eq. (2.10) gives, using Eqs. (2.37a) and (2.34i),

$$
\begin{aligned}
\underset{\text{esuq}}{U}(\mu_0) &= \frac{1}{\underset{\text{esuq}}{U}(\varepsilon_0) \cdot \underset{\text{esuq}}{U}(c)^2} = \frac{\text{dyne} \cdot \text{cm}^2}{\text{statcoulq}^2} \cdot \frac{\text{sec}^2}{\text{cm}^2} \\
&= \left(\frac{\text{erg}}{\text{statcoulq}} \right) \cdot \frac{\text{sec}^2}{\text{cm}} \cdot \frac{1}{\text{statcoulq}} \\
&= \frac{\text{statvoltq} \cdot \text{sec}^2}{\text{cm} \cdot \text{statcoulq}},
\end{aligned}
\tag{2.44}
$$

where $\text{erg} = \text{dyne} \cdot \text{cm}$ and $\text{statvoltq} = \text{erg}/\text{statcoulq}$ are used to get the final result. We can put this into a more compact form by defining two new esuq units, the statampq and the stathenryq:

$$
1 \text{ statampq} = \frac{\text{statcoulq}}{\text{sec}},
\tag{2.45a}
$$

$$
1 \text{ stathenryq} = \frac{\text{statvoltq} \cdot \text{sec}}{\text{statampq}}.
\tag{2.45b}
$$

Applying these definitions to Eq. (2.44) gives

$$
\underset{\text{esuq}}{U}(\mu_0) = \frac{\text{statvoltq} \cdot \text{sec}}{\text{statampq} \cdot \text{cm}} = \frac{\text{stathenryq}}{\text{cm}}.
\tag{2.45c}
$$

To find out the meaning of the new statampq unit, we apply $\underset{\text{esuq}}{U}$ to both sides of Eq. (2.7) to get

$$
\underset{\text{esuq}}{U}(I) = \frac{\underset{\text{esuq}}{U}(dQ)}{\underset{\text{esuq}}{U}(dt)} = \frac{\text{statcoulq}}{\text{sec}} = \text{statampq},
\tag{2.45d}
$$

showing that statampq is the esuq unit of current. To find the meaning of the stathenryq unit, we apply $\underset{\text{esuq}}{U}$ to both sides of Eq. (2.28a) to get

$$
\underset{\text{esuq}}{U}(L) = \frac{\underset{\text{esuq}}{U}(V) \cdot \underset{\text{esuq}}{U}(dt)}{\underset{\text{esuq}}{U}(dI)} = \frac{\text{statvoltq} \cdot \text{sec}}{\text{statampq}} = \text{stathenryq},
\tag{2.45e}
$$

showing stathenryq to be the esuq unit of induction. To find the numeric component of μ_0 in esuq units we consult Eqs. (2.11b) and (2.34g) to get

$$\underset{\text{esuq}}{\text{N}}(\mu_0) = c_{\text{cgs}}^{-2}. \tag{2.45f}$$

We can go through the same procedure using $\underset{\text{emuq}}{\text{U}}$ instead of $\underset{\text{esuq}}{\text{U}}$ in Eqs. (2.44) through (2.45e) to get

$$\underset{\text{emuq}}{\text{U}}(\mu_0) = \frac{\text{abvoltq} \cdot \text{sec}^2}{\text{cm} \cdot \text{abcoulq}}, \tag{2.46a}$$

which simplifies to

$$\underset{\text{emuq}}{\text{U}}(\mu_0) = \frac{\text{abvoltq} \cdot \text{sec}}{\text{abampq} \cdot \text{cm}} = \frac{\text{abhenryq}}{\text{cm}}, \tag{2.46b}$$

where

$$1\ \text{abampq} = \frac{\text{abcoulq}}{\text{sec}} \tag{2.46c}$$

and

$$1\ \text{abhenryq} = \frac{\text{abvoltq} \cdot \text{sec}}{\text{abampq}} \tag{2.46d}$$

are the emuq units of current and induction, respectively. Equations (2.12a) and (1.28b) give the numeric component of μ_0 in emuq units:

$$\underset{\text{emuq}}{\text{N}}(\mu_0) = 1. \tag{2.46e}$$

Readers familiar with mks units recognize that Eqs. (2.45c) and (2.46b) follow the mks rule of thumb explained in the discussion following Eq. (2.43c) above, since the esuq and emuq units for μ_0 are stathenryq/cm and abhenryq/cm, respectively, and the mks units for μ_0 are henry/m. To simplify the units of ε_0 and to show how they also match this rule of thumb, we define

$$1\ \text{statfaradq} = \frac{\text{statcoulq}}{\text{statvoltq}} \tag{2.47a}$$

and

$$1\ \text{abfaradq} = \frac{\text{abcoulq}}{\text{abvoltq}}. \tag{2.47b}$$

Applying $\underset{esuq}{U}$ and $\underset{emuq}{U}$ to Eq. (2.27a) then gives

$$\underset{esuq}{U}(C) = \frac{\underset{esuq}{U}(Q)}{\underset{esuq}{U}(V)} = \frac{statcoulq}{statvoltq} = statfaradq \qquad (2.47c)$$

and

$$\underset{emuq}{U}(C) = \frac{\underset{emuq}{U}(Q)}{\underset{emuq}{U}(V)} = \frac{abcoulq}{abvoltq} = abfaradq, \qquad (2.47d)$$

showing that statfaradq and abfaradq are the esuq and emuq units of capacitance. Now, using erg $=$ dyne $\cdot$ cm, Eq. (2.37a) can be written as

$$\underset{esuq}{U}(\varepsilon_0) = \frac{statcoulq^2}{erg \cdot cm} = \frac{statcoulq}{statvoltq \cdot cm} = \frac{statfaradq}{cm}, \qquad (2.48a)$$

where we use definitions (2.39b) and (2.47a) to get the final result. Equation (2.41b) can be treated similarly to get, using definitions (2.42b) and (2.47b),

$$\underset{emuq}{U}(\varepsilon_0) = \frac{abcoulq^2}{erg \cdot cm} = \frac{abcoulq}{abvoltq \cdot cm} = \frac{abfaradq}{cm}. \qquad (2.48b)$$

Because the units of ε_0 in the mks system are often given as farad/m, it is clear that Eqs. (2.48a,b) follow the mks rule of thumb.

Once the units of μ_0 are known, we have easy access to the units of all the magnetic physical quantities. Applying $\underset{esuq}{U}$ to Eqs. (2.2a, b) gives

$$dynes = \frac{\underset{esuq}{U}(p_H)^2}{\underset{esuq}{U}(\mu_0) \cdot cm^2}$$

or

$$\underset{esuq}{U}(p_H)^2 = dyne \cdot cm^2 \cdot \underset{esuq}{U}(\mu_0) = \left(\frac{dyne \cdot cm}{statcoulq}\right) \cdot statvoltq \cdot sec^2,$$

using Eq. (2.44) to eliminate $\underset{esuq}{U}(\mu_0)$. Since

$$\frac{dyne \cdot cm}{statcoulq} = \frac{erg}{statcoulq} = statvoltq,$$

the square of $\underset{esuq}{U}(p_H)$ must equal $statvoltq^2 \cdot sec^2$. Therefore, if we define

$$statweberq = statvoltq \cdot sec, \qquad (2.49)$$

we get that the magnetic pole strength in the esuq system is

$$\underset{esuq}{U}(p_H) = statweberq \qquad (2.50)$$

We now repeat this procedure, applying $\underset{emuq}{U}$ to Eqs. (2.2a, b) and using Eq. (2.46a) to get

$$\underset{emuq}{U}(p_H) = abweberq, \qquad (2.51a)$$

where we define

$$abweberq = abvoltq \cdot sec. \qquad (2.51b)$$

The unit of magnetic pole strength in the emuq system is the abweberq.
    Applying $\underset{esuq}{U}$ to Eq. (2.5) gives

$$\underset{esuq}{U}(H) = \frac{dynes}{\underset{esuq}{U}(p_H)} = \frac{dynes}{statweberq}. \qquad (2.52a)$$

From $erg = dyne \cdot cm$, $statvoltq = erg/statcoulq$, and $statweberq = statvoltq \cdot sec = (erg \cdot sec)/statcoulq$, we have that

$$\frac{dynes}{statweberq} = \left(\frac{erg}{cm}\right) \cdot \frac{statcoulq}{erg \cdot sec} = \frac{statampq}{cm},$$

where in the last step we use $statampq = statcoulq/sec$. Therefore, the esuq units of H can be written as

$$\underset{esuq}{U}(H) = \frac{statampq}{cm}. \qquad (2.52b)$$

This result can also be gotten directly from applying $\underset{esuq}{U}$ to both sides of Eq. (2.6). Following the same procedure with $\underset{emuq}{U}$, applying it to both sides of Eqs. (2.5) or (2.6), gives

$$\underset{emuq}{U}(H) = \frac{dynes}{abweberq} \qquad (2.52c)$$

or

$$\underset{emuq}{U}(H) = \frac{abampq}{cm}.$$ (2.52d)

The units of a permanent-magnet dipole moment m_H come from applying $\underset{esuq}{U}$ and $\underset{emuq}{U}$ to Eqs. (2.19a,b):

$$\underset{esuq}{U}(m_H) = statweberq \cdot cm,$$ (2.53a)

$$\underset{emuq}{U}(m_H) = abweberq \cdot cm.$$ (2.53b)

Applying $\underset{esuq}{U}$ and $\underset{emuq}{U}$ to the definition of the permanent-magnetic dipole density field in Eq. (2.19c) gives

$$\underset{esuq}{U}(M_H) = \frac{statweberq}{cm^2}$$ (2.54a)

and

$$\underset{emuq}{U}(M_H) = \frac{abweberq}{cm^2}.$$ (2.54b)

Turning to the definition of the magnetic induction B in Eq. (2.19d), we confirm that M_H and $\mu_0 H$ have the same units—as they must—by applying $\underset{esuq}{U}$ and $\underset{emuq}{U}$ to $\mu_0 H$. Equations (2.44) and (2.52b) show that

$$\underset{esuq}{U}(\mu_0 H) = \underset{esuq}{U}(\mu_0) \cdot \underset{esuq}{U}(H) = \frac{statvoltq \cdot sec^2}{cm \cdot statcoulq} \cdot \frac{statampq}{cm}$$
$$= \frac{statweberq}{cm^2},$$ (2.54c)

where the last step uses the definitions statampq = statcoulq/sec and statweberq = statvoltq · sec. Equations (2.46a) and (2.52d), and the definitions of abampq and abweberq, show that

$$\underset{emuq}{U}(\mu_0 H) = \frac{abvoltq \cdot sec^2}{cm \cdot abcoulq} \cdot \frac{abampq}{cm} = \frac{abweberq}{cm^2}.$$ (2.54d)

Comparison of Eqs. (2.54a, b) to (2.54c, d) shows that $\mu_0 H$ and M_H have the same units, so applying $\underset{esuq}{U}$ and $\underset{emuq}{U}$ to Eq. (2.19d) gives

$$\underset{esuq}{U}(B) = \frac{statweberq}{cm^2}$$ (2.55a)

and

$$\underset{\text{emuq}}{\text{U}}(B) = \frac{\text{abweberq}}{\text{cm}^2}.$$ (2.55b)

Once again, readers familiar with the mks system of units can see how it parallels the esuq and emuq units. The mks units of H are amp/m, as compared to the esuq and emuq units for H in Eqs. (2.52b, d), which are statampq/cm and abampq/cm, respectively. Similarly, the mks units for B are teslas, defined to be weber/m^2, as compared to the esuq and emuq units for B in Eqs. (2.55a, b), which are statweberq/cm^2 and abweberq/cm^2, respectively.

The esuq units are completed by applying $\underset{\text{esuq}}{\text{U}}$ to Eq. (2.25a) to get

$$\text{statvoltq} = \underset{\text{esuq}}{\text{U}}(R) \cdot \text{statampq}$$

or

$$\underset{\text{esuq}}{\text{U}}(R) = \text{statohmq},$$ (2.56a)

where we define

$$\text{statohmq} = \frac{\text{statvoltq}}{\text{statampq}}.$$ (2.56b)

The emuq units are completed by applying $\underset{\text{emuq}}{\text{U}}$ to the same equation to get

$$\underset{\text{emuq}}{\text{U}}(R) = \text{abohmq},$$ (2.56c)

where we define

$$\text{abohmq} = \frac{\text{abvoltq}}{\text{abampq}}.$$ (2.56d)

Clearly the units of resistance in the esuq and emuq systems are statohmq and abohmq, respectively.

Although, as already discussed after Eqs. (2.32a, c), it does not make sense to write

$$1\,\text{abcoul} \overset{?}{=} 3 \times 10^{10}\,\text{statcoul},$$

because, according to Eqs. (2.30a) and (2.31a), they have different dimensions; it does make sense to write

$$1\,\text{abcoulq} \cong 3 \times 10^{10}\,\text{statcoulq},$$

because both statcoulq and abcoulq are units of a newly recognized fourth fundamental dimension—electric charge. Consequently, we can write an equation relating statcoulq to abcoulq without violating any of the rules in Chapter 1. From Eq. (2.14i) and Eqs. (2.34a, b) we see that

$$\underset{\text{esuq}}{N}(Q) = c_{\text{cgs}} \cdot \underset{\text{emuq}}{N}(Q). \tag{2.57a}$$

So, from Rule I, it follows that

$$1 \text{ abcoulq} = c_{\text{cgs}} \cdot \text{statcoulq}. \tag{2.57b}$$

Tracing back the origin of this important result, we see that it stems from the fundamental choices made in Eqs. (2.11a, b) and (2.12a, b) to set up the esu and emu systems of units. In this sense, Eq. (2.57b) is entirely manmade and not the consequence of some natural symmetry. The choices made in Eqs. (2.11a, b) and Eqs. (2.12a, b) are not necessarily bad choices, but they are also not compellingly obvious. Any choice of ε_0 and μ_0 giving an $\varepsilon_0 \mu_0$ product equal to c^{-2}, as specified by Eq. (2.10), can be used to set up a system of units—we might consider, for example, choosing $\varepsilon_0 = \mu_0 = c^{-1}$. This choice is symmetric, making it perhaps more appealing to today's theoreticians; but historically speaking it was never pursued.

Equation (2.57b) is all that we need to relate every esuq unit to its dimensionally identical emuq counterpart. From Eqs. (2.57b), (2.42b), and (2.39b) we get

$$1 \text{ abvoltq} = \frac{\text{erg}}{\text{abcoulq}} = \frac{\text{erg}}{c_{\text{cgs}} \cdot \text{statcoulq}} = c_{\text{cgs}}^{-1} \cdot \text{statvoltq}. \tag{2.58a}$$

From Eqs. (2.46c) and (2.45a) we get

$$1 \text{ abampq} = \frac{\text{abcoulq}}{\text{sec}} = \frac{c_{\text{cgs}} \cdot \text{statcoulq}}{\text{sec}} = c_{\text{cgs}} \cdot \text{statampq}; \tag{2.58b}$$

and from Eqs. (2.46d), (2.58a), (2.58b), and (2.45b),

$$1 \text{ abhenryq} = \frac{\text{abvoltq} \cdot \text{sec}}{\text{abampq}} = \frac{c_{\text{cgs}}^{-1} \cdot \text{statvoltq} \cdot \text{sec}}{c_{\text{cgs}} \cdot \text{statampq}} = c_{\text{cgs}}^{-2} \cdot \text{stathenryq}. \tag{2.58c}$$

Equations (2.47b), (2.57b), (2.58a), and (2.47a) give

$$1 \text{ abfaradq} = \frac{\text{abcoulq}}{\text{abvoltq}} = \frac{c_{\text{cgs}} \cdot \text{statcoulq}}{c_{\text{cgs}}^{-1} \cdot \text{statvoltq}} = c_{\text{cgs}}^2 \cdot \text{statfaradq}; \tag{2.58d}$$

and Eqs. (2.51b), (2.58a), and (2.49) give

$$1 \text{ abweberq} = \text{abvoltq} \cdot \text{sec} = c_{\text{cgs}}^{-1} \cdot \text{statvoltq} \cdot \text{sec} = c_{\text{cgs}}^{-1} \cdot \text{statweberq}. \tag{2.58e}$$

Table 2.6 Relationships between the esuq and emuq units.

1 abampq $= c_{cgs} \cdot$ statampq
1 abcoulq $= c_{cgs} \cdot$ statcoulq
1 abweberq $= c_{cgs}^{-1} \cdot$ statweberq
1 abvoltq $= c_{cgs}^{-1} \cdot$ statvoltq
1 abfaradq $= c_{cgs}^{2} \cdot$ statfaradq
1 abhenryq $= c_{cgs}^{-2} \cdot$ stathenryq
1 abohmq $= c_{cgs}^{-2} \cdot$ statohmq

Equations (2.56d), (2.58a), (2.58b), and (2.56b) show that

$$1 \text{ abohmq} = \frac{\text{abvoltq}}{\text{abampq}} = \frac{c_{cgs}^{-1} \cdot \text{statvoltq}}{c_{cgs} \cdot \text{statampq}} = c_{cgs}^{-2} \cdot \text{statohmq}. \tag{2.58f}$$

Equations (2.57b) and (2.58a–f) are summarized in Table 2.6.

2.9 THE ESUQ AND EMUQ CONNECTION WITH THE ESU AND EMU SYSTEMS OF UNITS

The esuq and emuq units are a bridge between the esu and emu systems of units. Figure 2.7 shows the procedure for converting equations and formulas from esu to emu units. Following the solid line, we go from esu to esuq by recognizing charge as a fourth fundamental dimension, then convert to emuq units, and then drop down to the emu units by no longer recognizing charge as a separate dimension. The process is reversed when following the dotted line to convert equations from emu to esu units.

We have set up the four different systems of units to make these transformations simple, but it is important to realize that when dropping from the esuq, emuq upper level to the esu and emu lower level we are referring to two different ways of no longer recognizing charge as a separate dimension. To show how this works, we create an operator $\underset{mltq}{U}$ that recognizes the four fundamental dimensions of mass, length, time, and charge in any physical quantity b:

$$\underset{mltq}{U}(b) = \begin{array}{l} \text{dimensional formula of } b \text{ in} \\ \text{mass, length, time, and charge.} \end{array} \tag{2.59}$$

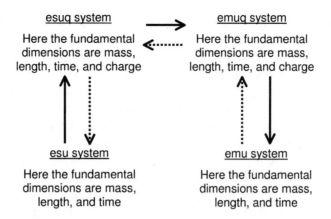

Figure 2.7 The solid arrows show the three-step transformation from esu units to emu units and the dotted arrows show the three-step transformation from emu units to esu units.

Operator $\underset{mltq}{U}$ is exactly the same as $\underset{mlt}{U}$ when $b = b_{MECH}$ is a strictly mechanical quantity [see discussion following Eq. (1.29c)]. When $b = b_{ELMAG}$ is an electromagnetic quantity, however, $\underset{mltq}{U}$ recognizes the fundamental dimension of charge exactly the way it is recognized in the esuq and emuq systems of units. Applying $\underset{mltq}{U}$ to Eqs. (2.1a, b), Coulomb's law for two point charges, gives

$$\underset{mltq}{U}(F) = \frac{\underset{mltq}{U}(Q_1) \cdot \underset{mltq}{U}(Q_1)}{\underset{mltq}{U}(\varepsilon_0) \cdot \underset{mltq}{U}(r)^2}$$

or

$$\underset{mltq}{U}(\varepsilon_0) = \frac{\text{charge}^2}{\text{mass} \cdot \text{length}^3 \cdot \text{time}^{-2}}. \qquad (2.60a)$$

Applying $\underset{mltq}{U}$ to Eq. (2.10), $\varepsilon_0\mu_0 = c^{-2}$, then gives

$$\underset{mltq}{U}(\mu_0) = \frac{\text{time}^2}{\text{length}^2} \cdot \frac{1}{\underset{mltq}{U}(\varepsilon_0)} = \frac{\text{mass} \cdot \text{length}}{\text{charge}^2}, \qquad (2.60b)$$

where Eq. (2.60a) is used in the last step of Eq. (2.60b). When dropping down to the esu system in Fig. 2.7, ε_0 must become a dimensionless quantity; so from Eq. (2.60a), charge loses its separate dimension by becoming equivalent to $\text{mass}^{1/2} \cdot \text{length}^{3/2} \cdot \text{time}^{-1}$. This is by no means the same as dropping down to the emu system. In the emu system, μ_0 must become dimensionless; so from Eq. (2.60b), charge loses its separate dimension by becoming equivalent to mass$^{1/2} \cdot$ length$^{1/2}$. These are clearly two distinct ways of no longer recognizing charge as a fundamental dimension. Using the terminology developed in Chapter 1, we see that the (statcoul, statcoulq) pair is a connecting unit between the upper level of Fig. 2.7 and the esu system of units, and the (abcoul, abcoulq) pair is a connecting unit between the upper level of Fig. 2.7 and the emu system of units. The third columns of Tables 2.4 and 2.5 give the esuq and emuq units of common electromagnetic physical quantities. In Table 2.4 we can pick out the connecting pairs between the esu system and the upper level of Fig. 2.7 by comparing unit expressions for the same physical quantity in the second and third columns. For example, the electric field E has units of statvolt/cm in the esu system and statvoltq/cm in the esuq system, so (statvolt/cm, statvoltq/cm) is a connecting pair between the esu and esuq systems. Similarly, the second and third columns of Table 2.5 show the connecting pairs between the emu system and the upper level of Fig. 2.7; the units of the permanent-magnet magnetic dipole moment m_H reveal, for example, that (abweber $\cdot$ cm, abweberq $\cdot$ cm) is a connecting pair between the emu and emuq systems.

Tables 2.6 and 2.7 provide the basic information needed to convert physical quantities and equations from esu units to emu units or from emu units to esu units. Following the convention used in Section 2.3, all variables with subscript "esu" are the numeric parts of physical quantities in the esu system of units, and all variables with the subscript "emu" are the numeric parts of physical quantities in the emu system of units. Variables with an "esuq" subscript are the numeric parts of physical quantities in the esuq system, and variables with an "emuq" subscript are the numeric parts of physical quantities in the emuq system. Equation (2.34a) shows that all physical variables in the esu and esuq systems automatically have the same numeric parts. Starting from the top of Table 2.7, we can write

$$A_{esu} = A_{esuq}$$
$$B_{esu} = B_{esuq}$$

etc.

Equation (2.34b) shows that the same holds true for all the physical variables in the emu, emuq systems.

$$A_{emu} = A_{emuq}$$
$$B_{emu} = B_{emuq}$$

etc.

Purely mechanical quantities, such as the speed of light, have the same numeric components in the esu, esuq, emu, emuq, and cgs systems of units, as indicated by Eqs. (2.34g, h). To keep it simple, when there is a choice of subscripts we choose the most well-known system of units. This means the numeric components of mechanical quantities get the cgs subscript when they are part of the esu, esuq, emu, or emuq systems; the numeric components of electromagnetic quantities get esu subscripts when they are part of the esu or esuq systems; and the numeric components of electromagnetic quantities get emu subscripts when they are part of the emu or emuq systems.

To show how to use Table 2.7, we convert the definition of the D field in esu units [see Eq. (2.20e) in Section 2.4],

$$\vec{D} = \vec{E} + 4\pi \vec{P},$$

to emu units. Split up into numeric components and the fundamental gm, cm, sec

Table 2.7 Numeric components of physical quantities in esu and emu units.

(magnetic vector potential) $A_{\text{emu}} = A_{\text{esu}} \cdot c_{\text{cgs}}$	(volume current density) $J_{\text{emu}} = J_{\text{esu}} \cdot c_{\text{cgs}}^{-1}$	(permeance) $\mathcal{P}_{\text{emu}} = \mathcal{P}_{\text{esu}} \cdot c_{\text{cgs}}^{2}$
(magnetic induction) $B_{\text{emu}} = B_{\text{esu}} \cdot c_{\text{cgs}}$	(surface current density) $(\mathcal{J}_S)_{\text{emu}} = (\mathcal{J}_S)_{\text{esu}} \cdot c_{\text{cgs}}^{-1}$	(charge) $Q_{\text{emu}} = Q_{\text{esu}} \cdot c_{\text{cgs}}^{-1}$
(capacitance) $C_{\text{emu}} = C_{\text{esu}} \cdot c_{\text{cgs}}^{-2}$	(inductance) $L_{\text{emu}} = L_{\text{esu}} \cdot c_{\text{cgs}}^{2}$	(resistance) $R_{\text{emu}} = R_{\text{esu}} \cdot c_{\text{cgs}}^{2}$
(electric displacement) $D_{\text{emu}} = D_{\text{esu}} \cdot c_{\text{cgs}}^{-1}$	(permanent-magnet dipole moment) $(m_H)_{\text{emu}} = (m_H)_{\text{esu}} \cdot c_{\text{cgs}}$	(reluctance) $\mathcal{R}_{\text{emu}} = \mathcal{R}_{\text{esu}} \cdot c_{\text{cgs}}^{-2}$
(electric field) $E_{\text{emu}} = E_{\text{esu}} \cdot c_{\text{cgs}}$	(current-loop magnetic dipole moment) $(m_I)_{\text{emu}} = (m_I)_{\text{esu}} \cdot c_{\text{cgs}}^{-1}$	(volume charge density) $(\rho_Q)_{\text{emu}} = (\rho_Q)_{\text{esu}} \cdot c_{\text{cgs}}^{-1}$
(dielectric constant) $\varepsilon_{\text{emu}} = \varepsilon_r \cdot c_{\text{cgs}}^{-2}, \varepsilon_{\text{esu}} = \varepsilon_r$	(permanent-magnet dipole density) $(M_H)_{\text{emu}} = (M_H)_{\text{esu}} \cdot c_{\text{cgs}}$	(resistivity) $(\rho_R)_{\text{emu}} = (\rho_R)_{\text{esu}} \cdot c_{\text{cgs}}^{2}$
(permittivity of free space) $(\varepsilon_0)_{\text{emu}} = c_{\text{cgs}}^{-2}, (\varepsilon_0)_{\text{esu}} = 1$	(current-loop magnetic dipole density) $(M_I)_{\text{emu}} = (M_I)_{\text{esu}} \cdot c_{\text{cgs}}^{-1}$	(elastance) $S_{\text{emu}} = S_{\text{esu}} \cdot c_{\text{cgs}}^{2}$
(magnetomotive force) $\mathcal{F}_{\text{emu}} = \mathcal{F}_{\text{esu}} \cdot c_{\text{cgs}}^{-1}$	(magnetic permeability) $\mu_{\text{emu}} = \mu_r, \mu_{\text{esu}} = \mu_r \cdot c_{\text{cgs}}^{-2}$	(surface charge density) $(S_Q)_{\text{emu}} = (S_Q)_{\text{esu}} \cdot c_{\text{cgs}}^{2}$
(magnetic flux) $(\Phi_B)_{\text{emu}} = (\Phi_B)_{\text{esu}} \cdot c_{\text{cgs}}$	(magnetic permeability of free space) $(\mu_0)_{\text{emu}} = 1, (\mu_0)_{\text{esu}} = c_{\text{cgs}}^{-2}$	(conductivity) $\sigma_{\text{emu}} = \sigma_{\text{esu}} \cdot c_{\text{cgs}}^{-2}$
(conductance) $G_{\text{emu}} = G_{\text{esu}} \cdot c_{\text{cgs}}^{-2}$	(magnetic pole strength) $(p_H)_{\text{emu}} = (p_H)_{\text{esu}} \cdot c_{\text{cgs}}$	(electric potential) $V_{\text{emu}} = V_{\text{esu}} \cdot c_{\text{cgs}}$
(magnetic field) $H_{\text{emu}} = H_{\text{esu}} \cdot c_{\text{cgs}}^{-1}$	(electric dipole moment) $p_{\text{emu}} = p_{\text{esu}} \cdot c_{\text{cgs}}^{-1}$	(magnetic scalar potential) $(\Omega_H)_{\text{emu}} = (\Omega_H)_{\text{esu}} \cdot c_{\text{cgs}}^{-1}$
(current) $I_{\text{emu}} = I_{\text{esu}} \cdot c_{\text{cgs}}^{-1}$	(electric dipole density) $P_{\text{emu}} = P_{\text{esu}} \cdot c_{\text{cgs}}^{-1}$	

units of the esu system, this equation becomes (see Table 2.4)

$$\vec{D}_{esu} \cdot \left(\frac{1}{sec} \cdot \frac{gm^{1/2}}{cm^{1/2}} \right) = \vec{E}_{esu} \cdot \left(\frac{1}{sec} \cdot \frac{gm^{1/2}}{cm^{1/2}} \right) + 4\pi \vec{P}_{esu} \cdot \left(\frac{1}{sec} \cdot \frac{gm^{1/2}}{cm^{1/2}} \right). \quad (2.61a)$$

To go from the esu to esuq units in Fig. 2.7, we recognize charge as a fourth fundamental dimension. According to Rule VIII, the first step in doing this is to use the meaning of the new charge dimension to rewrite Eq. (2.61a) so that it is balanced in both the invariant and connecting units. As it stands now, Eq. (2.61a) is only balanced in the invariant units gm, cm, and sec. From the electric displacement row, the electric field row, and the electric dipole density row of Table 2.4 we see that the connecting units of D, E, and P in the esu system are statcoul/cm^2, statvolt/cm, and statcoul/cm^2, respectively. Equation (2.61a) becomes

$$\vec{D}_{esu} \cdot \left(\frac{statcoul}{cm^2} \right) = \vec{E}_{esu} \cdot \left(\frac{statvolt}{cm} \right) + 4\pi \vec{P}_{esu} \cdot \left(\frac{statcoul}{cm^2} \right). \quad (2.61b)$$

At first glance, Eq. (2.61b) looks like its units do not match, but as long as we stay in the esu system we know that

$$\frac{statvolt}{cm} = \frac{1}{sec} \cdot \frac{gm^{1/2}}{cm^{1/2}} = \frac{statcoul}{cm^2},$$

so that Eq. (2.61b) has balanced units. Rule VIII, however, requires that Eq. (2.61b) obey Rules II, IV, and V—that is, it must have balanced units—in both its invariant and connecting units. For this to be true, we must multiply $\vec{E}_{esu}$ by (statcoul · statvolt^{-1} · cm^{-1}) to get

$$\vec{D}_{esu} \cdot \left(\frac{statcoul}{cm^2} \right) = \vec{E}_{esu} \cdot \left(\frac{statvolt}{cm} \right) \cdot \left(\frac{statcoul}{statvolt \cdot cm} \right)$$
$$+ 4\pi \vec{P}_{esu} \cdot \left(\frac{statcoul}{cm^2} \right). \quad (2.61c)$$

We can do this because in the esu system [see Table 2.4 or Eqs. (2.30a, c)]

$$\frac{statcoul}{statvolt \cdot cm} = \left(\frac{gm^{1/2} \cdot cm^{3/2}}{sec} \right) \cdot \frac{1}{cm} \cdot \left(\frac{sec}{gm^{1/2} \cdot cm^{1/2}} \right) = 1,$$

so all we have done is to multiply $\vec{E}_{esu}$ by 1. Equation (2.61c), unlike Eq. (2.61b), is balanced in both the connecting and invariant units, so we can recognize charge as a fourth fundamental dimension and write

$$
\begin{aligned}
\vec{D}_{esu} \cdot \left(\frac{statcoulq}{cm^2} \right) &= \vec{E}_{esu} \cdot \left(\frac{statvoltq}{cm} \right) \cdot \left(\frac{statcoulq}{statvoltq \cdot cm} \right) \\
&\quad + 4\pi \vec{P}_{esu} \cdot \left(\frac{statcoulq}{cm^2} \right) \\
&= \vec{E}_{esu} \cdot \left(\frac{statvoltq}{cm} \right) \cdot \left(\frac{statcoulq^2}{erg \cdot cm} \right) \\
&\quad + 4\pi \vec{P}_{esu} \cdot \left(\frac{statcoulq}{cm^2} \right),
\end{aligned}
\tag{2.61d}
$$

where the last step of Eq. (2.61d) uses statvoltq = erg/statcoulq. Note that our systems of units have been constructed so that recognizing charge as a fundamental dimension just means adding a "q" suffix to the unit names. Since 1 erg = gm · cm² · sec⁻², Eq. (2.61d) can be written as

$$
\begin{aligned}
\vec{D}_{esu} \cdot \left(\frac{statcoulq}{cm^2} \right) &= \vec{E}_{esu} \cdot \left(\frac{statvoltq}{cm} \right) \cdot \left(\frac{statcoulq^2 \cdot sec^2}{gm \cdot cm^3} \right) \\
&\quad + 4\pi \vec{P}_{esu} \cdot \left(\frac{statcoulq}{cm^2} \right).
\end{aligned}
\tag{2.61e}
$$

From Eq. (2.38),

$$
\begin{aligned}
1 \frac{statcoulq^2 \cdot sec^2}{gm \cdot cm^3} &= (\varepsilon_0)_{esuq} \frac{statcoulq^2 \cdot sec^2}{gm \cdot cm^3} \\
&= (\varepsilon_0)_{esu} \frac{statcoulq^2 \cdot sec^2}{gm \cdot cm^3}
\end{aligned}
\tag{2.61f}
$$

where we specify that the numeric components of physical quantities in the esu and esuq systems must be the same [and, of course, that $(\varepsilon_0)_{esuq}$ and $(\varepsilon_0)_{esu}$ are just different names for the number 1]. Equation (2.61f) is used to write Eq. (2.61e) as

$$
\begin{aligned}
\vec{D}_{esu} &\cdot \left(\frac{statcoulq}{cm^2} \right) \\
&= \vec{E}_{esu} \cdot \left(\frac{statvoltq}{cm} \right) \cdot \left[(\varepsilon_0)_{esu} \cdot \frac{statcoulq^2 \cdot sec^2}{gm \cdot cm^3} \right] \\
&\quad + 4\pi \vec{P}_{esu} \cdot \left(\frac{statcoulq}{cm^2} \right).
\end{aligned}
\tag{2.61g}
$$

Now Table 2.6 is used to make the second step from esuq to emuq units. Substituting abvoltq and abcoulq units for the statvoltq and statcoulq units gives

$$\vec{D}_{esu} \cdot \left(c_{cgs}^{-1} \cdot \frac{abcoulq}{cm^2} \right)$$

$$= \left(\vec{E}_{esu} \cdot c_{cgs} \cdot \frac{abvoltq}{cm} \right) \cdot \left[(\varepsilon_0)_{esu} \cdot c_{cgs}^{-2} \cdot \frac{abcoulq^2 \cdot sec^2}{gm \cdot cm^3} \right] \qquad (2.61h)$$

$$+ 4\pi \left(\vec{P}_{esu} \cdot c_{cgs}^{-1} \cdot \frac{abcoulq}{cm^2} \right).$$

Table 2.7 is next used to substitute emu-subscripted variables for the esu-subscripted variables.

$$\left(\vec{D}_{emu} \cdot \frac{abcoulq}{cm^2} \right)$$

$$= \left(\vec{E}_{emu} \cdot \frac{abvoltq}{cm} \right) \cdot \left[(\varepsilon_0)_{emu} \cdot \frac{abcoulq^2 \cdot sec^2}{gm \cdot cm^3} \right] \qquad (2.61i)$$

$$+ 4\pi \left(\vec{P}_{emu} \cdot \frac{abcoulq}{cm^2} \right).$$

We stop recognizing charge as a fundamental dimension, dropping to the emu system by removing the "q" suffix to get

$$\left(\vec{D}_{emu} \cdot \frac{abcoul}{cm^2} \right)$$

$$= \left(\vec{E}_{emu} \cdot \frac{abvolt}{cm} \right) \cdot \left[(\varepsilon_0)_{emu} \cdot \frac{sec^2}{cm^2} \right] + 4\pi \left(\vec{P}_{emu} \cdot \frac{abcoul}{cm^2} \right), \qquad (2.61j)$$

where we have used Eq. (2.31a) to make the substitution

$$\frac{abcoul^2 \cdot sec^2}{gm \cdot cm^3} = \frac{sec^2}{cm^2}.$$

Returning at last to variables representing physical quantities, Eq. (2.61j) can be written as

$$\vec{D} = \varepsilon_0 \vec{E} + 4\pi \vec{P} \qquad (2.61k)$$

or, using Eq. (2.12b) to replace ε_0 by c^2,

$$\vec{D} = \frac{1}{c^2} \vec{E} + 4\pi \vec{P}. \qquad (2.61l)$$

The procedure in Fig. 2.7 gives the correct answer even when 1 statcoulq/ (statvoltq · cm) is not recognized as ε_0 in the esuq system of units. Suppose we cancel the statvoltq units in the first step of Eq. (2.61d) to get

$$\vec{D}_{esu} \cdot \frac{statcoulq}{cm^2} = \vec{E}_{esu} \cdot \frac{statcoulq}{cm^2} + 4\pi \vec{P}_{esu} \cdot \frac{statcoulq}{cm^2}. \qquad (2.62a)$$

The expressions in Table 2.7 for the emu-subscripted numerics and the equality 1 abcoulq = c_{cgs} · statcoulq from Table 2.6 give

$$c_{cgs}\vec{D}_{emu} \cdot \frac{statcoulq}{cm^2} = c_{cgs}^{-1}\vec{E}_{emu} \cdot \frac{statcoulq}{cm^2} + 4\pi c_{cgs}\vec{P}_{emu} \cdot \frac{statcoulq}{cm^2}$$

or

$$\vec{D}_{emu} \cdot \frac{abcoulq}{cm^2} = c_{cgs}^{-2}\vec{E}_{emu} \cdot \frac{abcoulq}{cm^2} + 4\pi \vec{P}_{emu} \cdot \frac{abcoulq}{cm^2} \qquad (2.62b)$$

Dropping down to the emu system in Fig. 2.7 by removing the "q" suffix gives

$$\vec{D}_{emu} \cdot \frac{abcoul}{cm^2} = c_{cgs}^{-2}\vec{E}_{emu} \cdot \frac{abcoul}{cm^2} + 4\pi \vec{P}_{emu} \cdot \frac{abcoul}{cm^2}$$

or, using Table 2.3,

$$\vec{D}_{emu} \cdot \frac{gm^{1/2}}{cm^{3/2}} = c_{cgs}^{-2}\vec{E}_{emu} \cdot \frac{gm^{1/2}}{cm^{3/2}} + 4\pi \vec{P}_{emu} \cdot \frac{gm^{1/2}}{cm^{3/2}}. \qquad (2.62c)$$

From Table 2.5 we note that $gm^{1/2}/cm^{3/2}$ are the correct emu units for D and P but not for E, which must have units of $gm^{1/2} \cdot cm^{1/2}/sec^2$. Therefore we are forced to multiply the E_{emu} term by

$$\frac{sec^2}{cm^2} \cdot \frac{cm^2}{sec^2} = 1$$

to get

$$\vec{D}_{emu} \cdot \frac{gm^{1/2}}{cm^{3/2}} = \left(c_{cgs}^{-2} \cdot \frac{sec^2}{cm^2} \right) \cdot \vec{E}_{emu} \cdot \frac{gm^{1/2} \cdot cm^{1/2}}{sec^2} + 4\pi \vec{P}_{emu} \cdot \frac{gm^{1/2}}{cm^{3/2}}$$

or

$$\vec{D} = \frac{1}{c^2}\vec{E} + 4\pi \vec{P} \qquad (2.62d)$$

using the physical quantities c, D, E, and P. Equation (2.62d) is the same as Eq. (2.61l), which shows that we do not need to recognize 1statcoulq/(statvoltq · cm) as ε_0 to get the right answer.

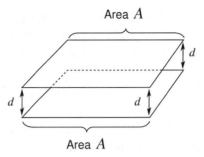

Area A

Figure 2.8 The two plates of this parallel-plate capacitor have the same shape and are placed one above the other, separated by a distance d.

The formula for the capacitance C of the parallel-plate capacitor in Fig. 2.8 is, in the emu system of units,

$$C = \frac{A}{4\pi \, dc^2}, \tag{2.63a}$$

where A is the area of each capacitor plate, d is the distance separating the plates, and c is the speed of light. Breaking this up into numeric parts and units gives, using Table 2.5,

$$C_{\text{emu}} \cdot \frac{\sec^2}{\text{cm}} = \frac{A_{\text{cgs}} \cdot \text{cm}^2}{4\pi \, (d_{\text{cgs}} \cdot \text{cm})\left(c_{\text{cgs}} \cdot \dfrac{\text{cm}}{\sec}\right)^2}. \tag{2.63b}$$

Consulting Rule VIII, we see that the first step is to use the physical meaning of capacitance to write C in terms of the connecting unit abfarad:

$$C_{\text{emu}} \cdot \text{abfarad} = \frac{A_{\text{cgs}} \cdot \text{cm}^2}{4\pi \, (d_{\text{cgs}} \cdot \text{cm})c_{\text{cgs}}^2} \cdot \frac{\sec^2}{\text{cm}^2}. \tag{2.63c}$$

Equation (2.63c) is balanced in the emu system, where 1 abfarad $= \sec^2/\text{cm}$, but does not yet obey the second part of Rule VIII, since it is not balanced in both the invariant units cm and sec and the connecting units (abfarad and abfaradq) to the upper level of Fig. 2.7. Multiplying the right-hand side by abfarad $\cdot$ cm/sec^2, which is 1 in the emu system, gives

$$C_{\text{emu}} \cdot \text{abfarad} = \frac{A_{\text{cgs}} \cdot \text{cm}^2}{4\pi \, (d_{\text{cgs}} \cdot \text{cm})} \cdot \left(\frac{1}{c_{\text{cgs}}^2} \frac{\sec^2}{\text{cm}^2}\right) \cdot \left(\frac{\text{abfarad} \cdot \text{cm}}{\sec^2}\right). \tag{2.63d}$$

This is clearly balanced in both the invariant and connecting units, so we can add the "q" suffix to transfer to the emuq system, recognizing charge as a fundamental dimension:

$$C_{\text{emu}} \cdot \text{abfaradq} = \frac{A_{\text{cgs}} \cdot \text{cm}^2}{4\pi \left(d_{\text{cgs}} \cdot \text{cm}\right)} \cdot \left(\frac{1}{c_{\text{cgs}}^2} \frac{\text{sec}^2}{\text{cm}^2}\right) \cdot \left(\frac{\text{abfaradq} \cdot \text{cm}}{\text{sec}^2}\right). \qquad (2.63\text{e})$$

From Table 2.6 we have $1\,\text{abfaradq} = c_{\text{cgs}}^2 \cdot \text{statfaradq}$, which gives

$$C_{\text{emu}} \cdot c_{\text{cgs}}^2 \cdot \text{statfaradq} = \frac{A_{\text{cgs}} \cdot \text{cm}^2}{4\pi \left(d_{\text{cgs}} \cdot \text{cm}\right)} \cdot \left(\frac{\text{statfaradq}}{\text{cm}}\right). \qquad (2.63\text{f})$$

Consulting Table 2.7, we note that $C_{\text{emu}} c_{\text{cgs}}^2 = C_{\text{esu}}$, so

$$C_{\text{esu}} \cdot \text{statfaradq} = \frac{A_{\text{cgs}} \cdot \text{cm}^2}{4\pi \left(d_{\text{cgs}} \cdot \text{cm}\right)} \cdot \frac{\text{statfaradq}}{\text{cm}} \qquad (2.63\text{g})$$

or

$$C_{\text{esu}} \cdot \text{statfarad} = \frac{A_{\text{cgs}} \cdot \text{cm}^2}{4\pi \left(d_{\text{cgs}} \cdot \text{cm}\right)} \cdot \frac{\text{statfarad}}{\text{cm}}, \qquad (2.63\text{h})$$

where in the last step we transfer from the esuq to the esu system in Fig. 2.7, no longer recognizing charge as a fundamental dimension by removing the "q" suffix. Table 2.4 shows that $1\,\text{statfarad} = \text{cm}$, so Eq. (2.63h) becomes

$$C_{\text{esu}} \cdot \text{statfarad} = \frac{A_{\text{cgs}} \cdot \text{cm}^2}{4\pi \left(d_{\text{cgs}} \cdot \text{cm}\right)}$$

or

$$C = \frac{A}{4\pi\ d}, \qquad (2.63\text{i})$$

when writing the formula in terms of the physical quantities C, A, and d.

2.10 Direct Conversion Between the ESU and EMU Systems of Units

Using esuq and emuq units to transform equations and formulas from the esu to emu system, or from the emu to esu system, is a straightforward process; but the job can also be done directly by combining Table 2.7 with the information in Table 2.8. The information in Table 2.8 comes from comparing the two columns of Table 2.3 and noticing that 1 statamp multiplied by sec/cm is the same as 1 abamp, 1 statvolt multiplied by cm/sec is the same as 1 abvolt, and so on.

Table 2.8 Unit relationships between the esu and emu systems.

$\text{abamp} = \dfrac{\text{sec}}{\text{cm}} \cdot \text{statamp}$
$\text{abcoul} = \dfrac{\text{sec}}{\text{cm}} \cdot \text{statcoul}$
$\text{abweber} = \dfrac{\text{cm}}{\text{sec}} \cdot \text{statweber}$
$\text{abvolt} = \dfrac{\text{cm}}{\text{sec}} \cdot \text{statvolt}$
$\text{abfarad} = \dfrac{\text{sec}^2}{\text{cm}^2} \cdot \text{statfarad}$
$\text{abhenry} = \dfrac{\text{cm}^2}{\text{sec}^2} \cdot \text{stathenry}$
$\text{abohm} = \dfrac{\text{cm}^2}{\text{sec}^2} \cdot \text{statohm}$

As an example of how to use Tables 2.7 and 2.8 to switch from emu to esu units, we consider the formula for B, Eq. (2.19e), in emu units. Broken up into numeric parts and units, this equation becomes

$$\vec{B}_{\text{emu}} \frac{1}{\text{sec}} \frac{\text{gm}^{1/2}}{\text{cm}^{1/2}} = \vec{H}_{\text{emu}} \frac{1}{\text{sec}} \frac{\text{gm}^{1/2}}{\text{cm}^{1/2}} + 4\pi \left(\vec{M}_H\right)_{\text{emu}} \frac{1}{\text{sec}} \frac{\text{gm}^{1/2}}{\text{cm}^{1/2}}. \qquad (2.64\text{a})$$

Consulting Table 2.5 for the ab-prefixed forms of emu units for B, H, and M_H, we write

$$\vec{B}_{\text{emu}} \frac{\text{abweber}}{\text{cm}^2} = \vec{H}_{\text{emu}} \frac{\text{abamp}}{\text{cm}} + 4\pi \left(\vec{M}_H\right)_{\text{emu}} \frac{\text{abweber}}{\text{cm}^2}. \qquad (2.64\text{b})$$

Using Table 2.7 to replace the emu-subscripted quantities by esu-subscripted quantities gives

$$\vec{B}_{\text{esu}} c_{\text{cgs}} \frac{\text{abweber}}{\text{cm}^2} = \vec{H}_{\text{esu}} c_{\text{cgs}}^{-1} \frac{\text{abamp}}{\text{cm}} + 4\pi \left(\vec{M}_H\right)_{\text{esu}} c_{\text{cgs}} \frac{\text{abweber}}{\text{cm}^2}. \qquad (2.64\text{c})$$

Multiplying all these terms by

$$\frac{\text{cm}}{\text{sec}} \cdot \frac{\text{sec}}{\text{cm}} = 1,$$

we have

$$\vec{B}_{esu}c_{cgs} \cdot \frac{cm}{sec} \cdot \frac{sec}{cm} \cdot \frac{abweber}{cm^2}$$

$$= \vec{H}_{esu}c_{cgs}^{-1} \cdot \frac{sec}{cm} \cdot \frac{cm}{sec} \cdot \frac{abamp}{cm} + 4\pi\left(\vec{M}_H\right)_{esu}c_{cgs} \cdot \frac{cm}{sec} \cdot \frac{sec}{cm} \cdot \frac{abweber}{cm^2},$$

or, consulting Table 2.8 to find that 1 statweber = (sec/cm) · abweber and 1 statamp = (cm/sec) · abamp, we get

$$\left(\vec{B}_{esu}\frac{statweber}{cm^2}\right) \cdot \left(c_{cgs}\frac{cm}{sec}\right)$$

$$= \left(\vec{H}_{esu}\frac{statamp}{cm}\right)\frac{1}{\left(c_{cgs}\dfrac{cm}{sec}\right)} \tag{2.64d}$$

$$+ 4\pi\left[\left(\vec{M}_H\right)_{esu}\frac{statweber}{cm^2}\right] \cdot \left(c_{cgs}\frac{cm}{sec}\right).$$

Writing this in terms of the physical quantities c, B, H, and M_H then gives

$$c\vec{B} = \frac{\vec{H}}{c} + 4\pi c\vec{M}_H \quad \text{or} \quad \vec{B} = \frac{\vec{H}}{c^2} + 4\pi\vec{M}_H \tag{2.64e}$$

in esu units. Since $\mu_0 = c^{-2}$ in the esu system, Eq. (2.64e) is the same as Eq. (2.19d), which is the expected result.

To show how this method can be used to go from the esu units to emu units, we take Eq. (2.63i) back into the emu system. Breaking both sides of the formula into numeric parts and units gives

$$C_{esu}\text{statfarad} = \frac{A_{cgs}cm^2}{4\pi d_{cgs}cm}, \tag{2.65a}$$

which becomes, using Tables 2.7 and 2.8,

$$C_{emu}c_{cgs}^2\frac{cm^2}{sec^2}\text{abfarad} = \frac{A_{cgs}cm^2}{4\pi d_{cgs}cm}. \tag{2.65b}$$

Written in terms of c, A, d, and the capacitance C in emu units, this becomes

$$C = \frac{A}{4\pi dc^2}. \tag{2.65c}$$

Equation (2.65c) is identical to Eq. (2.63a), as it should be.

2.11 THE B AND H FIELDS AT THE START OF THE TWENTIETH CENTURY

At the beginning of the twentieth century, physicists began to re-evaluate the roles of the B and H fields, with the magnetic induction B taken to be the fundamental magnetic field and the H field regarded as a sometimes-helpful auxiliary field. This was a natural change of opinion as it became more and more clear that isolated magnetic poles did not exist, that the behavior of permanent magnets could be explained in terms of microscopic current loops, and that moving electric charges interact directly with the B rather than the H field. A contributing factor was the just-proposed theory of special relativity, describing how a stationary charge's E field transforms into a combination of E and B fields when viewed by observers in motion with respect to the charge. We can easily show, in fact, that moving charges interact directly with the B field rather than the H field of a permanent magnet.

Figure 2.9 shows the south pole of a long, thin permanent magnet placed near an infinitely long wire carrying a constant current I. The pole strength of the south pole is $(-p_H)$ and the north pole is so far from the wire that its influence can be neglected. We model the current in the wire as a sequence of point charges moving at an average velocity v with an average linear density of λ_Q. The linear density λ_Q is constant along the wire and measured in terms of the charge per unit length, so

$$\frac{dQ}{dt} = I = \lambda_Q v. \tag{2.66a}$$

From Eqs. (2.5), (2.6), and (2.66a), we know that the total force on the magnet's south pole is

$$F_{SP} = p_H H_{WIRE} = \frac{2p_H I}{r} = \frac{2\lambda_Q v p_H}{r}, \tag{2.66b}$$

where H_{WIRE} is the magnetic field created by the current in the wire, $\vec{F}_{SP}$ points into the page, and r is the distance of the pole from the wire. We have not specialized these equations to one set of units, which means Eqs. (2.66a,b) are good in any of the four systems of units so far discussed. By the middle of the nineteenth century it was well established that a short wire segment of length $d\ell$ carrying a constant current I experiences a force proportional to

$$(Id\ell)H \sin\phi$$

in a magnetic field H making an angle ϕ with respect to the wire (see Fig. 2.9). Taking the constant of proportionality to be α, we have for the force dF_{WIRE} on the wire segment of length $d\ell$

$$dF_{WIRE} = \alpha(Id\ell)H \sin\phi. \tag{2.66c}$$

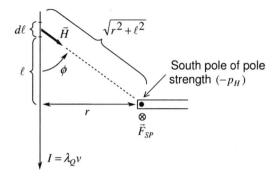

Figure 2.9 The south pole of a permanent magnet interacts with the constant current I in a long, straight wire. The pole strength of the south pole is $(-p_H)$.

From Eq. (2.3b) and Fig. 2.9 we know that

$$dF_{WIRE} = \alpha(Id\ell)\frac{p_H}{\mu_0} \cdot \frac{\sin\phi}{r^2 + \ell^2} = \left(\frac{\alpha\lambda_Q v p_H}{\mu_0}\right) \cdot \frac{rd\ell}{(r^2 + \ell^2)^{3/2}}, \qquad (2.66d)$$

where in the last step we have used Eq. (2.66a) and the definition of $\sin\phi$ to eliminate the current and the angle from the formula for dF_{WIRE}. The total force on the wire is

$$F_{WIRE} = \frac{\alpha\lambda_Q v p_H r}{\mu_0} \int_{-\infty}^{\infty} \frac{d\ell}{(r^2 + \ell^2)^{3/2}} = \left(\frac{\alpha\lambda_Q v p_H r}{\mu_0}\right) \cdot \left[\frac{\ell}{r^2\sqrt{\ell^2 + r^2}}\right]_{-\infty}^{\infty}$$

$$= \frac{2\alpha\lambda_Q v p_H}{\mu_0 r}. \qquad (2.66e)$$

By Newton's third law we know that

$$F_{WIRE} = F_{SP}; \qquad (2.66f)$$

that is, the magnitude of the current's total force on the pole must equal the magnitude of the pole's total force on the current. From Eqs. (2.66b), (2.66e), and (2.66f), we see that

$$\frac{2\lambda_Q v p_H}{r} = \frac{2\alpha\lambda_Q v p_H}{\mu_0 r}$$

or

$$\alpha = \mu_0.$$

Hence, using Eq. (2.66c) and the definition of B in empty space [see Eq. (2.19d)], it follows that

$$dF_{WIRE} = (Id\ell \sin\phi)(\mu_0 H) = (Id\ell \sin\phi)B.$$

Using Eq. (2.66a), this can be written as

$$dF_{WIRE} = (\lambda_Q d\ell)vB \sin\phi.$$

Since $\lambda_Q d\ell$ is the amount of moving charge dQ in a length $d\ell$ of the wire, we end up with

$$dF_{WIRE} = dQvB \sin\phi$$

or

$$d\vec{F}_{WIRE} = dQ \cdot (\vec{v} \times \vec{B}), \tag{2.66g}$$

where the definition of the vector cross product is used to write the formula for the vector force $d\vec{F}_{WIRE}$ in terms of the vector velocity $\vec{v}$, the moving charge $dQ = \lambda_Q d\ell$, and the $\vec{B}$ field. This shows what we wanted to prove, that moving charge interacts directly with the $\vec{B}$ field rather than the $\vec{H}$ field. Hence, if isolated magnetic poles do not exist and all magnetic fields are created by currents—either macroscopic or microscopic—it is only natural to regard the magnetic induction $\vec{B}$ as the fundamental magnetic field. The magnetic induction is then defined by the force it exerts on an isolated point charge Q moving with a vector velocity $\vec{v}$:

$$\vec{F} = Q\vec{E} + Q(\vec{v} \times \vec{B}). \tag{2.66h}$$

Equation (2.66h) is called the Lorentz force law, for the force $\vec{F}$ experienced by a point charge Q moving with a velocity $\vec{v}$ at a field point where the electric field is $\vec{E}$ and the magnetic induction is $\vec{B}$. Because neither ε_0, μ_0, nor c are part of the Lorentz force law, we expect it to have the same form in the esu, emu, esuq, and emuq systems of units—which is in fact the case. Equation (2.66h) is a more convincing fundamental equation than Eq. (2.2b), Coulomb's law for magnetic poles, because isolated charges, unlike isolated magnetic poles, are known to exist in the form of elementary particles such as the electron, proton, etc.

In Table 2.7 the vector potential $\vec{A}$, the current-loop magnetic dipole density $\vec{M}_I$, and the current-loop magnetic dipole moment $\vec{m}_I$ all reflect the interchanged roles of the B and H magnetic fields.

The B field in Eq. (2.19d) is constructed to be a zero-divergence field; as the primary magnetic field it makes sense to represent it by the curl of another vector field $\vec{A}$, called the vector potential:

$$\vec{B} = \text{curl}\vec{A} = \vec{\nabla} \times \vec{A}, \tag{2.67a}$$

because the divergence of the curl of any vector field is always zero;

$$\text{div}\left(\text{curl}\vec{A}\right) = \vec{\nabla} \cdot \left(\vec{\nabla} \times \vec{A}\right) = 0 = \vec{\nabla} \cdot \vec{B}. \tag{2.67b}$$

Equations (2.67a, b) have the same form in the esu, emu, esuq, and emuq systems of units. Many of the equations of quantum mechanics take on a simpler form when written in terms of the vector potential $\vec{A}$ instead of the magnetic induction $\vec{B}$.

Since $\vec{H}$ is now the auxiliary field, Eq. (2.19d) defines $\vec{H}$ rather than $\vec{B}$ and should be written as

$$\vec{H} = \frac{1}{\mu_0} \cdot \vec{B} - 4\pi \left(\frac{\vec{M}_H}{\mu_0}\right). \tag{2.68a}$$

Written this way, it is natural to define a new quantity

$$\vec{M}_I = \frac{1}{\mu_0} \cdot \vec{M}_H \tag{2.68b}$$

called the current-loop magnetic dipole density field, with

$$\vec{m}_I = \frac{1}{\mu_0} \cdot \vec{m}_H \tag{2.68c}$$

defined to be the current-loop magnetic dipole moment. Now Eq. (2.68a) can be written as

$$\vec{H} = \frac{1}{\mu_0} \cdot \vec{B} - 4\pi \vec{M}_I. \tag{2.68d}$$

In general, the distinctions drawn between m_I, m_H and M_I, M_H in Eqs. (2.68b, c) are irrelevant in emu units—in fact, they are irrelevant in any system where μ_0 is the dimensionless numeric 1 (such as the Gaussian and Heaviside-Lorentz systems discussed in Chapter 3). From this point on, when we convert equations containing the magnetic dipole moment or magnetic dipole density from systems where $\mu_0 = 1$ to other systems where $\mu_0 \neq 1$, we must first decide whether the magnetic dipole moment is m_I or m_H and whether the magnetic dipole density is M_I or M_H. This sounds like more of a problem than it really is; all we are really deciding is whether we want the corresponding formula in the other systems where $\mu_0 \neq 1$ to be in terms of m_I, M_I or m_H, M_H. It is good practice to make the choice immediately by attaching I or H subscripts to m and M before starting the conversion, because then there is never any doubt about which units and table entries to use. For this reason we will continue to label the magnetic dipole moment and magnetic dipole density with subscripts I and H, even when working in systems of units where $\mu_0 = 1$; this way it is always clear what these variables are intended to be when $\mu_0 \neq 1$.

The $\vec{m}_I$ magnetic dipole moment of any current loop in a plane is

$$\vec{m}_I = (I \cdot A)\hat{n}, \tag{2.68e}$$

where I is the constant current in the loop of area A, and $\hat{n}$ is the dimensionless unit vector perpendicular to the planar surface inside the loop (we note that the current must circulate counterclockwise around the base of $\hat{n}$ (as shown in Fig. 2.10). Equation (2.68e) has the same form in the esu, emu, esuq, and emuq systems of units.

Because many textbooks talk about the magnetic dipole moment $\vec{m}$ without clearly specifying its exact nature, even when working in systems where $\mu_0 \neq 1$, it is important to determine whether m_I or m_H is being referred to. One helpful indication is whether the torque $\vec{T}$ on an isolated magnetic dipole $\vec{m}$ is written as

$$\vec{T} = \vec{m} \times \vec{H} \quad \text{or} \quad \vec{T} = \vec{m} \times \vec{B}.$$

The first formula reveals $\vec{m}$ to be $\vec{m}_H$, so

$$\vec{T} = \vec{m}_H \times \vec{H}, \tag{2.69a}$$

and the second formula reveals $\vec{m}$ to be $\vec{m}_I$,

$$\vec{T} = \vec{m}_I \times \vec{B}. \tag{2.69b}$$

Many times these relationships are presented as a potential energy of orientation of the dipole in an external magnetic field. If the potential energy U for a magnetic dipole $\vec{m}$ is written as

$$U = -\vec{m} \cdot \vec{H},$$

then we know that $\vec{m} = \vec{m}_H$ and it is more precise to write this equation as

$$U = -\vec{m}_H \cdot \vec{H}. \tag{2.69c}$$

If the potential energy is written as

$$U = -\vec{m} \cdot \vec{B},$$

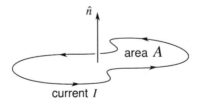

Figure 2.10 The current-loop magnetic dipole moment is $\vec{m}_I = I A\hat{n}$.

then we know that $\vec{m} = \vec{m}_I$ and it is more precise to write this equation as

$$U = -\vec{m}_I \cdot \vec{B}. \tag{2.69d}$$

2.12 ELECTROMAGNETIC CONCEPTS USED TO ANALYZE BULK MATTER

Most of the yet-to-be-specified quantities in Table 2.7 are straightforward extensions of concepts which have already been discussed. For example, the resistance of a wire, R_{WIRE}, is directly proportional to its length L_{WIRE} and inversely proportional to its cross-sectional area A_{WIRE}; therefore, the material of the wire can be assigned a constant resistivity

$$\rho_R = \frac{R_{WIRE} \cdot A_{WIRE}}{L_{WIRE}}. \tag{2.70a}$$

The conductivity of the wire material is

$$\sigma = \frac{1}{\rho_R}. \tag{2.70b}$$

Tables 2.4 and 2.5 give the units of resistivity and conductivity in the esu, esuq, emu, and emuq systems of units. We cannot give their practical units because the practical system of units described in Section 2.6 does not have an official unit of length.

The behavior of electric fields and charges inside matter can often be analyzed by treating the substance as a region of space characterized by a permittivity that is proportional to the permittivity of free space:

$$\varepsilon = \varepsilon_r \varepsilon_0, \tag{2.70c}$$

where ε_r is a dimensionless numeric. The physical quantity ε is often called the dielectric constant and ε_r is often called the relative dielectric constant. There is a similar equation for analyzing magnetic fields inside matter:

$$\mu = \mu_r \mu_0, \tag{2.70d}$$

with μ called the magnetic permeability of the substance and μ_r a dimensionless numeric that is sometimes called the relative magnetic permeability. Different materials have different values of ε_r and μ_r, although for most nonmetallic substances $\mu_r \cong 1$. For any given material the values of ε and μ, unlike the values of ε_0 and μ_0 in empty space, depend on the frequency of the electromagnetic phenomena being analyzed. Equations (2.70c, d) show that ε and μ have the same units as ε_0 and μ_0 in all systems of electromagnetic units.

The electric and magnetic susceptibilities, χ_e and χ_m respectively, are dimensionless scalars defined by the equations

$$\vec{P} = \chi_e \varepsilon_0 \vec{E} \tag{2.70e}$$

and

$$\vec{M}_H = \chi_m \mu_0 \vec{H}, \tag{2.70f}$$

when the material is such that $\vec{P}$ is parallel to $\vec{E}$ and $\vec{M}_H$ is parallel to $\vec{H}$. We note that when $\vec{P}$ is not parallel to $\vec{E}$, electric phenomena cannot be analyzed by treating the material as a region of space with a scalar permittivity of ε instead of ε_0; and when $\vec{M}_H$ is not parallel to $\vec{H}$, magnetic phenomena cannot be analyzed by treating the material as a region of space with a scalar permeability μ instead of μ_0. It should be pointed out that neither rationalization nor a change of units affects the values of ε_r and μ_r, which is perhaps no surprise since they are dimensionless; but the values of χ_e and χ_m, although they are also dimensionless and unaffected by a change of units, are indeed changed by rationalization (see Tables 3.6 and 3.14 in Chapter 3).

The volume charge density field ρ_Q is just the amount of charge per unit volume at any field point. This means the amount of charge dQ inside an infinitesimal volume dV at that field point is

$$dQ = \rho_Q \cdot dV. \tag{2.71a}$$

We can also define a surface charge density S_Q such that the amount of charge inside an infinitesimal area dA of a surface is

$$dQ = S_Q \cdot dA. \tag{2.71b}$$

The volume current density $\vec{J}$ describes the flow of charge. It points in the direction the charge is flowing and has a magnitude such that the infinitesimal electric current dI crossing an infinitesimal area dA at any field point is

$$dI = \left(\vec{J} \cdot \hat{n}\right) dA = \left(|\vec{J}| \cos\theta\right) dA. \tag{2.71c}$$

Here $\hat{n}$ is the dimensionless unit vector which is normal to dA and pointing in the direction for which dI is defined to be positive. The angle between $\vec{J}$ and $\hat{n}$ is θ. We can use the $\vec{J}$ field inside a wire such as the one shown in Fig. 2.11(a) to write the wire's total current I as

$$I = \int_{\text{over S}} dI = \int_{\text{over S}} \left(\vec{J} \cdot \hat{n}\right) dA, \tag{2.71d}$$

where S is the wire's cross-sectional surface. The surface current density $\mathcal{J}_S$ is a vector pointing in the direction of a current flowing on a surface and can be used to write the current dI flowing across a line element $\vec{d\ell}$ on that surface as

$$dI = \vec{J}_S \cdot \left(\vec{d\ell} \times \hat{n} \right) \qquad (2.71e)$$

In Fig. 2.11(b), we show that the total surface-current I flowing across a curve L can be written as

$$I = \int_{\text{overL}} dI = \int_{\text{overL}} \vec{J}_S \cdot \left(\vec{d\ell} \times \hat{n} \right) \qquad (2.71f)$$

Tables 2.4 and 2.5 give the units of ρ_Q, S_Q, $\vec{J}$, and $\vec{J}_S$ in the esu, esuq, emu, and emuq systems of units. Note that ρ_Q and $\vec{J}$ can describe the distribution and flow of charge even when no wires are present—for example when unbound electrons and protons interact in a plasma.

The remaining physical quantities in Table 2.7—the permeance, reluctance, magnetic flux, and magnetomotive force—are useful when analyzing the design of electromagnets and transformers, topics not usually covered in introductory textbooks of electricity and magnetism. It can be shown that an ideal electromagnet of the type sketched in Fig. 2.12 can be modelled as a magnetic circuit, with the $\vec{B}$ field playing the role of the $\vec{J}$ field in an ordinary electric circuit and the coil

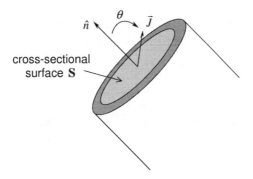

Figure 2.11a The current inside a wire can be represented by a volume current density $\vec{J}$ specified over a cross-sectional surface S.

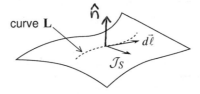

Figure 2.11b The current flowing along a surface can be represented by a surface current density $\mathcal{J}_S$ specified along a curve L.

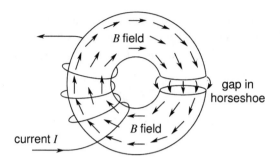

Figure 2.12 The current I generates a magnetic B field inside the iron or steel horseshoe.

of wire creating a magnetomotive force analogous to the total electric potential or voltage V of an electric circuit. The magnetomotive force is proportional to the electric current I in the coil of wire and always has the same electromagnetic units as an electric current. The iron or steel "horseshoe" around which the wire is coiled has a small gap. The B field is almost entirely zero outside the horseshoe except in the gap, where it tends to keep the approximately constant value it has inside the horseshoe. Since $\vec{B}$ is analogous to $\vec{J}$, the magnetic flux

$$\Phi_B = \int_{\substack{\text{cross–section} \\ \text{of horseshoe}}} (\vec{B} \cdot \hat{n}) dA \qquad (2.72a)$$

is analogous to electric current; and for magnetic circuits we can write an equation analogous to Ohm's law [see Eq. (2.25a) above]:

$$\mathcal{R} \cdot \Phi_B = \mathcal{F}. \qquad (2.72b)$$

In Eq. (2.72b), $\mathcal{F}$ is the magnetomotive force and $\mathcal{R}$ is the reluctance. The permeance $\mathcal{P}$ of the magnetic circuit is $1/\mathcal{R}$, so it is analogous to the conductance of an electric circuit. If the gap in Fig. 2.12 is small, then B is approximately constant in the gap and zero outside, so

$$\Phi_B \cong B \cdot A_{GAP}, \qquad (2.72c)$$

where A_{GAP} is the cross-sectional area of the gap. Given the reluctance (or permeance) and magnetomotive force of the electromagnet, Eqs. (2.72b, c) can be used to find the magnetic induction B inside the gap.

APPENDIX 2.A: MAGNETIC-FIELD MEASUREMENT IN THE EARLY NINETEENTH CENTURY

A small, thin compass needle suspended from its center of mass and allowed to swing freely will settle down to an equilibrium position parallel to the local magnetic field $\vec{H}$. Figure 2.A.1 shows that in this equilibrium position a vector drawn from the needle's south pole of pole strength $(-p_H) < 0$ to its north pole of pole strength $p_H > 0$ points in the direction of $\vec{H}$. We say that the needle has a permanent-magnet dipole moment

$$m_H = p_H \cdot L, \tag{2.A.1}$$

where L is the distance between the needle's north and south poles. A thin bar with constant mass density and constant cross-sectional area has a moment of inertia

$$\iota = \frac{M\ell^2}{12}, \tag{2.A.2}$$

where ℓ is the total length of the bar and M is the total mass of the bar. In general, ℓ, the length of a thin bar magnet, is slightly larger than L, the distance between the effective positions of the north and south magnetic poles.

When the needle in Fig. 2.A.1 is slightly disturbed, it undergoes small-angle oscillations about its equilibrium position. Because the needle is suspended from its center of mass, gravity cannot act to restore the needle to its equilibrium position, because for every mass element that experiences a drop in height there is a corresponding mass element that experiences a gain in height, leaving the overall gravitational potential energy constant. There is, however, a change in potential energy from the interaction of the needle with the local magnetic field $\vec{H}$, as shown in Fig. 2.A.2. For small-angle oscillations about equilibrium we define two angles,

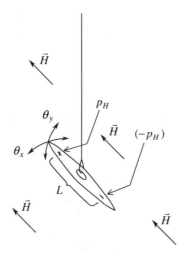

Figure 2.A.1 The small needle suspended in a constant magnetic field $\vec{H}$ is a permanent magnet with a north pole of pole strength p_H and a south pole of pole strength $(-p_H)$. It oscillates with a period which is inversely proportional to the square root of $|\vec{H}|$.

θ_x and θ_y, describing the swing of the needle's pole tip from its equilibrium angle (see Fig. 2.A.1). The equations of motion of the swinging needle are

$$\iota \frac{d^2\theta_x}{dt^2} = -m_H H \theta_x \qquad (2.A.3a)$$

and

$$\iota \frac{d^2\theta_y}{dt^2} = -m_H H \theta_y, \qquad (2.A.3b)$$

where $H = |\vec{H}|$ is the magnitude of the magnetic field at the position of the needle. These two simple-harmonic-oscillator equations have the general solution

$$\theta_x(t) = A_x \sin\left(\frac{2\pi t}{T}\right) + B_x \cos\left(\frac{2\pi t}{T}\right) \qquad (2.A.4a)$$

and

$$\theta_y(t) = A_y \sin\left(\frac{2\pi t}{T}\right) + B_y \cos\left(\frac{2\pi t}{T}\right), \qquad (2.A.4b)$$

where

$$T = 2\pi \sqrt{\frac{\iota}{m_H H}} \qquad (2.A.4c)$$

is the period of the needle's oscillation, and A_x, B_x, A_y, and B_y are arbitrary real constants determined by the type of initial disturbance experienced by the needle.

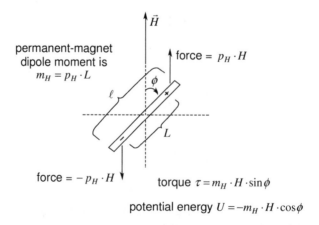

Figure 2.A.2 The poles of a long, thin bar magnet of length ℓ are separated by a distance L. For bar magnets L is slightly less than ℓ.

No matter what the values of A_x, B_x, A_y, and B_y are—that is, no matter what type of (small) disturbance starts the needle oscillating—we see that

$$\theta_x(t+T) = \theta_x(t)$$

and

$$\theta_y(t+T) = \theta_y(t).$$

This makes the period T of a small, swinging needle an easy way to measure the size of an unknown magnetic field $\vec{H}$. Consequently, the equilibrium orientation of the needle gives the direction of the $\vec{H}$ field, as in Fig. 2.A.1, and the magnitude of the $\vec{H}$ field is given by

$$H = \frac{4\pi^2 \iota}{m_H T^2}. \tag{2.A.5a}$$

From Eqs. (2.A.1), (2.A.2), and (2.A.5a) we have

$$H = \frac{4\pi^2 \left(\dfrac{M\ell^2}{12}\right)}{(p_H L)T^2} = \frac{\pi^2}{3} \cdot \frac{M\ell^2}{p_H L T^2}. \tag{2.A.5b}$$

Both ι and m_H are intrinsic properties of the needle in Eq. (2.A.5a). Therefore, we can create a known magnetic field H_K, measure the period of a suspended needle T_K in field H_K, move the needle to an unknown magnetic field H_U, measure the period T_U of the same needle in H_U, and calculate H_U from the known quantities H_K, T_K, and T_U:

$$H_U = H_K \left(\frac{T_K}{T_U}\right)^2. \tag{2.A.6}$$

APPENDIX 2.B: Dimensionless vector derivatives

There are three basic types of product defined for vectors: the product of a vector and a scalar, the dot product of two vectors, and the cross product of two vectors. Each vector product can be defined by specifying the behavior of the dimensionless unit vectors of a Cartesian coordinate system.

Figure 2.B gives the x, y, z scalar components of a vector physical quantity $\vec{b}$ in a three-dimensional Cartesian coordinate system. The three components of $\vec{b}$ are called b_x, b_y, and b_z, respectively, and all three scalars have the same physical

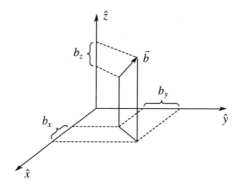

Figure 2.B The $\hat{x}, \hat{y}, \hat{z}$ components of vector $\vec{b}$ are b_x, b_y, b_z respectively.

dimensions and customarily are measured in the same units. The $\hat{x}, \hat{y}, \hat{z}$ dimensionless unit vectors can be used to write $\vec{b}$ as

$$\vec{b} = b_x\hat{x} + b_y\hat{y} + b_z\hat{z}. \tag{2.B.1a}$$

Equation (2.B.1a) is equivalent to the standard definition of a three-dimensional Cartesian vector. For example, in column-vector notation

$$\hat{x} = \begin{pmatrix} 1 \\ 0 \\ 0 \end{pmatrix}, \quad \hat{y} = \begin{pmatrix} 0 \\ 1 \\ 0 \end{pmatrix}, \quad \hat{z} = \begin{pmatrix} 0 \\ 0 \\ 1 \end{pmatrix},$$

so that

$$b_x\hat{x} = \begin{pmatrix} b_x \\ 0 \\ 0 \end{pmatrix}, \quad b_y\hat{y} = \begin{pmatrix} 0 \\ b_y \\ 0 \end{pmatrix}, \quad b_z\hat{z} = \begin{pmatrix} 0 \\ 0 \\ b_z \end{pmatrix}, \tag{2.B.1b}$$

and

$$\vec{b} = b_x\hat{x} + b_y\hat{y} + b_z\hat{z} = \begin{pmatrix} b_x \\ 0 \\ 0 \end{pmatrix} + \begin{pmatrix} 0 \\ b_y \\ 0 \end{pmatrix} + \begin{pmatrix} 0 \\ 0 \\ b_z \end{pmatrix} = \begin{pmatrix} b_x \\ b_y \\ b_z \end{pmatrix}. \tag{2.B.1c}$$

If we use row-vector notation

$$\hat{x} = (1\ 0\ 0), \quad \hat{y} = (0\ 1\ 0), \quad \hat{z} = (0\ 0\ 1),$$

we end up with

$$\vec{b} = (b_x\ b_y\ b_z).$$

We note that Eqs. (2.B.1a–c) assume some understanding of what happens when a scalar multiplies a vector, since b_x, b_y, b_z are scalars and $\hat{x}, \hat{y}, \hat{z}$ are vectors. Strictly speaking, however, Eqs. (2.B.1a–c) only specify what happens when scalars multiply dimensionless unit vectors. To define multiplication of any vector $\vec{b}$ by a real or complex scalar α we note that

$$\alpha\vec{b} = \alpha\left(b_x\hat{x} + b_y\hat{y} + b_z\hat{z}\right) = (\alpha b_x)\hat{x} + (\alpha b_y)\hat{y} + (\alpha b_z)\hat{z}, \qquad (2.B.2a)$$

which shows that the components b_x, b_y, and b_z become αb_x, αb_y, and αb_z when α multiplies $\vec{b}$. This matches our intuitive understanding of what happens when α multiplies $\vec{b}$. To see why this is so, note that the length of $\alpha\vec{b}$ is α times the length of $\vec{b}$:

$$\left|\alpha\vec{b}\right| = \sqrt{\alpha^2 b_x^2 + \alpha^2 b_y^2 + \alpha^2 b_z^2} = \alpha\sqrt{b_x^2 + b_y^2 + b_z^2} = \alpha\left|\vec{b}\right|;$$

while the direction of $\alpha\vec{b}$ remains the same as the direction of $\vec{b}$ because the ratios of the x, y, z components do not change:

$$\frac{\alpha b_x}{\alpha b_y} = \frac{b_x}{b_y}, \quad \frac{\alpha b_x}{\alpha b_z} = \frac{b_x}{b_z}, \quad \frac{\alpha b_y}{\alpha b_z} = \frac{b_y}{b_z}.$$

The product of a scalar and a vector is another vector, but the dot product of two vectors is a scalar. We define

$$\hat{x} \cdot \hat{x} = 1, \quad \hat{x} \cdot \hat{y} = \hat{x} \cdot \hat{z} = 0, \qquad (2.B.3a)$$

$$\hat{y} \cdot \hat{y} = 1, \quad \hat{y} \cdot \hat{x} = \hat{y} \cdot \hat{z} = 0, \qquad (2.B.3b)$$

$$\hat{z} \cdot \hat{z} = 1, \quad \hat{z} \cdot \hat{x} = \hat{z} \cdot \hat{y} = 0. \qquad (2.B.3c)$$

For any three vectors $\vec{a}, \vec{b}$, and $\vec{c}$ and any three real or complex scalars α, β, and γ, we require the dot product to distribute over vector addition:

$$\left(\alpha\vec{a}\right) \cdot \left(\beta\vec{b} + \gamma\vec{c}\right) = \left(\beta\vec{b} + \gamma\vec{c}\right) \cdot \left(\alpha\vec{a}\right) = (\alpha\beta)\left(\vec{b} \cdot \vec{a}\right) + (\alpha\gamma)\left(\vec{a} \cdot \vec{c}\right)$$
$$= (\alpha\beta)\left(\vec{a} \cdot \vec{b}\right) + (\alpha\gamma)\left(\vec{c} \cdot \vec{a}\right). \qquad (2.B.3d)$$

The first and third step in Eq. (2.B.3d) show that the dot product commutes: $\vec{a} \cdot \vec{b} = \vec{b} \cdot \vec{a}$ for any two vectors $\vec{a}$ and $\vec{b}$. Now we know enough to find the scalar value corresponding to the dot product of any two vectors

$$\vec{a} = a_x\hat{x} + a_y\hat{y} + a_z\hat{z} \quad \text{and} \quad \vec{b} = b_x\hat{x} + b_y\hat{y} + b_z\hat{z}.$$

We have, from Eqs. (2.B.3a–d), that

$$\vec{a} \cdot \vec{b} = \left(a_x \hat{x} + a_y \hat{y} + a_z \hat{z}\right) \cdot \left(b_x \hat{x} + b_y \hat{y} + b_z \hat{z}\right)$$

$$= a_x\left(b_x \hat{x} \cdot \hat{x} + b_y \hat{x} \cdot \hat{y} + b_z \hat{x} \cdot \hat{z}\right)$$

$$+ a_y\left(b_x \hat{y} \cdot \hat{x} + b_y \hat{y} \cdot \hat{y} + b_z \hat{y} \cdot \hat{z}\right) \tag{2.B.3e}$$

$$+ a_z\left(b_x \hat{z} \cdot \hat{x} + b_y \hat{z} \cdot \hat{y} + b_z \hat{z} \cdot \hat{z}\right) = a_x a_x + a_y a_y + a_z a_z.$$

This is, of course, the standard formula for the dot product of two vectors in x, y, z Cartesian coordinates. The dot product can also be used to specify the x, y, z components of any vector $\vec{a}$. From Eqs. (2.B.3a–c), we get

$$\vec{a} \cdot \hat{x} = \hat{x} \cdot \vec{a} = a_x, \tag{2.B.3f}$$

$$\vec{a} \cdot \hat{y} = \hat{y} \cdot \vec{a} = a_y, \tag{2.B.3g}$$

$$\vec{a} \cdot \hat{z} = \hat{z} \cdot \vec{a} = a_z. \tag{2.B.3h}$$

The cross product of two vectors is another vector. We define

$$\hat{x} \times \hat{x} = \hat{y} \times \hat{y} = \hat{z} \times \hat{z} = 0, \tag{2.B.4a}$$

$$\hat{x} \times \hat{y} = \hat{z} = -\left(\hat{y} \times \hat{x}\right), \tag{2.B.4b}$$

$$\hat{z} \times \hat{x} = \hat{y} = -\left(\hat{x} \times \hat{z}\right), \tag{2.B.4c}$$

$$\hat{y} \times \hat{z} = \hat{x} = -\left(\hat{z} \times \hat{y}\right). \tag{2.B.4d}$$

Interchanging the order of the cross product of any two vectors $\vec{a}$ and $\vec{b}$ gives

$$\vec{a} \times \vec{b} = -\left(\vec{b} \times \vec{a}\right). \tag{2.B.4e}$$

We note that the cross products of the dimensionless unit vectors in Eqs. (2.B.4a–d) obey this rule. For any three vectors $\vec{a}, \vec{b}$, and $\vec{c}$, and any three real or complex scalars α, β, and γ, we require the vector cross product to distribute over addition:

$$\left(\alpha \vec{a}\right) \times \left(\beta \vec{b} + \gamma \vec{c}\right) = \left(\alpha\beta\right)\left(\vec{a} \times \vec{b}\right) + \left(\alpha\gamma\right)\left(\vec{a} \times \vec{c}\right). \tag{2.B.4f}$$

Now we know enough to find the cross product of any two vectors $\vec{a}$ and $\vec{b}$. Equations (2.B.4a–f) give

$$
\begin{aligned}
\vec{a} \times \vec{b} &= \left(a_x\hat{x} + a_y\hat{y} + a_z\hat{z}\right) \times \left(b_x\hat{x} + b_y\hat{y} + b_z\hat{z}\right) \\
&= a_x\left(b_x\hat{x} \times \hat{x} + b_y\hat{x} \times \hat{y} + b_z\hat{x} \times \hat{z}\right) \\
&\quad + a_y\left(b_x\hat{y} \times \hat{x} + b_y\hat{y} \times \hat{y} + b_z\hat{y} \times \hat{z}\right) \\
&\quad + a_z\left(b_x\hat{z} \times \hat{x} + b_y\hat{z} \times \hat{y} + b_z\hat{z} \times \hat{z}\right) \\
&= (a_yb_z - a_zb_y)\hat{x} + (a_zb_x - a_xb_z)\hat{y} + (a_xb_y - a_yb_x)\hat{z}.
\end{aligned}
\tag{2.B.4g}
$$

Equation (2.B.4g) is the standard formula for the cross product of two vectors in x, y, z Cartesian coordinates.

There is a strong tradition in mathematical physics of defining a vector differential operator $\vec{\nabla}$, called the del operator, by

$$
\vec{\nabla} = \hat{x}\frac{\partial}{\partial x} + \hat{y}\frac{\partial}{\partial y} + \hat{z}\frac{\partial}{\partial z},
\tag{2.B.5a}
$$

and using it to examine the change of vector and scalar fields with respect to changes in the location, or field point, at which the fields are evaluated. Although Eq. (2.B.5a) is the customary way of writing $\vec{\nabla}$, it is equally correct to write

$$
\vec{\nabla} = \frac{\partial}{\partial x}\hat{x} + \frac{\partial}{\partial y}\hat{y} + \frac{\partial}{\partial z}\hat{z}
\tag{2.B.5b}
$$

because the location and size of the $\hat{x}, \hat{y}, \hat{z}$ dimensionless unit vectors do not change when the location of the field point changes. In effect, vectors $\hat{x}, \hat{y}, \hat{z}$ are three constant vector fields whose derivatives in the x, y, z directions are always zero. (Note that the same cannot be said for vectors such as $\hat{r}$, the dimensionless unit vector pointing from the origin of the coordinate system to the field point.)

There are three standard derivatives using $\vec{\nabla}$, with each standard derivative corresponding to one of the three types of vector product discussed above. A scalar field ϕ is defined to be a scalar function of position $\phi = \phi(x, y, z)$. The gradient of a scalar field ϕ corresponds to multiplying a vector (i.e., operator $\vec{\nabla}$) by a scalar (i.e., the field ϕ):

$$
\vec{\nabla}\phi = \left(\hat{x}\frac{\partial}{\partial x} + \hat{y}\frac{\partial}{\partial y} + \hat{z}\frac{\partial}{\partial z}\right)\phi = \hat{x}\frac{\partial\phi}{\partial x} + \hat{y}\frac{\partial\phi}{\partial y} + \hat{z}\frac{\partial\phi}{\partial z}.
\tag{2.B.6a}
$$

Unless ϕ is a constant field, we expect $\vec{\nabla}\phi$ to be a vector, making $\vec{\nabla}\phi$ a vector field. When ϕ is a physical quantity it must have units, which can be called uphi.

Measuring ϕ in units of uphi and the x, y, z coordinates in units of length called ulength gives, using the notation of Chapter 1,

$$
\begin{aligned}
\vec{\nabla}\phi &= \hat{x}\frac{\text{uphi}}{\text{ulength}}\frac{\partial\phi_{[\text{uphi}]}}{\partial x_{[\text{ulength}]}} + \hat{y}\frac{\text{uphi}}{\text{ulength}}\frac{\partial\phi_{[\text{uphi}]}}{\partial y_{[\text{ulength}]}} \\
&\quad + \hat{z}\frac{\text{uphi}}{\text{ulength}}\frac{\partial\phi_{[\text{uphi}]}}{\partial z_{[\text{ulength}]}} \\
&= \frac{\text{uphi}}{\text{ulength}}\left(\hat{x}\frac{\partial\phi_{[\text{uphi}]}}{\partial x_{[\text{ulength}]}} + \hat{y}\frac{\partial\phi_{[\text{uphi}]}}{\partial y_{[\text{ulength}]}} + \hat{z}\frac{\partial\phi_{[\text{uphi}]}}{\partial z_{[\text{ulength}]}}\right) \\
&= \left(\frac{1}{\text{ulength}}\vec{\nabla}_{[\text{ulength}]}\right)\left(\text{uphi}\cdot\phi_{[\text{uphi}]}\right) = \frac{1}{\text{ulength}}\vec{\nabla}_{[\text{ulength}]}\phi,
\end{aligned}
$$

(2.B.6b)

where

$$
\begin{aligned}
\vec{\nabla}_{[\text{ulength}]} &= \hat{x}\frac{\partial}{\partial x_{[\text{ulength}]}} + \hat{y}\frac{\partial}{\partial y_{[\text{ulength}]}} + \hat{z}\frac{\partial}{\partial z_{[\text{ulength}]}} \\
&= \frac{\partial}{\partial x_{[\text{ulength}]}}\hat{x} + \frac{\partial}{\partial y_{[\text{ulength}]}}\hat{y} + \frac{\partial}{\partial z_{[\text{ulength}]}}\hat{z}.
\end{aligned}
$$

(2.B.6c)

The divergence of a vector field $\vec{b}$ corresponds to the dot product of $\vec{\nabla}$ and $\vec{b}$. Equations (2.B.3f) through (2.B.3h) give

$$
\begin{aligned}
\vec{\nabla}\cdot\vec{b} &= \left(\frac{\partial}{\partial x}\hat{x} + \frac{\partial}{\partial y}\hat{y} + \frac{\partial}{\partial z}\hat{z}\right)\cdot\vec{b} \\
&= \left[\frac{\partial}{\partial x}(\hat{x}\cdot\vec{b}) + \frac{\partial}{\partial y}(\hat{y}\cdot\vec{b}) + \frac{\partial}{\partial z}(\hat{z}\cdot\vec{b})\right] \\
&= \frac{\partial b_x}{\partial x} + \frac{\partial b_y}{\partial x} + \frac{\partial b_z}{\partial x}.
\end{aligned}
$$

(2.B.6d)

This is the standard formula for the divergence of a vector field in x, y, z Cartesian coordinates. Measuring $\vec{b}$ in units of ub and the x, y, z coordinates in units of ulength gives

$$
\begin{aligned}
\vec{\nabla}\cdot\vec{b} &= \frac{\text{ub}}{\text{ulength}}\left(\frac{\partial(b_x)_{[\text{ub}]}}{\partial x_{[\text{ulength}]}} + \frac{\partial(b_y)_{[\text{ub}]}}{\partial x_{[\text{ulength}]}} + \frac{\partial(b_z)_{[\text{ub}]}}{\partial x_{[\text{ulength}]}}\right) \\
&= \frac{1}{\text{ulength}}\vec{\nabla}_{[\text{ulength}]}\cdot(\text{ub}\vec{b}_{[\text{ub}]}) = \frac{1}{\text{ulength}}\vec{\nabla}_{[\text{ulength}]}\cdot\vec{b}.
\end{aligned}
$$

(2.B.6e)

The curl of a vector field $\vec{b}$ corresponds to the cross product of $\vec{\nabla}$ and $\vec{b}$. Using Eqs. (2.B.6a–d) we get

$$
\begin{aligned}
\vec{\nabla} \times \vec{b} &= \left(\frac{\partial}{\partial x}\hat{x} + \frac{\partial}{\partial y}\hat{y} + \frac{\partial}{\partial z}\hat{z} \right) \times \left(\hat{x}b_x + \hat{y}b_y + \hat{z}b_z \right) \\
&= \left(\hat{z}\frac{\partial b_y}{\partial x} - \hat{y}\frac{\partial b_z}{\partial x} \right) + \left(-\hat{z}\frac{\partial b_x}{\partial y} + \hat{x}\frac{\partial b_z}{\partial y} \right) + \left(\hat{y}\frac{\partial b_x}{\partial z} - \hat{x}\frac{\partial b_y}{\partial z} \right) \quad \text{(2.B.6f)} \\
&= \hat{x}\left(\frac{\partial b_z}{\partial y} - \frac{\partial b_y}{\partial z} \right) + \hat{y}\left(\frac{\partial b_x}{\partial z} - \frac{\partial b_z}{\partial x} \right) + \hat{z}\left(\frac{\partial b_y}{\partial x} - \frac{\partial b_x}{\partial y} \right).
\end{aligned}
$$

This is the standard formula for the curl of a vector field in an x, y, z coordinate system. Measuring the x, y, z coordinates in units of ulength and $\vec{b}$ in units of ub gives

$$
\begin{aligned}
\vec{\nabla} \times \vec{b} &= \frac{ub}{u\text{length}}\hat{x}\left(\frac{\partial (b_z)_{[ub]}}{\partial y_{[u\text{length}]}} - \frac{\partial (b_y)_{[ub]}}{\partial z_{[u\text{length}]}} \right) \\
&\quad + \frac{ub}{u\text{length}}\hat{y}\left(\frac{\partial (b_x)_{[ub]}}{\partial z_{[u\text{length}]}} - \frac{\partial (b_z)_{[ub]}}{\partial x_{[u\text{length}]}} \right) \\
&\quad + \frac{ub}{u\text{length}}\hat{z}\left(\frac{\partial (b_y)_{[ub]}}{\partial x_{[u\text{length}]}} - \frac{\partial (b_x)_{[ub]}}{\partial y_{[u\text{length}]}} \right) \\
&= \frac{1}{u\text{length}}\vec{\nabla}_{[u\text{length}]} \times \left(ub\vec{b}_{[ub]} \right) = \frac{1}{u\text{length}}\vec{\nabla}_{[u\text{length}]} \times \vec{b}.
\end{aligned}
$$

(2.B.6g)

The units and dimensions of $\vec{\nabla}$ are clear for these three different types of vector derivative. In Eqs. (2.B.6b), (2.B.6e), and (2.B.6g), $\vec{\nabla}$ always has units of ulength^{-1} and has a "numeric part" given by Eq. (2.B.6c). Hence when we analyze equations containing $\vec{\nabla}\phi$, $\vec{\nabla} \cdot \vec{b}$, or $\vec{\nabla} \times \vec{b}$, we can always replace $\vec{\nabla}$ by $u$length$^{-1}\vec{\nabla}_{[u\text{length}]}$. For the cgs systems of electromagnetic units we define, as a matter of notational convenience, the operator $\vec{\nabla}_{\text{cgs}}$ by the equalities

$$
\begin{aligned}
\text{cm}^{-1}\vec{\nabla}_{\text{cgs}} &= \text{cm}^{-1}\left(\hat{x}\frac{\partial}{\partial x_{\text{cgs}}} + \hat{y}\frac{\partial}{\partial y_{\text{cgs}}} + \hat{z}\frac{\partial}{\partial z_{\text{cgs}}} \right) \\
&= \vec{\nabla} = \text{cm}^{-1}\vec{\nabla}_{[\text{cm}]} = \text{cm}^{-1}\left(\hat{x}\frac{\partial}{\partial x_{[\text{cm}]}} + \hat{y}\frac{\partial}{\partial y_{[\text{cm}]}} + \hat{z}\frac{\partial}{\partial z_{[\text{cm}]}} \right),
\end{aligned}
$$

(2.B.7a)

and for the mks systems of electromagnetic units we define

$$
\begin{aligned}
m^{-1}\vec{\nabla}_{\text{cgs}} &= m^{-1}\left(\hat{x}\frac{\partial}{\partial x_{\text{mks}}} + \hat{y}\frac{\partial}{\partial y_{\text{mks}}} + \hat{z}\frac{\partial}{\partial z_{\text{mks}}} \right) \\
&= \vec{\nabla} = m^{-1}\vec{\nabla}_{[\text{m}]} = m^{-1}\left(\hat{x}\frac{\partial}{\partial x_{[\text{m}]}} + \hat{y}\frac{\partial}{\partial y_{[\text{m}]}} + \hat{z}\frac{\partial}{\partial z_{[\text{m}]}} \right).
\end{aligned}
$$

(2.B.7b)

When ulength is the unit of length in any system of units having U and N operators which for the sake of specificity we call $\underset{vvv}{U}$ and $\underset{vvv}{N}$, then

$$\underset{vvv}{U}\left(\vec{\nabla}\phi\right) = \text{ulength}^{-1}\,\underset{vvv}{U}\left(\phi\right), \tag{2.B.8a}$$

$$\underset{vvv}{U}\left(\vec{\nabla}\cdot\vec{b}\right) = \text{ulength}^{-1}\,\underset{vvv}{U}\left(\vec{b}\right), \tag{2.B.8b}$$

$$\underset{vvv}{U}\left(\vec{\nabla}\times\vec{b}\right) = \text{ulength}^{-1}\,\underset{vvv}{U}\left(\vec{b}\right), \tag{2.B.8c}$$

and

$$\underset{vvv}{N}\left(\vec{\nabla}\phi\right) = \vec{\nabla}_{[\text{ulength}]}\left(\underset{vvv}{N}\left(\phi\right)\right), \tag{2.B.9a}$$

$$\underset{vvv}{N}\left(\vec{\nabla}\cdot\vec{b}\right) = \vec{\nabla}_{[\text{ulength}]}\cdot\left(\underset{vvv}{N}\left(\vec{b}\right)\right), \tag{2.B.9b}$$

$$\underset{vvv}{N}\left(\vec{\nabla}\times\vec{b}\right) = \vec{\nabla}_{[\text{ulength}]}\times\left(\underset{vvv}{N}\left(\vec{b}\right)\right). \tag{2.B.9c}$$

In particular, if we work with a set of electromagnetic dimensions based on the three fundamental dimensions of mass, length, and time, not recognizing charge as a new fundamental dimension, then

$$\underset{mlt}{U}\left(\vec{\nabla}\phi\right) = \text{length}^{-1}\,\underset{mlt}{U}\left(\phi\right), \tag{2.B.10a}$$

$$\underset{mlt}{U}\left(\vec{\nabla}\cdot\vec{b}\right) = \text{length}^{-1}\,\underset{mlt}{U}\left(\vec{b}\right), \tag{2.B.10b}$$

and

$$\underset{mlt}{U}\left(\vec{\nabla}\times\vec{b}\right) = \text{length}^{-1}\,\underset{mlt}{U}\left(\vec{b}\right). \tag{2.B.10c}$$

If we recognize charge as a new fundamental dimension, then

$$\underset{mltq}{U}\left(\vec{\nabla}\phi\right) = \text{length}^{-1}\,\underset{mltq}{U}\left(\phi\right), \tag{2.B.11a}$$

$$\underset{mltq}{U}\left(\vec{\nabla}\cdot\vec{b}\right) = \text{length}^{-1}\,\underset{mltq}{U}\left(\vec{b}\right), \tag{2.B.11b}$$

and

$$\underset{mltq}{U}\left(\vec{\nabla}\times\vec{b}\right) = \text{length}^{-1}\,\underset{mltq}{U}\left(\vec{b}\right). \tag{2.B.11c}$$

We have defined three standard derivatives—the gradient, the divergence and the curl—using the symbol $\vec{\nabla}$. We often use the gradient, divergence and curl to specify an additional two types of derivative by defining, for any scalar field ϕ, that

$$\nabla^2\phi = \vec{\nabla}\cdot(\vec{\nabla}\phi) \tag{2.B.12a}$$

and, for any vector field $\vec{b}$, that

$$\nabla^2\vec{b} = \vec{\nabla}\left(\vec{\nabla}\cdot\vec{b}\right) - \vec{\nabla}\times\left(\vec{\nabla}\times\vec{b}\right). \tag{2.B.12b}$$

The symbol ∇^2 is sometimes called the Laplacian. Note that operating on a scalar field with the Laplacian returns another scalar field, and operating on a vector field with the Laplacian returns another vector field. In a Cartesian coordinate system there is no great difficulty in showing that

$$\nabla^2\phi = \frac{\partial^2\phi}{\partial x^2} + \frac{\partial^2\phi}{\partial y^2} + \frac{\partial^2\phi}{\partial z^2}, \tag{2.B.12c}$$

and with somewhat more effort we find that

$$\nabla^2\vec{b} = \hat{x}\left(\frac{\partial^2 b_x}{\partial x^2} + \frac{\partial^2 b_x}{\partial y^2} + \frac{\partial^2 b_x}{\partial z^2}\right) + \hat{y}\left(\frac{\partial^2 b_y}{\partial x^2} + \frac{\partial^2 b_y}{\partial y^2} + \frac{\partial^2 b_y}{\partial z^2}\right)$$
$$+ \hat{z}\left(\frac{\partial^2 b_z}{\partial x^2} + \frac{\partial^2 b_z}{\partial y^2} + \frac{\partial^2 b_z}{\partial z^2}\right) = \hat{x}\nabla^2 b_x + \hat{y}\nabla^2 b_y + \hat{z}\nabla^2 b_z. \tag{2.B.12d}$$

The right-hand sides of Eqs. (2.B.12c) and (2.B.12d) show us how to get the units and dimensions of $\nabla^2\phi$ and $\nabla^2\vec{b}$. When analyzing equations containing $\nabla^2\phi$ or $\nabla^2\vec{b}$, we can always, when ulength is the unit of length, replace ∇^2 by $\text{ulength}^{-2}\nabla^2_{[\text{ulength}]}$. We can use the $\vec{\nabla}_{[\text{ulength}]}$ operator described above to define

$$\nabla^2_{[\text{ulength}]}\phi = \vec{\nabla}_{[\text{ulength}]}\cdot\left(\vec{\nabla}_{[\text{ulength}]}\phi\right)$$

for scalar fields ϕ and

$$\nabla^2_{[\text{ulength}]}\vec{b} = \vec{\nabla}_{[\text{ulength}]}\left(\vec{\nabla}_{[\text{ulength}]}\cdot\vec{b}\right) - \vec{\nabla}_{[\text{ulength}]}\times\left(\vec{\nabla}_{[\text{ulength}]}\times\vec{b}\right)$$

for vector fields $\vec{b}$. In a Cartesian coordinate system this becomes

$$\nabla^2_{[\text{ulength}]}\phi = \frac{\partial^2\phi}{\partial x^2_{[\text{ulength}]}} + \frac{\partial^2\phi}{\partial y^2_{[\text{ulength}]}} + \frac{\partial^2\phi}{\partial z^2_{[\text{ulength}]}}$$

and

$$\nabla^2_{[\text{ulength}]}\vec{b} = \hat{x}\nabla^2_{[\text{ulength}]}b_x + \hat{y}\nabla^2_{[\text{ulength}]}b_y + \hat{z}\nabla^2_{[\text{ulength}]}b_z,$$

respectively. When ulength is the unit of length in any system of units having U and N operators called $\underset{\text{vvv}}{U}$ and $\underset{\text{vvv}}{N}$, then

$$\underset{\text{vvv}}{U}\left(\nabla^2\phi\right) = \text{ulength}^{-2}\underset{\text{vvv}}{U}(\phi), \tag{2.B.12e}$$

$$\mathop{U}_{\text{vvv}}\left(\nabla^2\vec{b}\right) = \text{ulength}^{-2}\,\mathop{U}_{\text{vvv}}\left(\vec{b}\right),\qquad\qquad(2.\text{B}.12\text{f})$$

and

$$\mathop{N}_{\text{vvv}}\left(\nabla^2_{[\text{ulength}]}\phi\right) = \nabla^2_{[\text{ulength}]}\left(\mathop{N}_{\text{vvv}}(\phi)\right),\qquad\qquad(2.\text{B}.12\text{g})$$

$$\mathop{N}_{\text{vvv}}\left(\nabla^2_{[\text{ulength}]}\vec{b}\right) = \nabla^2_{[\text{ulength}]}\left(\mathop{N}_{\text{vvv}}\left(\vec{b}\right)\right).$$

UNITS ASSOCIATED WITH TWENTIETH-CENTURY ELECTROMAGNETIC THEORY

By the beginning of the twentieth century it had become customary to base electromagnetic theory on Maxwell's equations and the Lorentz force law. In principle, any classical electromagnetic formula can be derived from these equations; so, whenever a system of units changes the form of Maxwell's equations or the Lorentz force law, we can expect corresponding changes in the equations and formulas of classical electromagnetism.

Perhaps the first set of units to become popular in the twentieth century was the Gaussian system of units. Physicists, always interested in reducing theoretical clutter, liked these units because they removed both ε_0 and μ_0 from electromagnetic equations and formulas. The price paid for this advantage—extra factors of the speed of light in Maxwell's equations and the Lorentz force law—was thought to be well worth the resulting conceptual simplification of electromagnetic theory. The Gaussian system of units, like the esu and emu systems, only recognizes the three fundamental dimensions of mass, length, and time.

Although twentieth-century physicists liked Gaussian units, electrical engineers, after a few decades of indecision, chose the rationalized mks system of electromagnetic equations. This system, like the esuq and emuq systems discussed in Chapter 2, recognizes the existence of a fourth fundamental electromagnetic dimension in addition to the three traditional dimensions of mass, length, and time. Like the esuq and emuq systems, both ε_0 and μ_0 are kept as explicit constants; unlike the esuq and emuq systems, a process called "rationalization" is used to eliminate all factors of 4π from Maxwell's equations.

We start the discussion of twentieth-century electromagnetic units by combining the esu and emu units to create the Gaussian system and then move on to describe a rationalized cousin to the Gaussian units, called the Heaviside-Lorentz system, which has recently become popular with elementary particle physicists.[1] We show how to go from the Gaussian or Heaviside-Lorentz systems to the unrationalized and the rationalized mks systems. The unrationalized mks system, although not in use today and never very popular, is a helpful traditional system when converting equations to and from the rationalized mks system. Just as in Chapter 2, emphasis is placed on how to convert electromagnetic equations and formulas from one system of units to another, explaining as we go the diagrams and tables needed to make the change.

3.1 MAXWELL'S EQUATIONS

Much of today's electromagnetic theory can be derived from Maxwell's equations, a set of equations given their present form by James Clerk Maxwell in 1865. Although the relationships in Eqs. (3.1a), (3.1b), and (3.1c) had been used earlier in the nineteenth century, Maxwell was the first to introduce the term $\partial \vec{D}/\partial t$, called the displacement current, into Eq. (3.1d):

$$\vec{\nabla} \cdot \vec{D} = 4\pi \rho_Q, \tag{3.1a}$$

$$\vec{\nabla} \cdot \vec{B} = 0, \tag{3.1b}$$

$$\vec{\nabla} \times \vec{E} + \frac{\partial \vec{B}}{\partial t} = 0, \tag{3.1c}$$

$$\vec{\nabla} \times \vec{H} = 4\pi \vec{J} + \frac{\partial \vec{D}}{\partial t}, \tag{3.1d}$$

where

$$\vec{D} = \varepsilon_0 \vec{E} + 4\pi \vec{P} \tag{3.1e}$$

and

$$\vec{B} = \mu_0 \vec{H} + 4\pi \vec{M}_H. \tag{3.1f}$$

In Eqs. (3.1a–f), $\vec{E}$ is the electric field, $\vec{H}$ is the magnetic field, ρ_Q is the density of unbound charge, $\vec{J}$ is the volume current density, $\vec{P}$ is the electric dipole density, $\vec{M}_H$ is the permanent-magnet dipole density, μ_0 is the permeability of free space, and ε_0 is the permittivity of free space. The $\vec{\nabla}$ operator used in these equations is defined in Appendix 2.B. As has already been discussed in Section 2.11, during the twentieth century Eq. (3.1f) came to be written as

$$\vec{H} = \frac{\vec{B}}{\mu_0} - 4\pi \vec{M}_I, \tag{3.1g}$$

where

$$\vec{M}_I = \frac{\vec{M}_H}{\mu_0}. \tag{3.1h}$$

Here we implicitly make H the auxiliary field defined in terms of the fundamental B and M_I fields rather than, as in Eqs. (3.1e, f), implicitly making B the auxiliary field defined in terms of the fundamental H and M_H fields. After Maxwell introduced the $\partial \vec{D}/\partial t$ term into Eq. (3.1d), he pointed out that Eqs. (3.1a) through

(3.1f) now permitted the existence of electromagnetic radiation travelling through empty space with the velocity

$$c = \frac{1}{\sqrt{\varepsilon_0 \mu_0}} \tag{3.2}$$

[see Eq. (2.10)]. By the middle of the nineteenth century, physicists already knew the value of c from direct measurements of the velocity of light. Measuring the $\varepsilon_0 \mu_0$ product indirectly with electrical circuits and mechanical balances therefore verified Eq. (3.2), simultaneously confirming that the $\partial \vec{D}/\partial t$ term belongs in Eq. (3.1d) and that light is indeed electromagnetic radiation.

3.2 THE GAUSSIAN SYSTEM OF UNITS

Equations (3.1a–d) have the same form in both the esu and emu systems of units; and Eqs. (3.1e, f) can be put into either system by choosing the values given to ε_0 and μ_0. If $\varepsilon_0 = 1$ and $\mu_0 = c^{-2}$ are chosen, we put Eqs. (3.1a–f) into the esu system of units; and if $\varepsilon_0 = c^{-2}$ and $\mu_0 = 1$ are chosen, we put Eqs. (3.1a–f) into the emu system of units. A natural procedure to follow when deriving electromagnetic formulas is to start with Eqs. (3.1a–d), holding off on a formal choice of esu or emu units until it becomes clear from Eqs. (3.1e, f) which of the constants, ε_0 or μ_0, figures more prominently in the final results. As soon as this is known, esu or emu units are specified to make either $\varepsilon_0 = 1$ or $\mu_0 = 1$, eliminating the more troublesome constant from the problem. Sooner or later it becomes clear that this recipe amounts to measuring electrical quantities in esu units and magnetic quantities in emu units. The next obvious step is to start all the derivations this way, writing Maxwell's equations with esu units for the electric quantities $\vec{D}, \vec{E}, \vec{J}, \vec{P}$, and ρ_Q and emu units for the magnetic quantities $\vec{B}, \vec{H}, \vec{M}_I$, and $\vec{M}_H$. This idea for combining the esu and emu units into a single system came to be called the Gaussian cgs system of units. We can put Eqs. (3.1a, e), which contain only electric quantities, into the Gaussian system by thinking of them as having esu units; and we can put Eqs. (3.1b, f, g, h), which contain only magnetic quantities, into the Gaussian system by thinking of them as having emu units. This is simple enough, and one benefit of Gaussian units is already clear—because Eq. (3.1e) is in esu units while (3.1f, g, h) are in emu units, we have the best of both systems with $\varepsilon_0 = \mu_0 = 1$. All that remains is to see what happens to the two equations containing mixed electric and magnetic quantities—Eqs. (3.1c, d)—when they are converted to Gaussian units.

The $\vec{\nabla}$ operator has dimensions of length^{-1}, and Eq. (2.B.7a) from Appendix 2.B of Chapter 2 lets us write Eq. (3.1c) in esu units as (see also Table 2.4 in Chapter 2)

$$\left(cm^{-1} \vec{\nabla}_{cgs} \right) \times \left(\vec{E}_{esu} \frac{1}{sec} \frac{gm^{1/2}}{cm^{1/2}} \right) = -\left(sec^{-1} \frac{\partial}{\partial t_{cgs}} \right) \left(\vec{B}_{esu} \frac{gm^{1/2}}{cm^{3/2}} \right).$$

Table 2.7 is used to replace $\vec{B}_{esu}$ with $\vec{B}_{emu}$, giving

$$\left(cm^{-1}\vec{\nabla}_{cgs}\right) \times \left(\vec{E}_{esu}\frac{1}{sec}\frac{gm^{1/2}}{cm^{1/2}}\right) = -\left(sec^{-1}\frac{\partial}{\partial t_{cgs}}\right)\left(c_{cgs}^{-1}\vec{B}_{emu}\frac{gm^{1/2}}{cm^{3/2}}\right).$$

Multiplying the right-hand side by $\dfrac{sec}{cm} \cdot \dfrac{cm}{sec} = 1$, we get

$$\left(cm^{-1}\vec{\nabla}_{cgs}\right) \times \left(\vec{E}_{esu}\frac{statvolt}{cm}\right)$$
$$= -\left(sec^{-1}\frac{\partial}{\partial t_{cgs}}\right)\left(c_{cgs}\frac{cm}{sec}\right)^{-1}\left(\vec{B}_{emu}\frac{gm^{1/2}}{cm^{1/2}sec}\right) \qquad (3.3a)$$
$$= -\left(sec^{-1}\frac{\partial}{\partial t_{cgs}}\right)\left(c_{cgs}\frac{cm}{sec}\right)^{-1}\left(\vec{B}_{emu}\,gauss\right),$$

where Tables 2.4 and 2.5 are used to replace $sec^{-1} \cdot gm^{1/2} \cdot cm^{-1/2}$ by statvolt $\cdot$ cm^{-1} for E, and $sec^{-1} \cdot gm^{1/2} \cdot cm^{-1/2}$ by gauss = abweber $\cdot$ cm^{-2} for B. Now the electric quantity is in esu units and the magnetic quantity is in emu units, so Eq. (3.3a), using physical quantities, is

$$\vec{\nabla} \times \vec{E} = -\frac{1}{c}\frac{\partial \vec{B}}{\partial t}. \qquad (3.3b)$$

Applying the same procedure to Eq. (3.1d), we start off this time in emu units to get

$$\vec{\nabla} \times \vec{H} = 4\pi\left(\vec{J}_{emu}\frac{abcoul}{cm^2 sec}\right) + \left(sec^{-1}\frac{\partial}{\partial t_{cgs}}\right)\left(\vec{D}_{emu}\frac{abcoul}{cm^2}\right). \qquad (3.4a)$$

Nothing is going to happen to the left-hand side of Eq. (3.4a), because it is a magnetic quantity expressed in emu units, but the right-hand side is all electric quantities and has to be converted to esu units. From Table 2.7 we get

$$\vec{\nabla} \times \vec{H} = 4\pi\left(c_{cgs}^{-1}\vec{J}_{esu}\frac{abcoul}{cm^2 sec}\right) + \left(sec^{-1}\frac{\partial}{\partial t_{cgs}}\right)\left(c_{cgs}^{-1}\vec{D}_{esu}\frac{abcoul}{cm^2}\right).$$

Multiplying the right-hand side by $\dfrac{\text{sec}}{\text{cm}} \cdot \dfrac{\text{cm}}{\text{sec}} = 1$ gives

$$\vec{\nabla} \times \vec{H} = 4\pi \left(c_{\text{cgs}} \frac{\text{cm}}{\text{sec}} \right)^{-1} \left(\vec{J}_{\text{esu}} \frac{\left(\dfrac{\text{cm}}{\text{sec}} \text{abcoul} \right)}{\text{cm}^2 \text{sec}} \right)$$

$$+ \left(c_{\text{cgs}} \frac{\text{cm}}{\text{sec}} \right)^{-1} \left(\text{sec}^{-1} \frac{\partial}{\partial t_{\text{cgs}}} \right) \left(\vec{D}_{\text{esu}} \frac{\left(\dfrac{\text{cm}}{\text{sec}} \text{abcoul} \right)}{\text{cm}^2} \right) \qquad (3.4\text{b})$$

$$= 4\pi \left(c_{\text{cgs}} \frac{\text{cm}}{\text{sec}} \right)^{-1} \left(\vec{J}_{\text{esu}} \frac{\text{statcoul}}{\text{cm}^2 \text{sec}} \right)$$

$$+ \left(c_{\text{cgs}} \frac{\text{cm}}{\text{sec}} \right)^{-1} \left(\text{sec}^{-1} \frac{\partial}{\partial t_{\text{cgs}}} \right) \left(\vec{D}_{\text{esu}} \frac{\text{statcoul}}{\text{cm}^2} \right),$$

where in the last step Table 2.8 is used to replace $\left(\dfrac{\text{cm}}{\text{sec}} \right) \cdot$ abcoul by statcoul. Equation (3.4b) is now in the correct mixed esu and emu units to belong to the Gaussian system, so we can write it using physical quantities:

$$\vec{\nabla} \times \vec{H} = \frac{4\pi}{c} \vec{J} + \frac{1}{c} \frac{\partial \vec{D}}{\partial t}. \qquad (3.4\text{c})$$

We conclude that Maxwell's equations in Gaussian cgs units are

$$\vec{\nabla} \cdot \vec{D} = 4\pi \rho_Q, \qquad (3.5\text{a})$$

$$\vec{\nabla} \cdot \vec{B} = 0, \qquad (3.5\text{b})$$

$$\vec{\nabla} \times \vec{E} + \frac{1}{c} \frac{\partial \vec{B}}{\partial t} = 0, \qquad (3.5\text{c})$$

$$\vec{\nabla} \times \vec{H} = \frac{4\pi}{c} \vec{J} + \frac{1}{c} \frac{\partial \vec{D}}{\partial t}, \qquad (3.5\text{d})$$

where

$$\vec{D} = \vec{E} + 4\pi \vec{P} \qquad (3.5\text{e})$$

and

$$\vec{H} = \vec{B} - 4\pi \vec{M}_H = \vec{B} - 4\pi \vec{M}_I. \qquad (3.5\text{f})$$

We note that $\vec{M}_H = \vec{M}_I$ automatically in the Gaussian system because $\mu_0 = 1$ [see discussion following Eq. (2.68d)].

Table 3.1 shows which of the common electromagnetic physical quantities the Gaussian system treats as electric quantities and which it treats as magnetic quantities. Once an electromagnetic physical quantity b_{ELMAG} is classified as either electric or magnetic, we can say that the U operator for the Gaussian system is defined to be

$$\underset{gs}{U}(b_{ELMAG}) = \begin{cases} \underset{esu}{U}(b_{ELMAG}) & \text{if } b_{ELMAG} \text{ is electric} \\ \underset{emu}{U}(b_{ELMAG}) & \text{if } b_{ELMAG} \text{ is magnetic.} \end{cases} \tag{3.6a}$$

For mechanical physical quantities b_{MECH} we have, as in the esu and emu systems,

$$\underset{gs}{U}(b_{MECH}) = \underset{cgs}{U}(b_{MECH}). \tag{3.6b}$$

The N operators for the Gaussian system are

$$\underset{gs}{N}(b_{ELMAG}) = \begin{cases} \underset{esu}{N}(b_{ELMAG}) & \text{if } b_{ELMAG} \text{ is electric} \\ \underset{emu}{N}(b_{ELMAG}) & \text{if } b_{ELMAG} \text{ is magnetic,} \end{cases} \tag{3.6c}$$

with

$$\underset{gs}{N}(b_{MECH}) = \underset{cgs}{N}(b_{MECH}). \tag{3.6d}$$

The numeric parts of electromagnetic quantities in the Gaussian system are given the subscript "gs," so for electric quantities we have

$$\underset{gs}{N}(E) = E_{gs} = E_{esu},$$

$$\underset{gs}{N}(Q) = Q_{gs} = Q_{esu},$$

etc.

and for magnetic quantities we have

$$\underset{gs}{N}(B) = B_{gs} = B_{emu},$$

$$\underset{gs}{N}(M_I) = (M_I)_{gs} = (M_I)_{emu},$$

etc.

Table 3.1 gives the preferred names for the Gaussian units in boldface type; the "stat" and "ab" prefixed units, although strictly speaking just as correct as in the esu and emu systems, are often not the preferred usage (especially not for the magnetic quantities). Many physicists call the unit of charge in Gaussian units the

Table 3.1 Quantities measured in esu units in Gaussian system. (Preferred unit names are given in boldface.)

(capacitance) $$\underset{\text{gs}}{U}(C) = \underset{\text{esu}}{U}(C) = \text{statfarad} = \mathbf{cm}$$	(electric dipole moment) $$\underset{\text{gs}}{U}(p) = \underset{\text{esu}}{U}(p) = \mathbf{statcoul \cdot cm}$$ $$= \frac{\text{gm}^{1/2} \cdot \text{cm}^{5/2}}{\text{sec}}$$
(electric displacement) $$\underset{\text{gs}}{U}(D) = \underset{\text{esu}}{U}(D) = \frac{\mathbf{statcoul}}{\mathbf{cm^2}} = \frac{\text{gm}^{1/2}}{\text{cm}^{1/2} \cdot \text{sec}}$$	(electric dipole density) $$\underset{\text{gs}}{U}(P) = \underset{\text{esu}}{U}(P) = \frac{\mathbf{statcoul}}{\mathbf{cm^2}} = \frac{\text{gm}^{1/2}}{\text{cm}^{1/2} \cdot \text{sec}}$$
(electric field) $$\underset{\text{gs}}{U}(E) = \underset{\text{esu}}{U}(E) = \frac{\mathbf{statvolt}}{\mathbf{cm}} = \frac{\text{gm}^{1/2}}{\text{cm}^{1/2} \cdot \text{sec}}$$	(charge) $$\underset{\text{gs}}{U}(Q) = \underset{\text{esu}}{U}(Q) = \mathbf{statcoul} = \frac{\text{gm}^{1/2} \cdot \text{cm}^{3/2}}{\text{sec}}$$
(dielectric constant) $$\underset{\text{gs}}{U}(\varepsilon) = \underset{\text{esu}}{U}(\varepsilon) = 1$$	(resistance) $$\underset{\text{gs}}{U}(R) = \underset{\text{esu}}{U}(R) = \text{statohm} = \frac{\mathbf{sec}}{\mathbf{cm}}$$
(permittivity of free space) $$\underset{\text{gs}}{U}(\varepsilon_0) = \underset{\text{esu}}{U}(\varepsilon_0) = 1$$	(volume charge density) $$\underset{\text{gs}}{U}(\rho_Q) = \underset{\text{esu}}{U}(\rho_Q) = \frac{\mathbf{statcoul}}{\mathbf{cm^3}} = \frac{\text{gm}^{1/2}}{\text{cm}^{3/2} \cdot \text{sec}}$$
(conductance) $$\underset{\text{gs}}{U}(G) = \underset{\text{esu}}{U}(G) = \text{statohm}^{-1} = \frac{\mathbf{cm}}{\mathbf{sec}}$$	(resistivity) $$\underset{\text{gs}}{U}(\rho_R) = \underset{\text{esu}}{U}(\rho_R) = \text{statohm} \cdot \text{cm} = \mathbf{sec}$$
(current) $$\underset{\text{gs}}{U}(I) = \underset{\text{esu}}{U}(I) = \text{statamp} = \frac{\mathbf{statcoul}}{\mathbf{sec}}$$ $$= \frac{\text{gm}^{1/2} \cdot \text{cm}^{3/2}}{\text{sec}^2}$$	(elastance) $$\underset{\text{gs}}{U}(S) = \underset{\text{esu}}{U}(S) = \text{statfarad}^{-1} = \mathbf{cm^{-1}}$$
(volume current density) $$\underset{\text{gs}}{U}(J) = \underset{\text{esu}}{U}(J) = \frac{\mathbf{statcoul}}{\mathbf{cm^2 \cdot sec}} = \frac{\text{gm}^{1/2}}{\text{cm}^{1/2} \cdot \text{sec}^2}$$	(surface charge density) $$\underset{\text{gs}}{U}(S_Q) = \underset{\text{esu}}{U}(S_Q) = \frac{\mathbf{statcoul}}{\mathbf{cm^2}} = \frac{\text{gm}^{1/2}}{\text{cm}^{1/2} \cdot \text{sec}}$$
(surface current density) $$\underset{\text{gs}}{U}(\mathcal{J}_S) = \underset{\text{esu}}{U}(\mathcal{J}_S) = \frac{\mathbf{statcoul}}{\mathbf{cm \cdot sec}} = \frac{\text{gm}^{1/2} \cdot \text{cm}^{1/2}}{\text{sec}^2}$$	(conductivity) $$\underset{\text{gs}}{U}(\sigma) = \underset{\text{esu}}{U}(\sigma) = (\text{statohm} \cdot \text{cm})^{-1} = \mathbf{sec^{-1}}$$
(inductance) $$\underset{\text{gs}}{U}(L) = \underset{\text{esu}}{U}(L) = \text{stathenry} = \frac{\mathbf{sec^2}}{\mathbf{cm}}$$	(electric potential) $$\underset{\text{gs}}{U}(V) = \underset{\text{esu}}{U}(V) = \mathbf{statvolt} = \frac{\text{gm}^{1/2} \cdot \text{cm}^{1/2}}{\text{sec}}$$

esu instead of the statcoul; from our point of view these are just two different names for the same unit, so 1 esu = 1 statcoul. The reader should be careful not to confuse "esu," the Gaussian unit of charge, with "esu," the cgs electrostatic system of units.

When Gaussian units are broken down to gm, cm, and sec, we see that the same combination of powers is given different names depending on the physical quantity

Table 3.1 (Continued). Quantities measured in emu units in Gaussian system. (Preferred unit names are given in boldface.)

(magnetic vector potential) $$\underset{gs}{U}(A) = \underset{emu}{U}(A) = \frac{abweber}{cm} = \mathbf{gauss \cdot cm}$$ $$= \frac{gm^{1/2} \cdot cm^{1/2}}{sec}$$	(current-loop magnetic dipole density) $$\underset{gs}{U}(M_I) = \underset{emu}{U}(M_I) = abamp \cdot cm^{-1}$$ $$= \mathbf{maxwell \cdot cm^{-2}} = \frac{gm^{1/2}}{cm^{1/2} \cdot sec}$$
(magnetic induction) $$\underset{gs}{U}(B) = \underset{emu}{U}(B) = \frac{abweber}{cm^2} = \mathbf{gauss}$$ $$= \frac{gm^{1/2}}{cm^{1/2} \cdot sec}$$	(magnetic permeability) $$\underset{gs}{U}(\mu) = \underset{emu}{U}(\mu) = 1$$
(magnetomotive force) $$\underset{gs}{U}(\mathcal{F}) = \underset{emu}{U}(\mathcal{F}) = abamp = \mathbf{gilbert}$$ $$= \frac{gm^{1/2} \cdot cm^{1/2}}{sec}$$	(magnetic permeability of free space) $$\underset{gs}{U}(\mu_0) = \underset{emu}{U}(\mu_0) = 1$$
(magnetic flux) $$\underset{gs}{U}(\Phi_B) = \underset{emu}{U}(\Phi_B) = abweber = \mathbf{gauss \cdot cm^2}$$ $$= \frac{gm^{1/2} \cdot cm^{3/2}}{sec}$$	(magnetic pole strength) $$\underset{gs}{U}(p_H) = \underset{emu}{U}(p_H) = abweber$$ $$= \mathbf{maxwell} = \frac{gm^{1/2} \cdot cm^{3/2}}{sec}$$
(magnetic field) $$\underset{gs}{U}(H) = \underset{emu}{U}(H) = \frac{abamp}{cm} = \mathbf{oersted}$$ $$= \frac{gm^{1/2}}{cm^{1/2} \cdot sec}$$	(permeance) $$\underset{gs}{U}(\mathcal{P}) = \underset{emu}{U}(\mathcal{P}) = cm = \frac{abweber}{abamp}$$ $$= \frac{\mathbf{maxwell}}{\mathbf{gilbert}}$$
(permanent-magnet dipole moment) $$\underset{gs}{U}(m_H) = \underset{emu}{U}(m_H) = abweber \cdot cm$$ $$= \mathbf{maxwell \cdot cm} = \frac{gm^{1/2} \cdot cm^{5/2}}{sec}$$	(reluctance) $$\underset{gs}{U}(\mathcal{R}) = \underset{emu}{U}(\mathcal{R}) = cm^{-1} = \frac{abamp}{abweber}$$ $$= \frac{\mathbf{gilbert}}{\mathbf{maxwell}}$$
(current-loop magnetic dipole moment) $$\underset{gs}{U}(m_I) = \underset{emu}{U}(m_I) = abamp \cdot cm^2$$ $$= \mathbf{maxwell \cdot cm} = \frac{gm^{1/2} \cdot cm^{5/2}}{sec}$$	(magnetic scalar potential) $$\underset{gs}{U}(\Omega_H) = \underset{emu}{U}(\Omega_H) = abamp$$ $$= \mathbf{oersted \cdot cm} = \frac{gm^{1/2} \cdot cm^{1/2}}{sec}$$
(permanent-magnet dipole density) $$\underset{gs}{U}(M_H) = \underset{emu}{U}(M_H) = abweber \cdot cm^{-2}$$ $$= \mathbf{maxwell \cdot cm^{-2}} = \frac{gm^{1/2}}{cm^{1/2} \cdot sec}$$	

it is used with. For example, statcoul/cm^2, the unit of electric displacement D, gauss, the unit of magnetic induction B, and oersted, the unit of the magnetic field strength H, are all gm$^{1/2}$/cm$^{1/2}$/sec. Both the maxwell, the unit of the magnetic pole strength p_H, and the statcoul, the unit of electric charge, are gm$^{1/2} \cdot$ cm$^{3/2}$/sec. Some of the duplications listed in Table 3.1 are already present in the esu and emu systems. In the esu system $\varepsilon_0 = 1$, so from Eq. (3.1e) D, E, and P must have the same units; and in the emu system $\mu_0 = 1$, so from Eqs. (3.1f, g, h) B, H, M_H, and M_I must have the same units. The Gaussian system has both ε_0 and μ_0 equal to 1, so all these esu and emu unit duplications are kept. Going back to the unit symmetries listed in the first column of Table 2.2, we can write the symmetries listed there as

$$\underset{\text{esu}}{U}(b_{EL}) = \underset{\text{emu}}{U}(b_{MAG}), \tag{3.7a}$$

where b_{EL} and b_{MAG} are any pair of electric and magnetic physical quantities used in the same box of Table 2.2. Equation (3.6a) shows that, in the Gaussian system, Eq. (3.7a) becomes

$$\underset{\text{gs}}{U}(b_{EL}) = \underset{\text{gs}}{U}(b_{MAG}) \tag{3.7b}$$

for all the b_{EL}, b_{MAG} pairs in the first column of Table 2.2. Therefore, according to Table 2.2 and Eqs. (3.1e–h), the electromagnetic physical quantities D, E, P, B, H, M_H, and M_I must all have the same units in the Gaussian system. It is easy to see why so many of the Gaussian units end up as different names for the same combinations of gm, cm, and sec.

We now convert Eq. (2.6), the equation for the magnetic field of a long, thin wire carrying a current I,

$$H = \frac{2I}{r},$$

into the Gaussian system of units. Equation (2.6) has the same form in both the esu and emu systems of units. Taking it to be esu units, we write

$$\left(H_{\text{esu}} \frac{\text{statamp}}{\text{cm}} \right) = \frac{2(I_{\text{esu}} \text{statamp})}{(r_{\text{cgs}} \text{cm})}. \tag{3.8a}$$

Table 3.2 lists the information needed to go between the Gaussian and esu systems. Since $H_{\text{gs}} = c_{\text{cgs}}^{-1} H_{\text{esu}}$ and $I_{\text{gs}} = I_{\text{esu}}$, Eq. (3.8a) can be written as

$$c_{\text{cgs}} H_{\text{gs}} \frac{\text{statamp}}{\text{cm}} = \frac{2 I_{\text{gs}} \text{statamp}}{r_{\text{cgs}} \text{cm}}. \tag{3.8b}$$

Table 3.1 shows that the Gaussian system, not surprisingly, lists H as a magnetic quantity, so $H_{\text{gs}} = H_{\text{emu}}$, and H must be measured in units of oersted $=$

abamp/cm. From Table 3.3 (which has the same information as Table 2.8), 1 statamp = (cm/sec) · abamp, so Eq. (3.8b) becomes

$$\left(c_{\mathrm{cgs}}\frac{\mathrm{cm}}{\mathrm{sec}}\right)\left(H_{\mathrm{gs}}\frac{\mathrm{abamp}}{\mathrm{cm}}\right) = \frac{2I_{\mathrm{gs}}\mathrm{statamp}}{r_{\mathrm{cgs}}\mathrm{cm}}$$

or

$$H = \frac{2I}{rc} \tag{3.8c}$$

written in terms of the physical quantities $H, I, r,$ and c.

Equation (2.66h), the Lorentz force law, also gains a c in the denominator when converted to Gaussian units. Equation (2.66h), like Eq. (2.6), has the same form in both the esu and emu systems. This time we use the emu system (see Table 2.5) to break the equation up into its units and numeric components:

$$\vec{F}_{\mathrm{cgs}}\mathrm{dynes} = (Q_{\mathrm{emu}}\mathrm{abcoul})\left(\vec{E}_{\mathrm{emu}}\frac{\mathrm{abvolt}}{\mathrm{cm}}\right)$$
$$+ (Q_{\mathrm{emu}}\mathrm{abcoul})\left(\vec{v}_{\mathrm{cgs}}\frac{\mathrm{cm}}{\mathrm{sec}}\right) \times (\vec{B}_{\mathrm{emu}}\mathrm{gauss}). \tag{3.9a}$$

From Table 3.1 we note that $\vec{B}$ is already in its correct form for the Gaussian system, because it is a magnetic quantity in emu units. The $\vec{F}$ and $\vec{v}$ mechanical quantities are in cgs units, which are common to the esu, emu, and Gaussian systems, so they do not have to be changed either. Using Tables 3.3 and 3.4 to convert the $Q, \vec{E}$ electric quantities to the Gaussian system gives

$$\vec{F} = \left(c_{\mathrm{cgs}}^{-1}Q_{\mathrm{gs}}\right)\left(\frac{\mathrm{sec}}{\mathrm{cm}}\mathrm{statcoul}\right)\left(\vec{E}_{\mathrm{gs}}c_{\mathrm{cgs}}\right)\left(\frac{\mathrm{cm}}{\mathrm{sec}}\frac{\mathrm{statvolt}}{\mathrm{cm}}\right)$$
$$+ \left(c_{\mathrm{cgs}}^{-1}Q_{\mathrm{gs}}\right)\left(\frac{\mathrm{sec}}{\mathrm{cm}}\mathrm{statcoul}\right)\vec{v} \times \vec{B}, \tag{3.9b}$$

where $\vec{F}, \vec{v},$ and $\vec{B}$ are left as physical quantities because we know their units do not change. Since

$$c_{\mathrm{cgs}}\frac{\mathrm{cm}}{\mathrm{sec}} \cdot c_{\mathrm{cgs}}^{-1}\frac{\mathrm{sec}}{\mathrm{cm}} = 1 \quad \text{and} \quad c_{\mathrm{cgs}}^{-1}\frac{\mathrm{sec}}{\mathrm{cm}} = c^{-1},$$

Eq. (3.9b) can be written as

$$\vec{F} = (Q_{\mathrm{gs}}\mathrm{statcoul})\left(\vec{E}_{\mathrm{gs}}\frac{\mathrm{statvolt}}{\mathrm{cm}}\right) + (Q_{\mathrm{gs}}\mathrm{statcoul})\frac{\vec{v} \times \vec{B}}{c}$$
$$= Q\vec{E} + \frac{Q}{c}(\vec{v} \times \vec{B}), \tag{3.9c}$$

Table 3.2 Numeric components of physical quantities in esu and Gaussian units.

(magnetic vector potential) $A_{emu} = A_{gs} = A_{esu} \cdot c_{cgs}$	(volume current density) $J_{gs} = J_{esu}$	(permeance) $\mathcal{P}_{emu} = \mathcal{P}_{gs} = \mathcal{P}_{esu} \cdot c_{cgs}^2$
(magnetic induction) $B_{emu} = B_{gs} = B_{esu} \cdot c_{cgs}$	(surface current density) $(\mathcal{J}_S)_{gs} = (\mathcal{J}_S)_{esu}$	(charge) $Q_{gs} = Q_{esu}$
(capacitance) $C_{gs} = C_{esu}$	(inductance) $L_{gs} = L_{esu}$	(resistance) $R_{gs} = R_{esu}$
(electric displacement) $D_{gs} = D_{esu}$	(permanent-magnet dipole moment) $(m_H)_{emu} = (m_H)_{gs} = (m_H)_{esu} \cdot c_{cgs}$	(reluctance) $\mathcal{R}_{emu} = \mathcal{R}_{gs} = \mathcal{R}_{esu} \cdot c_{cgs}^{-2}$
(electric field) $E_{gs} = E_{esu}$	(current-loop magnetic dipole moment) $(m_I)_{emu} = (m_I)_{gs} = (m_I)_{esu} \cdot c_{cgs}^{-1}$	(volume charge density) $(\rho_Q)_{gs} = (\rho_Q)_{esu}$
(dielectric constant) $\varepsilon_{gs} = \varepsilon_{esu} = \varepsilon_r$	(permanent-magnet dipole density) $(M_H)_{emu} = (M_H)_{gs} = (M_H)_{esu} \cdot c_{cgs}$	(resistivity) $(\rho_R)_{gs} = (\rho_R)_{esu}$
(permittivity of free space) $(\varepsilon_0)_{gs} = (\varepsilon_0)_{esu} = 1$	(current-loop magnetic dipole density) $(M_I)_{emu} = (M_I)_{gs} = (M_I)_{esu} \cdot c_{cgs}^{-1}$	(elastance) $S_{gs} = S_{esu}$
(magnetomotive force) $\mathcal{F}_{emu} = \mathcal{F}_{gs} = \mathcal{F}_{esu} \cdot c_{cgs}^{-1}$	(magnetic permeability) $\mu_{emu} = \mu_{gs} = \mu_r = c_{cgs}^2 \cdot \mu_{esu}$	(surface charge density) $(S_Q)_{gs} = (S_Q)_{esu}$
(magnetic flux) $(\Phi_B)_{emu} = (\Phi_B)_{gs} = (\Phi_B)_{esu} \cdot c_{cgs}$	(magnetic permeability of free space) $(\mu_0)_{emu} = (\mu_0)_{gs} = 1 = c_{cgs}^2 \cdot (\mu_0)_{esu}$	(conductivity) $\sigma_{gs} = \sigma_{esu}$
(conductance) $G_{gs} = G_{esu}$	(magnetic pole strength) $(p_H)_{emu} = (p_H)_{gs} = (p_H)_{esu} \cdot c_{cgs}$	(electric potential) $V_{gs} = V_{esu}$
(magnetic field) $H_{emu} = H_{gs} = H_{esu} \cdot c_{cgs}^{-1}$	(electric dipole moment) $p_{gs} = p_{esu}$	(magnetic scalar potential) $(\Omega_H)_{emu} = (\Omega_H)_{gs} = (\Omega_H)_{esu} \cdot c_{cgs}^{-1}$
(current) $I_{gs} = I_{esu}$	(electric dipole density) $P_{gs} = P_{esu}$	

Table 3.3 Unit relationships for the esu, emu, and Gaussian systems.

$\text{abamp} = \dfrac{\text{sec}}{\text{cm}} \cdot \text{statamp}$
$\text{abcoul} = \dfrac{\text{sec}}{\text{cm}} \cdot \text{statcoul}$
$\text{abweber} = \dfrac{\text{cm}}{\text{sec}} \cdot \text{statweber}$
$\text{abvolt} = \dfrac{\text{cm}}{\text{sec}} \cdot \text{statvolt}$
$\text{abfarad} = \dfrac{\text{sec}^2}{\text{cm}^2} \cdot \text{statfarad}$
$\text{abhenry} = \dfrac{\text{cm}^2}{\text{sec}^2} \cdot \text{stathenry}$
$\text{abohm} = \dfrac{\text{cm}^2}{\text{sec}^2} \cdot \text{statohm}$

where we have recognized statcoul and statvolt/cm as the correct Gaussian units for the charge and electric field, respectively. If we had started off writing the Lorentz force law in esu units, then Q, $\vec{E}$, $\vec{F}$, and $\vec{v}$ would already have been in the correct units for the Gaussian system, giving us from Table 2.4

$$\vec{F} = Q\vec{E} + Q\vec{v} \times \left(\vec{B}_{\text{esu}} \frac{\text{statweber}}{\text{cm}^2} \right).$$

From Tables 3.2 and 3.3

$$\vec{B}_{\text{esu}} \frac{\text{statweber}}{\text{cm}^2} = c_{\text{cgs}}^{-1} \vec{B}_{\text{emu}} \frac{\text{sec}}{\text{cm}} \frac{\text{abweber}}{\text{cm}^2}$$

$$= c^{-1} \left(\vec{B}_{\text{emu}} \frac{\text{abweber}}{\text{cm}^2} \right) = c^{-1} (\vec{B}_{\text{gs}} \text{gauss}) = \frac{\vec{B}}{c},$$

which once again gives Eq. (3.9c). This form of the Lorentz force law can be remembered by noting that in the Gaussian system, $\vec{B}$ in gauss and $\vec{E}$ in statvolt/cm really have the same basic unit $\text{gm}^{1/2}/\text{cm}^{1/2}/\text{sec}$. Therefore, the velocity $\vec{v}$ must have its units cancelled by another velocity if Rule IV is to be satisfied. The only velocity "fundamental enough" to play this role (and the only velocity that keeps appearing in the Gaussian system of units) is the velocity of light—so it is no surprise to find $\vec{v}$ divided by c, giving us Eq. (3.9c).

Table 3.4 Numeric components of physical quantities in emu and Gaussian units.

(magnetic vector potential) $A_{gs} = A_{emu}$	(volume current density) $J_{esu} = J_{gs} = J_{emu} \cdot c_{cgs}$	(permeance) $\mathcal{P}_{gs} = \mathcal{P}_{emu}$
(magnetic induction) $B_{gs} = B_{emu}$	(surface current density) $(\mathcal{J}_S)_{esu} = (\mathcal{J}_S)_{gs} = (\mathcal{J}_S)_{emu} \cdot c_{cgs}$	(charge) $Q_{esu} = Q_{gs} = Q_{emu} \cdot c_{cgs}$
(capacitance) $C_{esu} = C_{gs} = C_{emu} \cdot c_{cgs}^2$	(inductance) $L_{esu} = L_{gs} = L_{emu} \cdot c_{cgs}^{-2}$	(resistance) $R_{esu} = R_{gs} = R_{emu} \cdot c_{cgs}^{-2}$
(electric displacement) $D_{esu} = D_{gs} = D_{emu} \cdot c_{cgs}$	(permanent-magnet dipole moment) $(m_H)_{gs} = (m_H)_{emu}$	(reluctance) $\mathcal{R}_{gs} = \mathcal{R}_{emu}$
(electric field) $E_{esu} = E_{gs} = E_{emu} \cdot c_{cgs}^{-1}$	(current-loop magnetic dipole moment) $(m_I)_{gs} = (m_I)_{emu}$	(volume charge density) $(\rho_Q)_{esu} = (\rho_Q)_{gs} = (\rho_Q)_{emu} \cdot c_{cgs}$
(dielectric constant) $\varepsilon_{esu} = \varepsilon_r = \varepsilon_{gs} = \varepsilon_{emu} \cdot c_{cgs}^2$	(permanent-magnet dipole density) $(M_H)_{gs} = (M_H)_{emu}$	(resistivity) $(\rho_R)_{esu} = (\rho_R)_{gs} = (\rho_R)_{emu} \cdot c_{cgs}^{-2}$
(permittivity of free space) $(\varepsilon_0)_{esu} = 1 = (\varepsilon_0)_{gs} = (\varepsilon_0)_{emu} \cdot c_{cgs}^2$	(current-loop magnetic dipole density) $(M_I)_{gs} = (M_I)_{emu}$	(elastance) $S_{esu} = S_{gs} = S_{emu} \cdot c_{cgs}^{-2}$
(magnetomotive force) $\mathcal{F}_{gs} = \mathcal{F}_{emu}$	(magnetic permeability) $\mu_{gs} = \mu_{emu} = \mu_r$	(surface charge density) $(S_Q)_{esu} = (S_Q)_{gs} = (S_Q)_{emu} \cdot c_{cgs}$
(magnetic flux) $(\Phi_B)_{gs} = (\Phi_B)_{emu}$	(magnetic permeability of free space) $(\mu_0)_{gs} = (\mu_0)_{emu} = 1$	(conductivity) $\sigma_{esu} = \sigma_{gs} = \sigma_{emu} \cdot c_{cgs}^2$
(conductance) $G_{esu} = G_{gs} = G_{emu} \cdot c_{cgs}^2$	(magnetic pole strength) $(p_H)_{gs} = (p_H)_{emu}$	(electric potential) $V_{esu} = V_{gs} = V_{emu} \cdot c_{cgs}^{-1}$
(magnetic field) $H_{gs} = H_{emu}$	(electric dipole moment) $p_{esu} = p_{gs} = p_{emu} \cdot c_{cgs}$	(magnetic scalar potential) $(\Omega_H)_{gs} = (\Omega_H)_{emu}$
(current) $I_{esu} = I_{gs} = I_{emu} \cdot c_{cgs}$	(electric dipole density) $P_{esu} = P_{gs} = P_{emu} \cdot c_{cgs}$	

Equation (2.68e) can be written in Gaussian units as (choosing the magnetic dipole to be $\vec{m}_I$ rather than $\vec{m}_H$)

$$\vec{m}_I = \left(\frac{IA}{c}\right)\hat{n}, \tag{3.10a}$$

where $\vec{m}_I, I, A, \hat{n}$ have the same meaning as in Eq. (2.68e) and c is the speed of light. We now convert Eq. (3.10a) back to esu and emu units to show how it is done. When changing Eq. (3.10a) to the emu system, we note that $\vec{m}_I$ is a magnetic quantity in the Gaussian system and so does not have to be changed. Breaking the right-hand side of Eq. (3.10a) into numeric components and units gives

$$\vec{m}_I = \left[\frac{(I_{gs}\text{statamp})(A_{cgs}\text{cm}^2)}{\left(c_{cgs}\dfrac{\text{cm}}{\text{sec}}\right)}\right]\hat{n}. \tag{3.10b}$$

Tables 3.3 and 3.4 show that

$$\vec{m}_I = \left[\frac{\left(c_{cgs}I_{emu}\dfrac{\text{cm}}{\text{sec}}\text{abamp}\right)(A_{cgs}\text{cm}^2)}{\left(c_{cgs}\dfrac{\text{cm}}{\text{sec}}\right)}\right]\hat{n}$$

or

$$\vec{m}_I = I\,A\hat{n}. \tag{3.10c}$$

This is, as expected, the same form as Eq. (2.68e). When converting Eq. (3.10a) to the esu system, we know that I is an electric quantity in the Gaussian system and so only the left-hand side of the equation has to be changed. Breaking the left-hand side up into numeric components and units gives, using Table 3.5,

$$(\vec{m}_I)_{gs}\text{maxwell}\cdot\text{cm} = \frac{IA}{c}\hat{n}.$$

From Table 3.5 we see that maxwell $\cdot$ cm = abamp $\cdot$ cm^2, so Tables 3.2 and 3.3 show that

$$(\vec{m}_I)_{esu}c_{cgs}^{-1}\text{abamp}\cdot\text{cm}^2 = \frac{IA}{c}\hat{n}$$

or

$$(\vec{m}_I)_{esu}c_{cgs}^{-1}\left(\frac{\text{sec}}{\text{cm}}\text{statamp}\right)\cdot\text{cm}^2 = \frac{IA}{c}\hat{n}. \tag{3.10d}$$

Table 3.5 Gaussian units for physical quantities with and without Heaviside-Lorentz rationalization.

Physical quantity	Gaussian units
(magnetic vector potential) $_h A$, A	$\underset{gs}{U}(_h A) = \underset{gs}{U}(A) = \dfrac{gm^{1/2} \cdot cm^{1/2}}{sec} = \dfrac{abweber}{cm} = gauss \cdot cm$
(magnetic induction) $_h B$, B	$\underset{gs}{U}(_h B) = \underset{gs}{U}(B) = \dfrac{gm^{1/2}}{sec \cdot cm^{1/2}} = \dfrac{abweber}{cm^2} = gauss$
(capacitance) $_h C$, C	$\underset{gs}{U}(_h C) = \underset{gs}{U}(C) = cm = statfarad$
(electric displacement) $_h D$, D	$\underset{gs}{U}(_h D) = \underset{gs}{U}(D) = \dfrac{gm^{1/2}}{cm^{1/2} \cdot sec} = \dfrac{statcoul}{cm^2}$
(electric field) $_h E$, E	$\underset{gs}{U}(_h E) = \underset{gs}{U}(E) = \dfrac{gm^{1/2}}{cm^{1/2} \cdot sec} = \dfrac{statvolt}{cm}$
(dielectric constant) ε	$\underset{gs}{U}(\varepsilon) = 1$
(permittivity of free space) ε_0	$\underset{gs}{U}(\varepsilon_0) = 1$
(magnetomotive force) $_h \mathcal{F}$, $\mathcal{F}$	$\underset{gs}{U}(_h \mathcal{F}) = \underset{gs}{U}(\mathcal{F}) = \dfrac{gm^{1/2} \cdot cm^{1/2}}{sec} = abamp = gilbert$
(magnetic flux) $_h \Phi_B$, Φ_B	$\underset{gs}{U}(_h \Phi_B) = \underset{gs}{U}(\Phi_B) = \dfrac{gm^{1/2} \cdot cm^{3/2}}{sec} = abweber = gauss \cdot cm^2$
(conductance) $_h G$, G	$\underset{gs}{U}(_h G) = \underset{gs}{U}(G) = \dfrac{cm}{sec} = statohm^{-1}$
(magnetic field) $_h H$, H	$\underset{gs}{U}(_h H) = \underset{gs}{U}(H) = \dfrac{gm^{1/2}}{cm^{1/2} \cdot sec} = oersted = \dfrac{abamp}{cm}$
(current) $_h I$, I	$\underset{gs}{U}(_h I) = \underset{gs}{U}(I) = \dfrac{gm^{1/2} \cdot cm^{3/2}}{sec^2} = \dfrac{statcoul}{sec} = statamp$
(volume current density) $_h J$, J	$\underset{gs}{U}(_h J) = \underset{gs}{U}(J) = \dfrac{gm^{1/2}}{cm^{1/2} \cdot sec^2} = \dfrac{statcoul}{cm^2 \cdot sec}$

Table 3.5 (Continued).

Physical quantity	Gaussian units
(surface current density) $_h\mathcal{J}_S, \mathcal{J}_S$	$\underset{gs}{U}(_h\mathcal{J}_S) = \underset{gs}{U}(\mathcal{J}_S) = \dfrac{gm^{1/2} \cdot cm^{1/2}}{sec^2} = \dfrac{statcoul}{cm \cdot sec}$
(inductance) $_hL, L$	$\underset{gs}{U}(_hL) = \underset{gs}{U}(L) = \dfrac{sec^2}{cm} = stathenry$
(permanent-magnet dipole moment) $_hm_H, m_H$	$\underset{gs}{U}(_hm_H) = \underset{gs}{U}(m_H) = \dfrac{gm^{1/2} \cdot cm^{5/2}}{sec}$ $= abweber \cdot cm = maxwell \cdot cm$
(current-loop magnetic dipole moment) $_hm_I, m_I$	$\underset{gs}{U}(_hm_I) = \underset{gs}{U}(m_I) = \dfrac{gm^{1/2} \cdot cm^{5/2}}{sec}$ $= abamp \cdot cm^2 = maxwell \cdot cm$
(permanent-magnet dipole density) $_hM_H, M_H$	$\underset{gs}{U}(_hM_H) = \underset{gs}{U}(M_H) = \dfrac{gm^{1/2}}{sec \cdot cm^{1/2}}$ $= \dfrac{abweber}{cm^2} = \dfrac{maxwell}{cm^2}$
(current-loop magnetic dipole density) $_hM_I, M_I$	$\underset{gs}{U}(_hM_I) = \underset{gs}{U}(M_I) = \dfrac{gm^{1/2}}{sec \cdot cm^{1/2}}$ $= \dfrac{abamp}{cm} = \dfrac{maxwell}{cm^2}$
(magnetic permeability) μ	$\underset{gs}{U}(\mu) = 1$
(magnetic permeability of free space) μ_0	$\underset{gs}{U}(\mu_0) = 1$
(magnetic pole strength) $_hp_H, p_H$	$\underset{gs}{U}(_hp_H) = \underset{gs}{U}(p_H) = \dfrac{gm^{1/2} \cdot cm^{3/2}}{sec}$ $= abweber = maxwell$
(electric dipole moment) $_hp, p$	$\underset{gs}{U}(_hp) = \underset{gs}{U}(p) = \dfrac{gm^{1/2} \cdot cm^{5/2}}{sec} = statcoul \cdot cm$
(electric dipole density) $_hP, P$	$\underset{gs}{U}(_hP) = \underset{gs}{U}(P) = \dfrac{gm^{1/2}}{cm^{1/2} \cdot sec} = \dfrac{statcoul}{cm^2}$
(permeance) $\mathcal{P}$	$\underset{gs}{U}(\mathcal{P}) = cm = \dfrac{abweber}{abamp} = \dfrac{maxwell}{gilbert}$

Table 3.5 (Continued).

Physical quantity	Gaussian units
(charge) $_hQ, Q$	$\underset{gs}{U}(_hQ) = \underset{gs}{U}(Q) = \dfrac{gm^{1/2} \cdot cm^{3/2}}{sec} = statcoul$
(resistance) $_hR, R$	$\underset{gs}{U}(_hR) = \underset{gs}{U}(R) = \dfrac{sec}{cm} = statohm$
(reluctance) $\mathcal{R}$	$\underset{gs}{U}(\mathcal{R}) = cm^{-1} = \dfrac{abamp}{abweber} = \dfrac{gilbert}{maxwell}$
(volume charge density) $_h\rho_Q, \rho_Q$	$\underset{gs}{U}(_h\rho_Q) = \underset{gs}{U}(\rho_Q) = \dfrac{gm^{1/2}}{cm^{3/2} \cdot sec} = \dfrac{statcoul}{cm^3}$
(resistivity) $_h\rho_R, \rho_R$	$\underset{gs}{U}(_h\rho_R) = \underset{gs}{U}(\rho_R) = sec = statohm \cdot cm$
(elastance) $_hS, S$	$\underset{gs}{U}(_hS) = \underset{gs}{U}(S) = cm^{-1} = statfarad^{-1}$
(surface charge density) $_hS_Q, S_Q$	$\underset{gs}{U}(_hS_Q) = \underset{gs}{U}(S_Q) = \dfrac{gm^{1/2}}{cm^{1/2} \cdot sec} = \dfrac{statcoul}{cm^2}$
(conductivity) $_h\sigma, \sigma$	$\underset{gs}{U}(_h\sigma) = \underset{gs}{U}(\sigma) = sec^{-1} = \dfrac{statohm^{-1}}{cm}$
(electric potential) $_hV, V$	$\underset{gs}{U}(_hV) = \underset{gs}{U}(V) = \dfrac{gm^{1/2} \cdot cm^{1/2}}{sec} = statvolt$
(magnetic scalar potential) $_h\Omega_H, \Omega_H$	$\underset{gs}{U}(_h\Omega_H) = \underset{gs}{U}(\Omega_H) = \dfrac{gm^{1/2} \cdot cm^{1/2}}{sec} = abamp = oersted \cdot cm$

Table 2.4 shows that the correct esu units for $\vec{m}_I$ are statamp $\cdot$ cm^2, so Eq. (3.10d) written in esu units is

$$\frac{\vec{m}_I}{c} = \frac{IA}{c}\hat{n},$$

which again reduces to Eq. (3.10c),

$$\vec{m}_I = IA\hat{n}.$$

We see that it is as easy to return to esu or emu units from Gaussian units as it is to convert to Gaussian units in the first place.

The Gaussian system of units, unlike the esu and emu systems of units, is in widespread use today, especially in theoretical electromagnetic analysis. From the physicist's point of view there are many advantages to the Gaussian system. The annoying constants ε_0 and μ_0 have been eliminated from electromagnetic theory, so from Eqs. (3.5e, f)

$$\vec{D} = \vec{E} \quad \text{and} \quad \vec{B} = \vec{H},$$

when

$$\vec{P} = \vec{M}_H = \vec{M}_I = 0.$$

Therefore, in empty space the $\vec{D}$ field becomes the same as the $\vec{E}$ field and the $\vec{H}$ field becomes the same as the $\vec{B}$ field. This is a significant conceptual simplification for today's physicists, who often do their electromagnetic analysis in vacuum—either on a microscopic scale between and inside atoms or on a macroscopic scale in the space surrounding localized electromagnetic disturbances. When doing relativistic calculations, physicists can even switch to a system of units where the speed of light is 1 (see Section 1.10), thereby eliminating all three constants ε_0, μ_0, and c from Maxwell's equations. In this sense, the Gaussian system combines the best aspects of both the esu and emu systems of units. On the other hand, the electrical engineer, who rarely needs to use relativity theory, might remark after examining the Gaussian system that we pay for eliminating ε_0 and μ_0 by introducing the speed of light c into many formulas—such as Eqs. (3.5c), (3.8c), (3.9c), and (3.10a)—where it previously did not exist. Engineers analyzing electric fields in matter notice that they have not even really eliminated the permittivity and permeability in return for all the extra factors of c. The constants $\varepsilon = \varepsilon_r \varepsilon_0$ and $\mu = \mu_r \mu_0$ used to describe the interaction of electric and magnetic fields with bulk matter are, since $\varepsilon_0 = \mu_0 = 1$, replaced by just ε_r and μ_r in the Gaussian system. Variables ε_r and μ_r are dimensionless, but they are all too often not equal to 1. Formulas that previously contained ε and μ still contain ε_r and μ_r, and it is just as much trouble to keep track of ε_r and μ_r as it is to keep track of ε and μ. From this point of view, introducing all these factors of c has just complicated practical electromagnetic calculations, so it is not surprising that electrical engineers have never really had much use for Gaussian units.

3.3 RATIONALIZATION AND THE HEAVISIDE-LORENTZ SYSTEM

One concept that did eventually catch on with electrical engineers—although not in the form first proposed—is the rationalization of electromagnetic equations. The idea of rationalizing the electromagnetic equations was first suggested by Oliver

Heaviside in 1882. He objected to an "eruption of 4π's" in equations having no obvious connection with spherical or circular geometry.[2] Equation (2.63i) in Chapter 2 is an example of what Heaviside was objecting to; this equation for the capacitance C of the parallel-plate capacitor shown in Fig. 2.8 is (in esu or Gaussian units)

$$C = \frac{A}{4\pi d}, \tag{3.11a}$$

even though there is nothing circular or spherical about the capacitor's geometry. Oliver Heaviside suggested rescaling the units used to measure electrical quantities by powers of 4π to eliminate these constants from formulas such as Eq. (3.11a). If we take Eq. (3.11a) to be in reality an equation between purely numeric quantities, which in our notation would be (using esu or Gaussian units)

$$C_{gs} = C_{[\text{statfarad}]} = \frac{A_{cgs}}{4\pi\, d_{cgs}} = \frac{A_{[\text{cm}^2]}}{4\pi d_{[\text{cm}]}}, \tag{3.11b}$$

then by Rule I we could decide to measure C in units of statfarad/(4π) rather than statfarad to get

$$C_{[\text{statfarad}/(4\pi)]} = 4\pi\, C_{[\text{statfarad}]}. \tag{3.11c}$$

From Eqs. (3.11b) and (3.11c) we then have the desired result

$$C_{[\text{statfarad}/(4\pi)]} = \frac{A_{[\text{cm}^2]}}{d_{[\text{cm}]}}. \tag{3.11d}$$

Note that all the variables in Eqs. (3.11b) through (3.11d), unlike the variables in Eq. (3.11a), are dimensionless numerics; and, in fact, during the nineteenth and early twentieth century there was often no hard and fast distinction drawn between these two types of variable. From our point of view, however, Eq. (3.11a) involves physical quantities C, A, and d, while Eqs. (3.11b) through (3.11d) involve the numerical components of physical quantities $C_{[\text{statfarad}]}$, $A_{[\text{cm}^2]}$, and $d_{[\text{cm}]}$. If we start with Eq. (3.11a) split up into units and numeric components to get, using the equality statfarad $=$ cm,

$$C_{[\text{statfarad}]}\text{statfarad} = C_{[\text{statfarad}]}\text{cm} = \frac{A_{[\text{cm}^2]}\text{cm}^2}{4\pi d_{[\text{cm}]}\text{cm}}, \tag{3.12a}$$

and then decide to measure C in the new unit

$$1\ \text{newfarad} = \text{statfarad}/(4\pi), \tag{3.12b}$$

we end up with

$$(4\pi C_{[\text{statfarad}]})\text{newfarad} = \frac{A_{[\text{cm}^2]}\text{cm}^2}{4\pi d_{[\text{cm}]}\text{cm}}.$$

But, from Eq. (3.11c) this becomes

$$C_{[\text{statfarad}/(4\pi)]}\text{newfarad} = \frac{A_{[\text{cm}^2]}\text{cm}^2}{4\pi d_{[\text{cm}]}\text{cm}}, \qquad (3.12c)$$

or, again using variables representing physical quantities,

$$C = \frac{A}{4\pi d}.$$

Equation (3.12c) is clearly no different from Eq. (3.11a); the dimensionless factor of 4π refuses to disappear. As long as variables like C, A, and d are thought of as a numeric multiplied by a unit, equations like Eq. (3.11a) will keep their form no matter how we stretch or shrink the units that are used to measure its variables. Although Oliver Heaviside could, back in the nineteenth century, regard his suggestion as a proposal to measure electromagnetic quantities in a new set of units, we cannot agree with him.

From our point of view, the new set of electromagnetic equations first proposed by Oliver Heaviside and later popularized, at the beginning of the twentieth century, by H. A. Lorentz, is in reality a system of rescaled physical quantities, which we call the Heaviside-Lorentz system of electromagnetic quantities. Many people have realized that, according to today's understanding of what is meant by a change of units, Oliver Heaviside and H. A. Lorentz were proposing a rescaled set of electromagnetic quantities rather than a change of units;[3] however, reference to a change of units is thoroughly entrenched in the literature. In fact, all systems of electromagnetic equations that have been rescaled to eliminate factors of 4π, and not just the Heaviside-Lorentz system of rescaled equations, are often referred to as electromagnetic equations in rationalized units, or electromagnetic equations written using a system of rationalized units. To avoid breaking sharply with this tradition we prefer to describe these new electromagnetic formulas as belonging to a rationalized system, such as the Heaviside-Lorentz rationalized system, while leaving the nature of the system unspecified.

Table 3.6 contains all the information needed to convert electromagnetic formulas to the Heaviside Lorentz system. We give the rationalized electromagnetic quantities the prefix "h;" the Table 3.6 shows that only ε_0, μ_0, ε_r, μ_r, $\mathcal{P}$, and $\mathcal{R}$ are not changed in this system of rationalization. We see from the table that $_hC = 4\pi C$, so now Eq. (3.11a) becomes

$$_hC = \frac{A}{d}, \qquad (3.13)$$

Table 3.6 Heaviside-Lorentz rationalization for electromagnetic physical quantities.

(magnetic vector potential, rationalized) $_hA = A/\sqrt{4\pi}$	(magnetic permeability, unchanged) μ
(magnetic induction, rationalized) $_hB = B/\sqrt{4\pi}$	(relative magnetic permeability, unchanged) μ_r
(capacitance, rationalized) $_hC = 4\pi C$	(magnetic permeability of free space, unchanged) μ_0
(electric displacement, rationalized) $_hD = D/\sqrt{4\pi}$	(magnetic pole strength, rationalized) $_hp_H = p_H \cdot \sqrt{4\pi}$
(electric field, rationalized) $_hE = E/\sqrt{4\pi}$	(electric dipole moment, rationalized) $_hp = p \cdot \sqrt{4\pi}$
(dielectric constant, unchanged) ε	(electric dipole density, rationalized) $_hP = P \cdot \sqrt{4\pi}$
(relative dielectric constant, unchanged) ε_r	(permeance, unchanged) $\mathcal{P}$
(permittivity of free space, unchanged) ε_0	(charge, rationalized) $_hQ = Q \cdot \sqrt{4\pi}$
(magnetomotive force, rationalized) $_h\mathcal{F} = \mathcal{F}/\sqrt{4\pi}$	(resistance, rationalized) $_hR = R/(4\pi)$
(magnetic flux, rationalized) $_h\Phi_B = \Phi_B/\sqrt{4\pi}$	(reluctance, unchanged) $\mathcal{R}$
(conductance, rationalized) $_hG = 4\pi G$	(volume charge density, rationalized) $_h\rho_Q = \rho_Q \cdot \sqrt{4\pi}$
(magnetic field, rationalized) $_hH = H/\sqrt{4\pi}$	(resistivity, rationalized) $_h\rho_R = \rho_R/(4\pi)$
(current, rationalized) $_hI = I \cdot \sqrt{4\pi}$	(elastance, rationalized) $_hS = S/(4\pi)$
(volume current density, rationalized) $_hJ = J \cdot \sqrt{4\pi}$	(surface charge density, rationalized) $_hS_Q = S_Q \cdot \sqrt{4\pi}$
(surface current density, rationalized) $_h\mathcal{J}_S = \mathcal{J}_S \cdot \sqrt{4\pi}$	(conductivity, rationalized) $_h\sigma = 4\pi\sigma$
(inductance, rationalized) $_hL = L/(4\pi)$	(electric potential, rationalized) $_hV = V/\sqrt{4\pi}$
(permanent-magnet dipole moment, rationalized) $_hm_H = m_H \cdot \sqrt{4\pi}$	(magnetic scalar potential, rationalized) $_h\Omega_H = \Omega_H/\sqrt{4\pi}$
(current-loop magnetic dipole moment, rationalized) $_hm_I = m_I \cdot \sqrt{4\pi}$	(electric susceptibility, rationalized) $_h\chi_e = 4\pi\chi_e$
(permanent-magnet dipole density, rationalized) $_hM_H = M_H \cdot \sqrt{4\pi}$	(magnetic susceptibility, rationalized) $_h\chi_m = 4\pi\chi_m$
(current-loop magnetic dipole density, rationalized) $_hM_I = M_I \cdot \sqrt{4\pi}$	

eliminating the inconvenient factor of 4π. The reader is warned that in the literature there is usually no prefix, subscript, or superscript of any sort used to distinguish the rescaled from the nonrescaled variables in rationalized systems of electromagnetic quantities.

Because the equalities in Table 3.6 are between physical quantities, the U and N operators for any system of units can be applied to them. Looking, for example, at the electric field, we have

$$_h\vec{E} = \frac{1}{\sqrt{4\pi}}\vec{E}. \tag{3.14a}$$

From Eqs. (1.25a, b) and (1.28a, b) of Chapter 1 it follows that for any pair of U and N operators

$$\mathrm{U}\left(_h\vec{E}\right) = \mathrm{U}\left(\vec{E}\right) \tag{3.14b}$$

and

$$\mathrm{N}\left(_h\vec{E}\right) = \frac{1}{\sqrt{4\pi}}\mathrm{N}\left(\vec{E}\right). \tag{3.14c}$$

Equation (3.14b) explicitly shows that Heaviside-Lorentz rationalization can be used with any system of electromagnetic units, because only the numeric parts of electromagnetic physical quantities are changed. In emuq units,

$$\underset{\mathrm{emuq}}{\mathrm{U}}\left(_h\vec{E}\right) = \underset{\mathrm{emuq}}{\mathrm{U}}\left(\vec{E}\right) = \frac{\mathrm{abvoltq}}{\mathrm{cm}};$$

in Gaussian or esu units,

$$\underset{\mathrm{gs}}{\mathrm{U}}\left(_h\vec{E}\right) = \underset{\mathrm{esu}}{\mathrm{U}}\left(_h\vec{E}\right) = \underset{\mathrm{gs}}{\mathrm{U}}\left(\vec{E}\right) = \underset{\mathrm{esu}}{\mathrm{U}}\left(\vec{E}\right) = \frac{\mathrm{statvolt}}{\mathrm{cm}} = \frac{\mathrm{gm}^{1/2}}{\mathrm{cm}^{1/2}\cdot\mathrm{sec}};$$

and so on. When textbooks refer to the Heaviside-Lorentz system, however, they are applying Heaviside-Lorentz rationalization to the Gaussian system of units. This means that the U and N operators for the Heaviside-Lorentz system are just the U and N operators for the Gaussian system.

$$\underset{\substack{\mathrm{Heaviside}\\\mathrm{Lorentz}}}{\mathrm{U}} = \underset{\mathrm{gs}}{\mathrm{U}}, \tag{3.15a}$$

$$\underset{\substack{\mathrm{Heaviside}\\\mathrm{Lorentz}}}{\mathrm{N}} = \underset{\mathrm{gs}}{\mathrm{N}}. \tag{3.15b}$$

Instead of defining a new pair of U and N operators for the Heaviside-Lorentz system, we just use $\underset{\mathrm{gs}}{\mathrm{U}}$ and $\underset{\mathrm{gs}}{\mathrm{N}}$ to describe the behavior of the rescaled electromagnetic physical quantities.

Converting Gaussian electromagnetic equations to the Heaviside-Lorentz system is an exercise in algebraic substitution; there is no need to split equations into units and numeric components. To see how this works, we convert Coulomb's law for electric charges and magnetic poles to the Heaviside-Lorentz system. Coulomb's law for electric charges involves only electric quantities and is the same in the esu and Gaussian systems of units. Writing Eq. (2.1b) in Chapter 2 with $\varepsilon_0 = 1$, we have

$$\vec{F}_{12} = \frac{Q_1 Q_2}{r^2} \hat{r}_{12}.$$

Substituting the rescaled values $_h Q_1 = Q_1 \sqrt{4\pi}$ and $_h Q_2 = Q_2 \sqrt{4\pi}$ gives

$$\vec{F}_{12} = \frac{(_h Q_1)(_h Q_2)}{4\pi r^2} \hat{r}_{12}. \tag{3.16a}$$

Coulomb's law for magnetic poles involves only magnetic quantities and is the same in emu and Gaussian units. Substituting $(_h p_H)_1 = (p_H)_1 \sqrt{4\pi}$ and $(_h p_H)_2 = (p_H)_2 \sqrt{4\pi}$ into Eq. (2.2b) in Chapter 2 with $\mu_0 = 1$ takes us from the emu or Gaussian system to the Heaviside-Lorentz system:

$$\vec{F}_{12} = \frac{(_h p_H)_1 (_h p_H)_2}{4\pi r^2} \hat{r}_{12}. \tag{3.16b}$$

Clearly the factors of 4π that have been removed from one set of electromagnetic formulas re-appear in other sets of electromagnetic formulas. Oliver Heaviside and H. A. Lorentz knew this, of course; what they intended was to put the factors of 4π into little-used formulas while removing them from frequently used formulas. Thus, we can infer that by the end of the nineteenth century formulas like Eq. (3.11a) had become much more useful than Coulomb's law for electric charges or magnetic poles. We note that Eqs. (3.5a–f), Maxwell's equations in Gaussian units, become

$$\vec{\nabla} \cdot {}_h\vec{D} = {}_h\rho_Q, \tag{3.17a}$$

$$\vec{\nabla} \cdot {}_h\vec{B} = 0, \tag{3.17b}$$

$$\vec{\nabla} \times {}_h\vec{E} + \frac{1}{c}\frac{\partial {}_h\vec{B}}{\partial t} = 0, \tag{3.17c}$$

$$\vec{\nabla} \times {}_h\vec{H} = \frac{1}{c}{}_h\vec{J} + \frac{1}{c}\frac{\partial {}_h\vec{D}}{\partial t}, \tag{3.17d}$$

with

$${}_h\vec{D} = {}_h\vec{E} + {}_h\vec{P} \tag{3.17e}$$

and

$$_h\vec{H} = {_h\vec{B}} - {_h\vec{M}_I},\tag{3.17f}$$

when Table 3.6 is used to replace ρ_Q, $\vec{E}$, $\vec{D}$, $\vec{B}$, $\vec{H}$, $\vec{J}$, $\vec{P}$, and M_I by $_h\rho_Q$, $_h\vec{E}$, $_h\vec{D}$, $_h\vec{B}$, $_h\vec{H}$, $_h\vec{J}$, $_h\vec{P}$, and $_h M_I$. By the end of the nineteenth century Maxwell's equations, rather than Coulomb's laws, were the basis of electromagnetic field theory, and Eqs. (3.17a–f) are obviously simpler than Eqs. (3.1a–f) or Eqs. (3.5a–f). Indeed, all a relativistic physicist has to do is move to a set of units where the speed of light is equal to 1 to get the ultimate "user-friendly" set of equations:

$$\vec{\nabla} \cdot {_h\vec{D}} = {_h\rho_Q},\tag{3.18a}$$

$$\vec{\nabla} \cdot {_h\vec{B}} = 0,\tag{3.18b}$$

$$\vec{\nabla} \times {_h\vec{E}} + \frac{\partial\, {_h\vec{B}}}{\partial t} = 0,\tag{3.18c}$$

$$\vec{\nabla} \times {_h\vec{H}} = {_h\vec{J}} + \frac{\partial\, {_h\vec{D}}}{\partial t},\tag{3.18d}$$

where

$$_h\vec{D} = {_h\vec{E}} + {_h\vec{P}}\tag{3.18e}$$

and

$$_h\vec{H} = {_h\vec{B}} - {_h\vec{M}_I}.\tag{3.18f}$$

In this case there are no extraneous constants whatsoever to clutter up the algebra. In empty space with $_h\vec{P} = 0$ and $_h\vec{M} = 0$ the $\vec{D}$ and $\vec{H}$ fields become identical to the $\vec{E}$ and $\vec{B}$ fields, and Eqs. (3.18a–f) reduce to

$$\vec{\nabla} \cdot {_h\vec{E}} = {_h\rho_Q},\tag{3.19a}$$

$$\vec{\nabla} \cdot {_h\vec{B}} = 0,\tag{3.19b}$$

$$\vec{\nabla} \times {_h\vec{E}} + \frac{\partial\, {_h\vec{B}}}{\partial t} = 0,\tag{3.19c}$$

$$\vec{\nabla} \times {_h\vec{B}} = {_h\vec{J}} + \frac{\partial\, {_h\vec{E}}}{\partial t},\tag{3.19d}$$

with $_h\rho_Q$ and $_h\vec{J}$ usually taken to represent the density and flow of small point charges in a vacuum. From a mathematical point of view, Eqs. (3.19a–d) are an attractively simple set of field equations, and today there are large communities of physicists who routinely use the Heaviside-Lorentz system when doing relativistic analysis of elementary particles and electromagnetic fields.

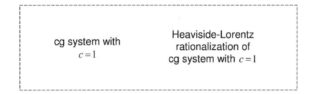

THREE FUNDAMENTAL DIMENSIONS
[mass, length, time]

esu cgs
system

emu cgs
system

Gaussian cgs
system

Heaviside-Lorentz
rationalization of the
Gaussian cgs system
(usually called just the
Heaviside-Lorentz
cgs system)

TWO FUNDAMENTAL DIMENSIONS
[mass-energy, space-time]

cg system with
$c = 1$

Heaviside-Lorentz
rationalization of
cg system with $c = 1$

Figure 3.1 Here are six electromagnetic systems which do not recognize charge as a fundamental dimension.

3.4 GAUSSIAN AND HEAVISIDE-LORENTZ SYSTEMS WITH c = 1 AND ℏ = c = 1

Since the Gaussian and Heaviside-Lorentz systems are often used with units where $c = 1$, it is worth taking time to examine exactly what this means. Remembering that the full name for Gaussian units is the Gaussian cgs system of units, we return to the cgc set of units constructed in Section 1.10 and consider the transformation from Gaussian cgs units to Gaussian cgc units. Looking at Table 3.5, we can pick any standard electromagnetic quantity, such as the magnetic induction $\vec{B}$, and write

$$\vec{B} = \vec{B}_{\text{gs}}\text{gauss} = \vec{B}_{\text{gs}}\frac{\text{gm}^{1/2}}{\text{cm}^{1/2}\text{sec}}. \tag{3.20a}$$

From Eq. (1.56a) this becomes in cgc units

$$\vec{B}_{\text{gs}}\frac{\text{gm}^{1/2}}{\text{cm}^{1/2}\text{sec}} = \vec{B}_{\text{gs}}c_{\text{cgs}}^{-1}\frac{\text{gm}^{1/2}}{\text{cm}^{1/2}\text{cmtime}_3}. \tag{3.20b}$$

We know that the gs subscript in $\vec{B}_{gs}$ stands for the numeric part of $\vec{B}$ in cgs Gaussian units

$$\vec{B}_{gs} = \underset{gs}{N}(\vec{B}).$$

Equation (1.56a) shows the unit of magnetic induction in Gaussian cgc units must be larger than the gauss by a factor of c_{cgs}:

$$1 \text{ gauss} = 1\left(\frac{gm^{1/2}}{cm^{1/2}sec}\right) = c_{cgs}^{-1}\frac{gm^{1/2}}{cm^{1/2}cmtime_3},$$

so from Rule I the numeric component of $\vec{B}$ in Gaussian cgc units must be

$$B_{gscgc} = c_{cgs}^{-1}B_{gs}. \tag{3.20c}$$

Now, Eqs. (3.20a, b) become

$$\vec{B} = \vec{B}_{gscgc}\frac{gm^{1/2}}{cm^{1/2}cmtime_3}. \tag{3.20d}$$

Still following the procedure of Section 1.10, we stop recognizing the separate dimension of time to get

$$cmtime_3 \rightarrow cmtime_2 = cm.$$

This takes us from the Gaussian cgc system to the Gaussian cg system. Equation (3.20d) then becomes, representing the Gaussian cg system by the subscript "gscg":

$$\vec{B} = \vec{B}_{gscg}\frac{gm^{1/2}}{cm^{3/2}}, \tag{3.20e}$$

where, following the precedent of Eq. (1.58a), we require the numeric components of all physical quantities in the Gaussian cg system to have the same value as in the Gaussian cgc system:

$$\vec{B}_{gscg} = \vec{B}_{gscgc}. \tag{3.20f}$$

Generalizing the rules given in equations of Section 1.10, we recognize that the U and N operators for the Gaussian cg system are basically the same as the $\underset{cg}{U}$ and $\underset{cg}{N}$ operators defined in Eqs. (1.58a, d). For any physical quantity b, with

$$\underset{gs}{U}(b) = gm^{\alpha}cm^{\beta}sec^{\gamma} \tag{3.21a}$$

for real exponents α, β, and γ, we define

$$\underset{\text{gscg}}{N}(b) = c_{\text{cgs}}^{\gamma} \underset{\text{gs}}{N}(b) \tag{3.21b}$$

and

$$\underset{\text{gscg}}{U}(b) = \left(\frac{\text{cm}}{\text{sec}}\right)^{\gamma} \underset{\text{gs}}{U}(b) = \text{gm}^{\alpha}\text{cm}^{\beta+\gamma}. \tag{3.21c}$$

Table 3.7 comes from applying Eqs. (3.21a–c) to our standard list of electromagnetic physical quantities.

Returning to the $\vec{B}$ field specified in Eq. (3.20a), we can transform to the Heaviside-Lorentz system with $c = 1$ by consulting Table 3.6 to get

$$_h\vec{B} = \frac{1}{\sqrt{4\pi}}\vec{B}. \tag{3.22a}$$

This equation is between physical quantities, so Eq. (3.20e) can be used to write

$$_h\vec{B} = \left(\frac{\vec{B}_{\text{gscg}}}{\sqrt{4\pi}}\right)\frac{\text{gm}^{1/2}}{\text{cm}^{3/2}} = \left(\frac{\vec{B}_{\text{gs}}}{c_{\text{cgs}}\sqrt{4\pi}}\right)\frac{\text{gm}^{1/2}}{\text{cm}^{3/2}}, \tag{3.22b}$$

where the last step comes from Eqs. (3.20c, f). Repeating this conversion for all the Heaviside-Lorentz field quantities in (3.18a–f) and (3.19a–d) gives

$$_h\vec{E} = \left(\frac{\vec{E}_{\text{gs}}}{c_{\text{cgs}}\sqrt{4\pi}}\right)\frac{\text{gm}^{1/2}}{\text{cm}^{3/2}}, \tag{3.23a}$$

$$_h\vec{D} = \left(\frac{\vec{D}_{\text{gs}}}{c_{\text{cgs}}\sqrt{4\pi}}\right)\frac{\text{gm}^{1/2}}{\text{cm}^{3/2}}, \tag{3.23b}$$

$$_h\rho_Q = \left[\frac{(\rho_Q)_{\text{gs}}\sqrt{4\pi}}{c_{\text{cgs}}}\right]\frac{\text{gm}^{1/2}}{\text{cm}^{5/2}}, \tag{3.23c}$$

$$_h\vec{H} = \left(\frac{\vec{H}_{\text{gs}}}{c_{\text{cgs}}\sqrt{4\pi}}\right)\frac{\text{gm}^{1/2}}{\text{cm}^{3/2}}, \tag{3.23d}$$

$$_h\vec{J} = \left(\frac{\vec{J}_{\text{gs}}\sqrt{4\pi}}{c_{\text{cgs}}^{2}}\right)\frac{\text{gm}^{1/2}}{\text{cm}^{5/2}}, \tag{3.23e}$$

$$_h\vec{P} = \left(\frac{\vec{P}_{\text{gs}}\sqrt{4\pi}}{c_{\text{cgs}}}\right)\frac{\text{gm}^{1/2}}{\text{cm}^{3/2}}, \tag{3.23f}$$

and

$$_h\vec{M}_I = \left[\frac{(\vec{M}_I)_{\text{gs}}\sqrt{4\pi}}{c_{\text{cgs}}}\right]\frac{\text{gm}^{1/2}}{\text{cm}^{3/2}}, \tag{3.23g}$$

Table 3.7 The Gaussian cg system (with $c = 1$) where time is no longer recognized as a fundamental dimension. Space, time both have units of cm and mass, energy both have units of gm.

Physical quantity	Gaussian cg units (gscg system)	Conversion of numeric components from gs to gscg systems
(magnetic vector potential) A	$\underset{\text{gscg}}{\text{U}}(A) = \dfrac{\text{gm}^{1/2}}{\text{cm}^{1/2}}$	$A_{\text{gscg}} = A_{\text{gs}} \cdot c_{\text{cgs}}^{-1}$
(magnetic induction) B	$\underset{\text{gscg}}{\text{U}}(B) = \dfrac{\text{gm}^{1/2}}{\text{cm}^{3/2}}$	$B_{\text{gscg}} = B_{\text{gs}} \cdot c_{\text{cgs}}^{-1}$
(capacitance) C	$\underset{\text{gscg}}{\text{U}}(C) = cm$	$C_{\text{gscg}} = C_{\text{gs}}$
(electric displacement) D	$\underset{\text{gscg}}{\text{U}}(D) = \dfrac{\text{gm}^{1/2}}{\text{cm}^{3/2}}$	$D_{\text{gscg}} = D_{\text{gs}} \cdot c_{\text{cgs}}^{-1}$
(electric field) E	$\underset{\text{gscg}}{\text{U}}(E) = \dfrac{\text{gm}^{1/2}}{\text{cm}^{3/2}}$	$E_{\text{gscg}} = E_{\text{gs}} \cdot c_{\text{cgs}}^{-1}$
(permittivity of free space) ε_0	$\underset{\text{gscg}}{\text{U}}(\varepsilon_0) = 1$	$(\varepsilon_0)_{\text{gscg}} = (\varepsilon_0)_{\text{gs}} = 1$
(magnetomotive force) $\mathcal{F}$	$\underset{\text{gscg}}{\text{U}}(\mathcal{F}) = \dfrac{\text{gm}^{1/2}}{\text{cm}^{1/2}}$	$\mathcal{F}_{\text{gscg}} = \mathcal{F}_{\text{gs}} \cdot c_{\text{cgs}}^{-1}$
(magnetic flux) Φ_B	$\underset{\text{gscg}}{\text{U}}(\Phi_B) = \text{gm}^{1/2} \cdot \text{cm}^{1/2}$	$(\Phi_B)_{\text{gscg}} = (\Phi_B)_{\text{gs}} \cdot c_{\text{cgs}}^{-1}$
(conductance) G	$\underset{\text{gscg}}{\text{U}}(G) = 1$	$G_{\text{gscg}} = G_{\text{gs}} \cdot c_{\text{cgs}}^{-1}$
(magnetic field) H	$\underset{\text{gscg}}{\text{U}}(H) = \dfrac{\text{gm}^{1/2}}{\text{cm}^{3/2}}$	$H_{\text{gscg}} = H_{\text{gs}} \cdot c_{\text{cgs}}^{-1}$
(current) I	$\underset{\text{gscg}}{\text{U}}(I) = \dfrac{\text{gm}^{1/2}}{\text{cm}^{1/2}}$	$I_{\text{gscg}} = I_{\text{gs}} \cdot c_{\text{cgs}}^{-2}$
(volume current density) J	$\underset{\text{gscg}}{\text{U}}(J) = \dfrac{\text{gm}^{1/2}}{\text{cm}^{5/2}}$	$J_{\text{gscg}} = J_{\text{gs}} \cdot c_{\text{cgs}}^{-2}$
(surface current density) $\mathcal{J}_S$	$\underset{\text{gscg}}{\text{U}}(\mathcal{J}_S) = \dfrac{\text{gm}^{1/2}}{\text{cm}^{3/2}}$	$(\mathcal{J}_S)_{\text{gscg}} = (\mathcal{J}_S)_{\text{gs}} \cdot c_{\text{cgs}}^{-2}$

Table 3.7 (Continued).

Physical quantity	Gaussian cg units (gscg system)	Conversion of numeric components from gs to gscg systems
(inductance) L	$\underset{\text{gscg}}{U}(L) = \text{cm}$	$L_{\text{gscg}} = L_{\text{gs}} \cdot c_{\text{cgs}}^2$
(permanent-magnet dipole moment) m_H	$\underset{\text{gscg}}{U}(m_H) = \text{gm}^{1/2} \cdot \text{cm}^{3/2}$	$(m_H)_{\text{gscg}} = (m_H)_{\text{gs}} \cdot c_{\text{cgs}}^{-1}$
(current-loop magnetic dipole moment) m_I	$\underset{\text{gscg}}{U}(m_I) = \text{gm}^{1/2} \cdot \text{cm}^{3/2}$	$(m_I)_{\text{gscg}} = (m_I)_{\text{gs}} \cdot c_{\text{cgs}}^{-1}$
(permanent-magnet dipole density) M_H	$\underset{\text{gscg}}{U}(M_H) = \dfrac{\text{gm}^{1/2}}{\text{cm}^{3/2}}$	$(M_H)_{\text{gscg}} = (M_H)_{\text{gs}} \cdot c_{\text{cgs}}^{-1}$
(current-loop magnetic dipole density) M_I	$\underset{\text{gscg}}{U}(M_I) = \dfrac{\text{gm}^{1/2}}{\text{cm}^{3/2}}$	$(M_I)_{\text{gscg}} = (M_I)_{\text{gs}} \cdot c_{\text{cgs}}^{-1}$
(magnetic permeability of free space) μ_0	$\underset{\text{gscg}}{U}(\mu_0) = 1$	$(\mu_0)_{\text{gscg}} = (\mu_0)_{\text{gs}} = 1$
(magnetic pole strength) p_H	$\underset{\text{gscg}}{U}(p_H) = \text{gm}^{1/2} \cdot \text{cm}^{1/2}$	$(p_H)_{\text{gscg}} = (p_H)_{\text{gs}} \cdot c_{\text{cgs}}^{-1}$
(electric dipole moment) p	$\underset{\text{gscg}}{U}(p) = \text{gm}^{1/2} \cdot \text{cm}^{3/2}$	$p_{\text{gscg}} = p_{\text{gs}} \cdot c_{\text{cgs}}^{-1}$
(electric dipole density) P	$\underset{\text{gscg}}{U}(P) = \dfrac{\text{gm}^{1/2}}{\text{cm}^{3/2}}$	$P_{\text{gscg}} = P_{\text{gs}} \cdot c_{\text{cgs}}^{-1}$
(permeance) $\mathcal{P}$	$\underset{\text{gscg}}{U}(\mathcal{P}) = \text{cm}$	$\mathcal{P}_{\text{gscg}} = \mathcal{P}_{\text{gs}}$
(charge) Q	$\underset{\text{gscg}}{U}(Q) = \text{gm}^{1/2} \cdot \text{cm}^{1/2}$	$Q_{\text{gscg}} = Q_{\text{gs}} \cdot c_{\text{cgs}}^{-1}$
(resistance) R	$\underset{\text{gscg}}{U}(R) = 1$	$R_{\text{gscg}} = R_{\text{gs}} \cdot c_{\text{cgs}}$
(reluctance) $\mathcal{R}$	$\underset{\text{gscg}}{U}(\mathcal{R}) = \text{cm}^{-1}$	$\mathcal{R}_{\text{gscg}} = \mathcal{R}_{\text{gs}}$

Table 3.7 (Continued).

Physical quantity	Gaussian cg units (gscg system)	Conversion of numeric components from gs to gscg systems
(volume charge density) ρ_Q	$\underset{\text{gscg}}{\text{U}}\,(\rho_Q) = \dfrac{\text{gm}^{1/2}}{\text{cm}^{5/2}}$	$(\rho_Q)_{\text{gscg}} = (\rho_Q)_{\text{gs}} \cdot c_{\text{cgs}}^{-1}$
(resistivity) ρ_R	$\underset{\text{gscg}}{\text{U}}\,(\rho_R) = \text{cm}$	$(\rho_R)_{\text{gscg}} = (\rho_R)_{\text{gs}} \cdot c_{\text{cgs}}$
(elastance) S	$\underset{\text{gscg}}{\text{U}}\,(S) = \text{cm}^{-1}$	$S_{\text{gscg}} = S_{\text{gs}}$
(surface charge density) S_Q	$\underset{\text{gscg}}{\text{U}}\,(S_Q) = \dfrac{\text{gm}^{1/2}}{\text{cm}^{3/2}}$	$(S_Q)_{\text{gscg}} = (S_Q)_{\text{gs}} \cdot c_{\text{cgs}}^{-1}$
(conductivity) σ	$\underset{\text{gscg}}{\text{U}}\,(\sigma) = \text{cm}^{-1}$	$\sigma_{\text{gscg}} = \sigma_{\text{gs}} \cdot c_{\text{cgs}}^{-1}$
(electric potential) V	$\underset{\text{gscg}}{\text{U}}\,(V) = \dfrac{\text{gm}^{1/2}}{\text{cm}^{1/2}}$	$V_{\text{gscg}} = V_{\text{gs}} \cdot c_{\text{cgs}}^{-1}$
(magnetic scalar potential) Ω_H	$\underset{\text{gscg}}{\text{U}}\,(\Omega_H) = \dfrac{\text{gm}^{1/2}}{\text{cm}^{1/2}}$	$(\Omega_H)_{\text{gscg}} = (\Omega_H)_{\text{gs}} \cdot c_{\text{cgs}}^{-1}$

in units where the speed of light is equal to 1, space and time are measured in cm, and mass and energy are measured in gm.

If we want to measure mass-energy in MeV, Eq. (1.68b) is used to replace gm by $\text{MeV}_2 = 10^6\,\text{eV}_2$. As an example of how to proceed, we choose a physical quantity from Table 3.7, for example the electric field $\vec{E}$, and write

$$\vec{E} = \vec{E}_{\text{gscg}} \frac{\text{gm}^{1/2}}{\text{cm}^{3/2}} = \left(\frac{\vec{E}_{\text{gs}}}{c_{\text{cgs}}}\right)\left(\frac{c_{\text{cgs}}^2 \text{MeV}_2}{10^{13} e_{[\text{coul}]}}\right)^{1/2} \text{cm}^{-3/2}$$

$$= \left(\frac{\vec{E}_{\text{gs}}}{\sqrt{10^{13} e_{[\text{coul}]}}}\right) \frac{(\text{MeV}_2)^{1/2}}{\text{cm}^{3/2}}. \tag{3.24a}$$

In the Heaviside-Lorentz system Table 3.6 and Eq. (3.24a) give

$$_h\vec{E} = \frac{1}{\sqrt{4\pi}}\vec{E} = \left[\frac{\vec{E}_{\text{gs}}}{\sqrt{(4\pi \cdot 10^{13}) e_{[\text{coul}]}}}\right] \frac{(\text{MeV}_2)^{1/2}}{\text{cm}^{3/2}}. \tag{3.24b}$$

The straightforward substitution of MeV_2 for gm shown in Eq. (3.24a) can be used to convert any electromagnetic physical quantity in Table 3.7 to units where mass and energy are measured in MeV_2, and Table 3.6 then specifies how to convert physical quantities to the Heaviside-Lorentz system.

If we want units where $\hbar = c = 1$, Eq. (1.90e) states that

$$1\,\mathrm{gm}_1 = \left(\frac{c_{\mathrm{cgs}}}{\hbar_{\mathrm{cgs}}}\right)\mathrm{cm}^{-1}, \tag{3.25a}$$

where the subscript "1" on gm reminds us that mass and energy are no longer being recognized as a separate dimension from space and time. Although this result comes from a system of units set up to make Boltzmann's constant as well as $\hbar$ and c equal to 1, the value of Boltzmann's constant comes from the unit of temperature, which is irrelevant here. The scaling and dimension chosen for the units for mass, length, and time that make $\hbar = c = 1$ in Chapter 1 have the same effect now, allowing us to use Eq. (3.25a) with confidence.

Table 3.7 shows how to separate the $\vec{E}$ field into numeric parts and units in the Gaussian cg system given $\vec{E}_{\mathrm{gs}}$, the numeric value of $\vec{E}$ in the Gaussian cgs system.

$$\vec{E} = \frac{\vec{E}_{\mathrm{gs}}}{c_{\mathrm{cgs}}}\frac{\mathrm{gm}^{1/2}}{\mathrm{cm}^{3/2}}.$$

We stop recognizing the separate dimension of mass-energy by replacing gm with gm_1, which gives

$$\vec{E} = \left(\frac{\vec{E}_{\mathrm{gs}}}{c_{\mathrm{cgs}}}\right)\frac{(\mathrm{gm}_1)^{1/2}}{\mathrm{cm}^{3/2}} = \left(\frac{\vec{E}_{\mathrm{gs}}}{c_{\mathrm{cgs}}}\right)\left(\frac{c_{\mathrm{cgs}}}{\hbar_{\mathrm{cgs}}}\mathrm{cm}^{-1}\right)^{1/2}\mathrm{cm}^{-3/2}$$
$$= \left(\frac{\vec{E}_{\mathrm{gs}}}{c_{\mathrm{cgs}}^{1/2}\hbar_{\mathrm{cgs}}^{1/2}}\right)\mathrm{cm}^{-2}. \tag{3.25b}$$

In the Heaviside-Lorentz system this becomes

$$_h\vec{E} = \left(\frac{\vec{E}_{\mathrm{gs}}}{c_{\mathrm{cgs}}^{1/2}\hbar_{\mathrm{cgs}}^{1/2}\sqrt{4\pi}}\right)\mathrm{cm}^{-2}. \tag{3.25c}$$

We can check these results by noting from Eq. (3.25a) that mass has the dimension of length^{-1} in units where $\hbar = c = 1$. Table 3.7 then shows that charge is dimensionless, so Eq. (2.3a) of Chapter 2—with $\varepsilon_0 = 1$ since we are in Gaussian units—shows the $\vec{E}$ field must have dimensions of length^{-2}, matching the right-hand side of Eq. (3.25c). Following this same procedure for the other field quantities in Eqs. (3.23b–g) gives

$$_h\vec{D} = \left(\frac{\vec{D}_{\mathrm{gs}}}{c_{\mathrm{cgs}}^{1/2}\hbar_{\mathrm{cgs}}^{1/2}\sqrt{4\pi}}\right)\mathrm{cm}^{-2}, \tag{3.25d}$$

$$_h\vec{B} = \left(\frac{\vec{B}_{\text{gs}}}{c_{\text{cgs}}^{1/2}\hbar_{\text{cgs}}^{1/2}\sqrt{4\pi}}\right)\text{cm}^{-2},\tag{3.25e}$$

$$_h\vec{H} = \left(\frac{\vec{H}_{\text{gs}}}{c_{\text{cgs}}^{1/2}\hbar_{\text{cgs}}^{1/2}\sqrt{4\pi}}\right)\text{cm}^{-2},\tag{3.25f}$$

$$_h\rho_Q = \left[\frac{(\rho_Q)_{\text{gs}}\sqrt{4\pi}}{c_{\text{cgs}}^{1/2}\hbar_{\text{cgs}}^{1/2}}\right]\text{cm}^{-3},\tag{3.25g}$$

$$_h\vec{J} = \left(\frac{\vec{J}_{\text{gs}}\sqrt{4\pi}}{c_{\text{cgs}}^{3/2}\hbar_{\text{cgs}}^{1/2}}\right)\text{cm}^{-3},\tag{3.25h}$$

$$_h\vec{P} = \left(\frac{\vec{P}_{\text{gs}}\sqrt{4\pi}}{c_{\text{cgs}}^{1/2}\hbar_{\text{cgs}}^{1/2}}\right)\text{cm}^{-2},\tag{3.25i}$$

and

$$_h\vec{M}_I = \left[\frac{(\vec{M}_I)_{\text{gs}}\sqrt{4\pi}}{c_{\text{cgs}}^{1/2}\hbar_{\text{cgs}}^{1/2}}\right]\text{cm}^{-2}.\tag{3.25j}$$

Not only do the $\vec{E}, \vec{D}, \vec{B}$, and $\vec{H}$ fields now all have the same units, keeping life simple; but a quick check of Table 3.7 also confirms what Eqs. (3.25c–j) suggest is true, that all the standard electromagnetic physical quantities end up with integer powers of the length unit when we start with the Gaussian cgs system and make $\hbar = c = 1$.

3.5 EQUIVALENCE OF THE ESU, EMU, AND GAUSSIAN SYSTEMS WHEN c = 1

Maxwell's equations and the Lorentz force law provide a complete description of classical electromagnetic phenomena.[4] From the work done in Section 1.10 we know that physical equations can be transformed to units where $c = 1$ simply by replacing c with 1 everywhere it appears. Equations (3.1a–d) are Maxwell's equations in either the esu or emu system depending on whether we choose $\varepsilon_0 = 1, \mu_0 = c^{-2}$ or $\mu_0 = 1, \varepsilon_0 = c^{-2}$. When $c = 1$, both choices are equivalent; and the resulting equations become identical to Maxwell's equations in the Gaussian system, Eqs. (3.5a–f), with $c = 1$. Equation (2.66h), the Lorentz force law, has the same form in esu or emu units; and Eq. (3.9c), the Lorentz force law in the Gaussian system, becomes the same as Eq. (2.66h) when $c = 1$. Since Maxwell's equations and the Lorentz force law are the foundation for all electromagnetic theory, we expect all electromagnetic equations written using physical quantities to have the same identical form when converted from the esu, emu, or Gaussian systems to corresponding systems of units where $c = 1$. We can convert to such a system by using Rule I to replace sec by cmtime$_3$, which then becomes

$\text{cmtime}_2 = \text{cm}$ when time is no longer recognized as having a separate dimension. This replaces sec by cm in the units of all physical quantities. Comparing Tables 2.4, 2.5, and 3.5, we see that the esu, emu, and Gaussian cgs units of the electromagnetic physical quantities become identical to the units of the electromagnetic physical quantities in Table 3.7, the Gaussian cg system, when sec is replaced by cm. Hence, when the esu, emu, and Gaussian cgs systems are converted in this way to units where $c = 1$, not only do all the electromagnetic formulas and equations become identical but also all the electromagnetic units become identical. Therefore, the numerical components of all the electromagnetic physical quantities become identical as well, and we end up with the same system of units—there is no longer any difference between the esu, emu, and Gaussian systems. The Gaussian cg (gscg) system of units can thus be labeled as just the centimeter-gram (cg) system of units, because it is the unique system of units we end up with when converting from the three sets of electromagnetic cgs units where charge is not recognized as having a separate dimension to the corresponding cg units where $c = 1$.

As a check on our reasoning, we pick some physical quantity, such as the current I, and use Table 2.5 to break it up into a numeric component and units in the emu system:

$$I = I_{\text{emu}} \frac{\text{gm}^{1/2}\text{cm}^{1/2}}{\text{sec}}. \tag{3.26a}$$

Following the procedure of Section 1.10 of Chapter 1, we measure time in units of $\text{cmtime}_3 = c_{\text{cgs}}^{-1}\text{sec}$ to get

$$I = \left(\frac{I_{\text{emu}}}{c_{\text{cgs}}}\right) \frac{\text{gm}^{1/2}\text{cm}^{1/2}}{\text{cmtime}_3}. \tag{3.26b}$$

When time is no longer recognized as a separate dimension, $\text{cmtime}_3 \to \text{cmtime}_2 = \text{cm}$ and the current becomes

$$I = \left(\frac{I_{\text{emu}}}{c_{\text{cgs}}}\right) \frac{\text{gm}^{1/2}}{\text{cm}^{1/2}}. \tag{3.26c}$$

This is the formula for I in the emu system converted to the cg system. Now, let us see what happens when we convert to the cg system from the esu or Gaussian systems. Since I is an electric quantity, it has the same units in both the esu and Gaussian systems. Keeping in mind that for this reason $I_{\text{gs}} = I_{\text{esu}}$, we use either Table 2.4 or 3.5 to write

$$I = I_{\text{gs}} \frac{\text{gm}^{1/2}\text{cm}^{3/2}}{\text{sec}^2}. \tag{3.27a}$$

Converting this expression to units of cmtime$_3$ gives

$$I = \left(\frac{I_{\mathrm{gs}}}{c_{\mathrm{cgs}}^2}\right) \frac{\mathrm{gm}^{1/2}\mathrm{cm}^{3/2}}{(\mathrm{cmtime}_3)^2}, \tag{3.27b}$$

which becomes

$$I = \left(\frac{I_{\mathrm{gs}}}{c_{\mathrm{cgs}}^2}\right) \frac{\mathrm{gm}^{1/2}}{\mathrm{cm}^{1/2}} \tag{3.27c}$$

when time is no longer recognized as a separate dimension. Comparing Eqs. (3.26c) and (3.27c), we note that for the two formulas to represent the same system of units, I must have the same numeric component:

$$\frac{I_{\mathrm{emu}}}{c_{\mathrm{cgs}}} = \frac{I_{\mathrm{gs}}}{c_{\mathrm{cgs}}^2},$$

which, since $I_{\mathrm{esu}} = I_{\mathrm{gs}}$, can be written as

$$I_{\mathrm{emu}} = I_{\mathrm{gs}}c_{\mathrm{cgs}}^{-1} = I_{\mathrm{esu}}c_{\mathrm{cgs}}^{-1}. \tag{3.28}$$

This is exactly what Tables 2.7 and 3.4 show to be the case. As a second check, we pick a magnetic quantity, i.e., the magnetic pole strength p_H. This is the same in emu and Gaussian units, so $(p_H)_{\mathrm{emu}} = (p_H)_{\mathrm{gs}}$ and from Tables 2.5 or 3.5 we have

$$p_H = (p_H)_{\mathrm{gs}}\frac{\mathrm{gm}^{1/2}\mathrm{cm}^{3/2}}{\sec}. \tag{3.29a}$$

Converting to cg units the same way as before, we get

$$p_H = \left[\frac{(p_H)_{\mathrm{gs}}}{c_{\mathrm{cgs}}}\right]\mathrm{gm}^{1/2}\mathrm{cm}^{1/2}. \tag{3.29b}$$

We repeat this conversion starting from the esu system. Table 2.4 shows that

$$p_H = (p_H)_{\mathrm{esu}}\mathrm{gm}^{1/2}\mathrm{cm}^{1/2}. \tag{3.29c}$$

We need do nothing at all to convert this to cg units because there are no powers of sec in Eq. (3.29c). For p_H to be represented by the same system of units in Eqs. (3.29b, c), p_H must have the same numeric component. Recognizing that $(p_H)_{\mathrm{emu}} = (p_H)_{\mathrm{gs}}$, we write

$$(p_H)_{\mathrm{esu}} = \frac{(p_H)_{\mathrm{gs}}}{c_{\mathrm{cgs}}} = \frac{(p_H)_{\mathrm{emu}}}{c_{\mathrm{cgs}}}. \tag{3.29d}$$

Tables 2.7 and 3.2 show that in fact Eq. (3.29d) is true, once again demonstrating that the esu, emu, and Gaussian systems become the same system of units when we make the speed of light $c = 1$ without recognizing charge as a fundamental dimension. We conclude that there can only be one centimeter-gram system of units for all mechanical and electromagnetic physical quantities, which means the $\underset{cg}{U}$ and $\underset{cg}{N}$ operators of the cg system presented in Chapter 1 can be identified with the $\underset{gscg}{U}$ and $\underset{gscg}{N}$ operators of the gscg system defined above in Eq. (3.21b, c):

$$\underset{cg}{U} = \underset{gscg}{U}, \tag{3.30a}$$

$$\underset{cg}{N} = \underset{gscg}{N}. \tag{3.30b}$$

3.6 RATIONALIZED AND UNRATIONALIZED MKS SYSTEMS

The most widespread system of electromagnetic units in use today is the mks system of units; when used with its rationalized system of electromagnetic quantities it is often referred to as SI, for *Systeme International*, units. Although Giovanni Giorgi introduced it in 1901 using its rationalized form, it was clear from the beginning that both the rationalized and unrationalized mks systems were in many ways an improvement over cgs esu and cgs emu units. From the start the rationalized system was more popular; and although in 1938 the International Electrical Congress recognized the existence of both the unrationalized and rationalized systems, in 1950 the International Electrical Congress officially adopted the rationalized system of mks units for communication of scientific and engineering data. In its rationalized form the mks system has become practically universal in all fields of electrical engineering, and almost all introductory courses of electromagnetic theory use these units.

For purposes of exposition, it is easier to talk first about the unrationalized mks system, because it follows easily and immediately from what has gone before simply by deciding, as with the esuq and emuq units of Chapter 2, to recognize the independent dimension of electric charge. In fact, both the esuq and emuq units, with the ε_0 and μ_0 constants explicitly present, have equations that are always identical in form to equations using the unrationalized mks system. When we compare the esuq, emuq, and unrationalized mks systems, we return to the world of classical mechanics, where the form of the equations does not change when the system of units changes—only now there are four fundmental dimensions mass, length, time and charge, rather than just the three fundamental dimensions mass, length, and time.

The great appeal of either the unrationalized or rationalized mks systems comes from their connection to the practical units coul, volt, amp, ohm, weber, henry, and farad discussed in Chapter 2. Since we regard rationalization as a redefinition of the physical quantities rather than as a change of units—and because there are no other mks electromagnetic units in use today—we can without confusion call the U and

N operators for both the rationalized and unrationalized mks systems $\underset{\text{mks}}{\text{U}}$ and $\underset{\text{mks}}{\text{N}}$. In the both the rationalized and unrationalized mks systems, the numerical parts of the physical quantities used in circuit theory—such as charge, electric potential, electric current, resistance, inductance, and capacitance —would be the same as when measured in nineteenth-century practical units.* If $\underset{\text{prac}}{\text{N}}$ is the N operator for the practical system of units introduced in Section 2.6 and $b_{CIRCUIT}$ is any physical quantity used in circuit theory, we have

$$\underset{\text{mks}}{\text{N}}\,(b_{CIRCUIT}) = \underset{\text{prac}}{\text{N}}\,(b_{CIRCUIT}). \qquad (3.31a)$$

For example, applying $\underset{\text{mks}}{\text{N}}$ to any circuit-theory potential V gives

$$\underset{\text{mks}}{\text{N}}\,(V) = \underset{\text{prac}}{\text{N}}\,(V) = \text{same dimensionless number of "volts;"}$$

applying $\underset{\text{mks}}{\text{N}}$ to any circuit-theory current I gives

$$\underset{\text{mks}}{\text{N}}\,(I) = \underset{\text{prac}}{\text{N}}\,(I) = \text{same dimensionless number of "amperes;"}$$

applying $\underset{\text{mks}}{\text{N}}$ to any circuit-theory charge Q gives

$$\underset{\text{mks}}{\text{N}}\,(Q) = \underset{\text{prac}}{\text{N}}\,(Q) = \text{same dimensionless number of "coulombs;"}$$

and so on. The change comes in what "volt," "ampere," "coulomb," etc., mean in the mks system. In Eq. (2.24a), the practical unit ampere is defined as one tenth the emu current unit—that is, it is one tenth of an abamp—and as such must have dimensions

$$\frac{\text{mass}^{1/2} \cdot \text{length}^{1/2}}{\text{time}}$$

since the units of an abamp are $\text{gm}^{1/2} \cdot \text{cm}^{1/2}/\text{sec}$. Similarly, in Eq. (2.26c) a practical coul is defined to be one $\text{amp} \cdot \text{sec}$ or one tenth the emu charge unit. Thus, its dimensions must be

$$\text{mass}^{1/2} \cdot \text{length}^{1/2},$$

because the units of abcoul, the emu charge unit, are $\text{gm}^{1/2} \cdot \text{cm}^{1/2}$. In the mks system, however, we recognize the separate dimension of charge; therefore "coulomb"

---

* The alert reader will notice that for this statement to be true the rationalization used in the SI system must be fundamentally different from Heaviside-Lorentz rationalization; see Section 3.8 below to discover how the mks type of rationalization works.

as an mks unit must have dimensions of charge; "ampere" as an mks unit must have dimensions of charge/time; and so on. Recognizing the separate dimension of charge gives all the mks circuit units different dimensions from the practical units, so

$$\underset{\text{mks}}{U}(b_{CIRCUIT}) \neq \underset{\text{prac}}{U}(b_{CIRCUIT}), \tag{3.31b}$$

even though the mks circuit-theory units (coulomb, volt, amp, ohm, weber, henry, and farad) have the same names as the practical circuit-theory units. Comparing the entries of Table 3.8 to the definitions of the practical units given in Chapter 2 [see Eqs. (2.25b) through (2.29)], we note that the unit equalities between the mks units are identical to the unit equalities of the practical units. Because neither the names of the units nor their inter-relationships change, it is easy to overlook the change in what they mean. Strictly speaking, we should give the mks and practical units different names, just as we gave all the units of the esuq and emuq systems the suffix "q" to distinguish them from the units of the esu and emu systems. In practice there is no confusion because there is no longer any reason to refer to the system of practical units. From now on we assume that coulomb, volt, amp, ohm, weber, henry, and farad refer to mks units recognizing the fourth fundamental dimension charge.

Table 3.9 gives the standard units for electromagnetic physical quantities in the mks system. Quantities labeled with an "f" prefix are part of the rationalized mks system which will be discussed later. Comparison with the esuq and emuq columns of Tables 2.4 and 2.5 of Chapter 2 shows the close correspondence between the esuq, emuq, and mks units; to go from esuq or emuq in Tables 2.4 or 2.5 to mks in Table 3.9 all that need be done is to remove the "stat" or "ab" prefix, drop the "q" suffix, and replace cm by m. In fact, we can formally define the $\underset{\text{mks}}{U}$ and $\underset{\text{mks}}{N}$ operators in terms of either the $\underset{\text{esuq}}{U}, \underset{\text{esuq}}{N}$ or $\underset{\text{emuq}}{U}, \underset{\text{emuq}}{N}$ operators of Chapter 2.

The first step in setting up these definitions is to use the definitions of the composite esuq or emuq units such as statvoltq and abvoltq; statampq and abampq; stathenryq and abhenryq; statfaradq and abfaradq; etc., presented in Section 2.8 to write all the esuq and emuq units as powers of the fundamental units gm, cm, sec,

Table 3.8 Units of the rationalized and unrationalized mks systems.

1 amp = coul/sec
1 volt = joule/coul
1 weber = volt · sec = joule · sec/coul
1 ohm = volt/amp = joule · sec/coul^2
1 henry = weber/amp = volt · sec/amp = joule · sec^2/coul^2
1 farad = coul/volt = coul^2/joule
1 siemens = mho = ohm^{-1}

Table 3.9 mks units for physical quantities with and without Fessenden rationalization.

Physical quantity	mks units
A (magnetic vector potential)	$\mathrm{U}_{\mathrm{mks}}(A) = \dfrac{\text{weber}}{\text{m}}$
B (magnetic induction)	$\mathrm{U}_{\mathrm{mks}}(B) = \dfrac{\text{weber}}{\text{m}^2} = \text{tesla}$
C (capacitance)	$\mathrm{U}_{\mathrm{mks}}(C) = \text{farad}$
$_fD,\ D$ (electric displacement)	$\mathrm{U}_{\mathrm{mks}}(D) = \dfrac{\text{coul}}{\text{m}^2}$
E (electric field)	$\mathrm{U}_{\mathrm{mks}}(E) = \dfrac{\text{volt}}{\text{m}}$
$_f\varepsilon,\ \varepsilon$ (dielectric constant)	$\mathrm{U}_{\mathrm{mks}}(_f\varepsilon) = \mathrm{U}_{\mathrm{mks}}(\varepsilon) = \dfrac{\text{farad}}{\text{m}}$
$_f\varepsilon_0,\ \varepsilon_0$ (permittivity of free space)	$\mathrm{U}_{\mathrm{mks}}(_f\varepsilon_0) = \mathrm{U}_{\mathrm{mks}}(\varepsilon_0) = \dfrac{\text{farad}}{\text{m}}$
$_f\mathcal{F},\ \mathcal{F}$ (magnetomotive force)	$\mathrm{U}_{\mathrm{mks}}(_f\mathcal{F}) = \mathrm{U}_{\mathrm{mks}}(\mathcal{F}) = \text{amp}$
Φ_B (magnetic flux)	$\mathrm{U}_{\mathrm{mks}}(\Phi_B) = \text{weber}$
G (conductance)	$\mathrm{U}_{\mathrm{mks}}(G) = \text{ohm}^{-1} = \text{mho} = \text{siemens}$
$_fH,\ H$ (magnetic field)	$\mathrm{U}_{\mathrm{mks}}(_fH) = \mathrm{U}_{\mathrm{mks}}(H) = \dfrac{\text{amp}}{\text{m}}$
I (current)	$\mathrm{U}_{\mathrm{mks}}(I) = \text{amp}$
J (volume current density)	$\mathrm{U}_{\mathrm{mks}}(J) = \dfrac{\text{coul}}{\text{m}^2 \cdot \text{sec}}$
$\mathcal{J}_S$ (surface current density)	$\mathrm{U}_{\mathrm{mks}}(\mathcal{J}_S) = \dfrac{\text{coul}}{\text{m} \cdot \text{sec}}$
L (inductance)	$\mathrm{U}_{\mathrm{mks}}(L) = \text{henry}$
$_fm_H,\ m_H$ (permanent-magnet dipole moment)	$\mathrm{U}_{\mathrm{mks}}(_fm_H) = \mathrm{U}_{\mathrm{mks}}(m_H) = \text{weber} \cdot \text{m}$
m_I (current-loop magnetic dipole moment)	$\mathrm{U}_{\mathrm{mks}}(m_I) = \text{amp} \cdot \text{m}^2$
$_fM_H,\ M_H$ (permanent-magnet dipole density)	$\mathrm{U}_{\mathrm{mks}}(_fM_H) = \mathrm{U}_{\mathrm{mks}}(M_H) = \dfrac{\text{weber}}{\text{m}^2} = \text{tesla}$

Table 3.9 (Continued).

Physical quantity	mks units
M_I (current-loop magnetic dipole density)	$\underset{\text{mks}}{\text{U}}(M_I) = \dfrac{\text{amp}}{\text{m}}$
$f\mu, \mu$ (magnetic permeability)	$\underset{\text{mks}}{\text{U}}(f\mu) = \underset{\text{mks}}{\text{U}}(\mu) = \dfrac{\text{henry}}{\text{m}}$
$f\mu_0, \mu_0$ (magnetic permeability of free space)	$\underset{\text{mks}}{\text{U}}(f\mu) = \underset{\text{mks}}{\text{U}}(\mu) = \dfrac{\text{henry}}{\text{m}}$
$f p_H, p_H$ (magnetic pole strength)	$\underset{\text{mks}}{\text{U}}(f p_H) = \underset{\text{mks}}{\text{U}}(p_H) = \text{weber}$
p (electric dipole moment)	$\underset{\text{mks}}{\text{U}}(p) = \text{coul} \cdot \text{m}$
P (electric dipole density)	$\underset{\text{mks}}{\text{U}}(P) = \dfrac{\text{coul}}{\text{m}^2}$
$f\mathcal{P}, \mathcal{P}$ (permeance)	$\underset{\text{mks}}{\text{U}}(f\mathcal{P}) = \underset{\text{mks}}{\text{U}}(\mathcal{P}) = \dfrac{\text{weber}}{\text{amp}}$
Q (charge)	$\underset{\text{mks}}{\text{U}}(Q) = \text{coul}$
R (resistance)	$\underset{\text{mks}}{\text{U}}(R) = \text{ohm}$
$f\mathcal{R}, \mathcal{R}$ (reluctance)	$\underset{\text{mks}}{\text{U}}(f\mathcal{R}) = \underset{\text{mks}}{\text{U}}(\mathcal{R}) = \dfrac{\text{amp}}{\text{weber}}$
ρ_Q (volume charge density)	$\underset{\text{mks}}{\text{U}}(\rho_Q) = \dfrac{\text{coul}}{\text{m}^3}$
ρ_R (resistivity)	$\underset{\text{mks}}{\text{U}}(\rho_R) = \text{ohm} \cdot \text{m}$
S (elastance)	$\underset{\text{mks}}{\text{U}}(S) = \text{farad}^{-1}$
S_Q (surface charge density)	$\underset{\text{mks}}{\text{U}}(S_Q) = \dfrac{\text{coul}}{\text{m}^2}$
σ (conductivity)	$\underset{\text{mks}}{\text{U}}(\sigma) = \dfrac{\text{ohm}^{-1}}{\text{m}}$
V (electric potential)	$\underset{\text{mks}}{\text{U}}(V) = \text{volt}$
$f\Omega_H, \Omega_H$ (magnetic scalar potential)	$\underset{\text{mks}}{\text{U}}(f\Omega_H) = \underset{\text{mks}}{\text{U}}(\Omega_H) = \text{amp}$

and charge (either statcoulq or abcoulq). For example, we can write the esuq unit of electric potential as

$$1 \text{ statvoltq} = 1 \frac{\text{erg}}{\text{statcoulq}} = \text{gm} \cdot \text{cm}^2 \cdot \text{sec}^{-2} \cdot \text{statcoulq}^{-1};$$

we can write the emuq unit of capacitance as

$$1 \text{ abfaradq} = \frac{\text{abcoulq}}{\text{abvoltq}} = \frac{\text{abcoulq}^2}{\text{erg}} = \text{gm}^{-1} \cdot \text{cm}^{-2} \cdot \text{sec}^2 \cdot \text{abcoulq}^2;$$

and so on. If any electromagnetic physical quantity b has a mechanical component, we can use the techniques of Chapter 1 to convert that into powers of gm, cm, sec in the cgs system to get, as a final result,

$$\underset{\text{esuq}}{U}(b) = \text{gm}^\alpha \text{cm}^\beta \text{sec}^\gamma \text{statcoulq}^\delta \qquad (3.32a)$$

or

$$\underset{\text{emuq}}{U}(b) = \text{gm}^\alpha \text{cm}^\beta \text{sec}^\gamma \text{abcoulq}^\delta, \qquad (3.32b)$$

where the numbers α, β, γ, and δ come from breaking down the mechanical and electromagnetic units of b into powers of gm, cm, sec, and the unit of charge. We can also get α, β, γ, and δ from dimensional analysis of b; and in fact, the $\underset{\text{mltq}}{U}$ operator introduced in Chapter 2 [see Eqs. (2.59) through (2.60b)] can be used to find them:

$$\underset{\text{mltq}}{U}(b) = \text{mass}^\alpha \text{length}^\beta \text{time}^\gamma \text{charge}^\delta. \qquad (3.32c)$$

This is, by the way, as good a time as any to point out that α, β, γ, and δ always turn out to be integers when charge is recognized as a fourth fundamental dimension and b is one of the electromagnetic physical quantities listed in Tables 3.1 to 3.9.

Having found integers α, β, γ, and δ, we are ready for the second step; now operator $\underset{\text{mks}}{U}$ can be defined by

$$\underset{\text{mks}}{U}(b) = \left(\frac{\text{kg}}{\text{gm}}\right)^\alpha \left(\frac{\text{m}}{\text{cm}}\right)^\beta \left(\frac{\text{coul}}{\text{statcoulq}}\right)^\delta \underset{\text{esuq}}{U}(b) \qquad (3.33a)$$

or by

$$\underset{\text{mks}}{U}(b) = \left(\frac{\text{kg}}{\text{gm}}\right)^\alpha \left(\frac{\text{m}}{\text{cm}}\right)^\beta \left(\frac{\text{coul}}{\text{abcoulq}}\right)^\delta \underset{\text{emuq}}{U}(b). \qquad (3.33b)$$

Equations (3.33a, b) have the same meaning, and either one can chosen as the definition of U_{mks}. Tables 3.10 and 3.11 state the equalities between mks units and the esuq, emuq units; all we need from them now is that in the mks system

Table 3.10 Relationships between the mks and esuq units.

$$1\,\mathrm{coul} = \frac{c_{\mathrm{cgs}}}{10} \cdot \mathrm{statcoulq}$$

$$1\,\mathrm{amp} = \frac{c_{\mathrm{cgs}}}{10} \cdot \mathrm{statampq}$$

$$1\,\mathrm{weber} = \frac{10^8}{c_{\mathrm{cgs}}} \cdot \mathrm{statweberq}$$

$$1\,\mathrm{volt} = \frac{10^8}{c_{\mathrm{cgs}}} \cdot \mathrm{statvoltq}$$

$$1\,\mathrm{farad} = \frac{c_{\mathrm{cgs}}^2}{10^9} \cdot \mathrm{statfaradq}$$

$$1\,\mathrm{henry} = \frac{10^9}{c_{\mathrm{cgs}}^2} \cdot \mathrm{stathenryq}$$

$$1\,\mathrm{ohm} = \frac{10^9}{c_{\mathrm{cgs}}^2} \cdot \mathrm{statohmq}$$

Table 3.11 Relationships between the mks and emuq units.

$$1\,\mathrm{coul} = 10^{-1} \cdot \mathrm{abcoulq}$$

$$1\,\mathrm{amp} = 10^{-1} \cdot \mathrm{abampq}$$

$$1\,\mathrm{weber} = 10^8 \cdot \mathrm{abweberq}$$

$$1\,\mathrm{volt} = 10^8 \cdot \mathrm{abvoltq}$$

$$1\,\mathrm{farad} = 10^{-9} \cdot \mathrm{abfaradq}$$

$$1\,\mathrm{henry} = 10^9 \cdot \mathrm{abhenryq}$$

$$1\,\mathrm{ohm} = 10^9 \cdot \mathrm{statohmq}$$

$$1 \text{ coul} = 10^{-1} \text{abcoulq} = \left(\frac{c_{cgs}}{10} \right) \text{statcoulq.} \tag{3.34a}$$

We also need to know that

$$1 \text{ kg} = 10^3 \text{ gm} \tag{3.34b}$$

and

$$1 \text{ m} = 10^2 \text{ cm.} \tag{3.34c}$$

Equations (3.34a–c) and Rule I can be used to write $\underset{\text{mks}}{N}$ for any physical quantity b as

$$\underset{\text{mks}}{N}(b) = \left(10^{-3}\right)^{\alpha} \left(10^{-2}\right)^{\beta} (10)^{\delta} \underset{\text{emuq}}{N}(b) \tag{3.34d}$$

or

$$\underset{\text{mks}}{N}(b) = \left(10^{-3}\right)^{\alpha} \left(10^{-2}\right)^{\beta} \left(\frac{10}{c_{cgs}} \right)^{\delta} \underset{\text{esuq}}{N}(b). \tag{3.34e}$$

Both of these have the same meaning and either one can be used to define $\underset{\text{mks}}{N}$. We also now say that for any electromagnetic or mechanical physical quantity b,

$$\underset{\text{mks}}{N}(b) = b_{\text{mks}}. \tag{3.34f}$$

3.7 CONVERSION OF EQUATIONS TO AND FROM THE UNRATIONALIZED MKS SYSTEM

To convert equations from esu or emu units to the unrationalized mks system, we need only convert them to esuq or emuq units, respectively, following the procedure developed in Chapter 2 and being careful to recognize the ε_0, μ_0 constants when they appear. Although the conversion from esu to esuq and from emu to emuq was thoroughly discussed in Chapter 2, there is no harm in giving a few more examples.

In esu units the attractive force ϕ per unit length between two long, parallel wires separated by a distance r and each carrying an electric current I flowing in the same direction is

$$\phi = \frac{2I^2}{rc^2}. \tag{3.35a}$$

Splitting Eq. (3.35a) into numeric components and units gives

$$\phi_{cgs}\left(\frac{dynes}{cm}\right) = \frac{2[I_{esu}\,statamp]^2}{(r_{cgs}cm)\left(c_{cgs}\dfrac{cm}{sec}\right)^2}. \tag{3.35b}$$

Because 1 dyne $= gm \cdot cm \cdot sec^{-2}$ and, from Table 2.3, 1 statamp $= gm^{1/2} \cdot cm^{3/2} \cdot sec^{-2}$, Eq. (3.35b) is balanced as written. We can confirm this by operating on both sides with $\underset{esu}{U}$ to get

$$\frac{gm}{sec^2} = \frac{gm \cdot cm^3/sec^4}{cm^3/sec^2} \quad or \quad \frac{gm}{sec^2} = \frac{gm}{sec^2}.$$

Before charge can be recognized as a new fundamental dimension, taking us from the esu to the esuq system of units, we must, according to Rule VIII, make Eq. (3.35b) balanced in both the invariant and connecting units. The connecting unit of Eq. (3.35b) is the (statamp, statampq) pair, so we multiply the right-hand side by

$$\frac{dyne}{statamp^2} \cdot \frac{cm^2}{sec^2} = \frac{(gm \cdot cm \cdot sec^{-2}) \cdot cm^2}{gm \cdot cm^3 \cdot sec^{-4} \cdot sec^2} = 1.$$

There is nothing subtle about how this combination is picked; we just use the ratio of units needed to make both sides balance without breaking up the connecting unit statamp. This gives

$$\phi_{cgs}\left(\frac{dyne}{cm}\right) = \frac{2(I_{esu}\,statamp)^2}{(r_{cgs}cm)\left(c_{cgs}\dfrac{cm}{sec}\right)^2} \cdot \left(\frac{dyne \cdot cm^2}{statamp^2 \cdot sec^2}\right). \tag{3.35c}$$

Equation (3.35c) satisfies Rule VIII, so we now recognize charge as a new fundamental dimension and write

$$\phi_{cgs}\left(\frac{dyne}{cm}\right) = \frac{2(I_{esu}\,statampq)^2}{(r_{cgs}cm)\left(c_{cgs}\dfrac{cm}{sec}\right)^2} \cdot \left(\frac{dyne \cdot cm^2}{statampq^2 \cdot sec^2}\right). \tag{3.35d}$$

Typically, in systems of units recognizing charge as a new fundamental dimension, the extra clump of units that is not connected in an obvious way to a physical quantity turns into ε_0, μ_0, or some combination of the two. From Eq. (2.37a) and the definition statampq $=$ statcoulq/sec, we have

$$\underset{esuq}{U}(\varepsilon_0) = \frac{statcoulq^2}{dyne \cdot cm^2} = \frac{statampq^2 \cdot sec^2}{dyne \cdot cm^2}.$$

Since $\underset{\text{esuq}}{\text{N}}\,(\varepsilon_0) = 1$ from Eq. (2.37b) and all physical quantities have the same numeric parts in the esu and esuq systems [see Eq. (2.34a)], Eq. (3.35d) now becomes

$$\phi = \frac{2I^2}{rc^2} \cdot \frac{1}{\varepsilon_0} \tag{3.35e}$$

or, using $\varepsilon_0\mu_0 = c^{-2}$ from Eq. (2.10),

$$\phi = \frac{2I^2\mu_0}{r}. \tag{3.35f}$$

Equation (3.35f) has the correct form for the esuq system of units, and since all variables in the equation represent physical quantities we know that this is also the correct form for the unrationalized mks system. We could have gone directly to Eq. (3.35f) by recognizing from Eq. (2.45c) that

$$\underset{\text{esuq}}{\text{U}}\,(\mu_0) = \frac{\text{stathenryq}}{\text{cm}} = \frac{(\text{erg/statcoulq}) \cdot \text{sec}}{\text{statampq} \cdot \text{cm}} = \frac{\text{dyne}}{\text{statampq}^2},$$

where we have used the equalities

$$\text{stathenryq} = \frac{\text{statvoltq} \cdot \text{sec}}{\text{statampq}}, \quad \text{statvoltq} = \frac{\text{erg}}{\text{statcoulq}}, \quad \text{and} \quad \text{erg} = \text{dyne} \cdot \text{cm}.$$

Since $\underset{\text{esuq}}{\text{N}}\,(\mu_0) = c_{\text{cgs}}^{-2}$ [see Eq. (2.45f)], the extra clump of units

$$\left(c_{\text{cgs}}\frac{\text{cm}}{\text{sec}}\right)^{-2} \cdot \frac{\text{dyne} \cdot \text{cm}^2}{\text{statampq}^2 \cdot \text{sec}^2}$$

becomes

$$\underset{\text{esuq}}{\text{N}}\,(\mu_0) \cdot \frac{\text{dyne}}{\text{statampq}^2} = \mu_0,$$

turning Eq. (3.35d) directly into Eq. (3.35f).

Even though we know from first principles that all equations in the esuq, emuq, and unrationalized mks systems have the same form, there is no harm in demonstrating how this works for this particular example.

Writing Eq. (3.35f) in esuq numeric components and units gives

$$\left(\phi_{\text{cgs}}\frac{\text{dyne}}{\text{cm}}\right) = \frac{2(I_{\text{esu}}\text{statampq})^2}{(r_{\text{cgs}}\text{cm})}\left[(\mu_0)_{\text{esu}}\frac{\text{stathenryq}}{\text{cm}}\right], \tag{3.36a}$$

where we use Eq. (2.34a) from Chapter 2 to write $I_{esu} = I_{esuq}$ and $(\mu_0)_{esu} = (\mu_0)_{esuq}$, and the units of μ_0 in the esuq system are written as stathenryq/cm. Substituting from Table 3.10 gives

$$\left(\phi_{cgs}\frac{10^{-5}\text{newton}}{10^{-2}\text{m}}\right) = \frac{2(I_{esu}10 \cdot c_{cgs}^{-1} \cdot \text{amp})^2}{(r_{cgs}10^{-2}\text{m})}\left[(\mu_0)_{esu}\frac{10^{-9}c_{cgs}^2\text{henry}}{10^{-2}\text{m}}\right], \quad (3.36b)$$

where we have used Table 1.3 to convert from the cgs mechanical units dyne and cm to the mks mechanical units newton and m. From Rule I we know that

$$\phi_{cgs}\frac{10^{-5}}{10^{-2}} = \phi_{cgs}10^{-3} = \phi_{mks}, \quad r_{mks} = 10^{-2}r_{cgs},$$

and from Table 3.12 we have

$$I_{mks} = 10c_{cgs}^{-1}I_{esu}, \quad (\mu_0)_{mks} = (\mu_0)_{esu}10^{-7}c_{cgs}^2;$$

so Eq. (3.36b) becomes

$$\left(\phi_{mks}\frac{\text{newton}}{\text{m}}\right) = \frac{2(I_{mks}\text{amp})^2}{(r_{mks}\text{m})}\left[(\mu_0)_{mks}\frac{\text{henry}}{\text{m}}\right] \quad (3.36c)$$

or

$$\phi = \frac{2I^2\mu_0}{r},$$

the same as Eq. (3.35f). One point worth mentioning, perhaps, is that the equality

$$(\mu_0)_{mks} = (\mu_0)_{esu}10^{-7}c_{cgs}^2$$

gives us, since

$$(\mu_0)_{esu} = \underset{esu}{\text{N}}(\mu_0) = c_{cgs}^{-2},$$

that

$$(\mu_0)_{mks} = 10^{-7} \quad (3.36d)$$

or

$$\mu_0 = 10^{-7}\text{henry/m} \quad (3.36e)$$

exactly in the unrationalized mks system. This result has some historical importance, because, after much discussion over the first half of the twentieth century, it was decided that the unrationalized mks system is uniquely specified by the requirement that $\mu_0 = 10^{-7}$ henry/m; or, what amounts to the same thing, that

$$\underset{\text{mks}}{\text{N}}\,(\mu_0) = 10^{-7}.$$

It is just as easy to convert equations to the unrationalized mks system from emu units as it is from esu units. The formula for the Poynting vector describing the energy flux of an electromagnetic radiation field in a vacuum can be written as

$$\vec{S} = \frac{1}{4\pi}\vec{E} \times \vec{B} \tag{3.37a}$$

in emu units. Splitting this up into units and numeric components gives

$$\vec{S}_{\text{cgs}}\frac{\text{ergs}}{\text{cm}^2\text{sec}} = \frac{1}{4\pi}\left(\vec{E}_{\text{emu}}\frac{\text{abvolt}}{\text{cm}}\right) \times \left(\vec{B}_{\text{emu}}\frac{\text{abweber}}{\text{cm}^2}\right), \tag{3.37b}$$

where we represent the emu unit of magnetic induction as abweber $\cdot$ cm^{-2} rather than the more customary gauss (see Table 2.5). From Table 2.3 we note that

$$\frac{\text{abvolt}}{\text{cm}} \cdot \frac{\text{abweber}}{\text{cm}^2} = \frac{\text{gm}^{1/2} \cdot \text{cm}^{1/2}}{\text{sec}^2} \cdot \frac{\text{gm}^{1/2}}{\text{cm}^{1/2} \cdot \text{sec}} = \frac{\text{gm} \cdot \text{cm}^2 \cdot \text{sec}^{-2}}{\text{cm}^2 \cdot \text{sec}} = \frac{\text{erg}}{\text{cm}^2 \cdot \text{sec}},$$

showing that Eq. (3.37b) has balanced units in the emu system. It cannot, however, satisfy Rule VIII until it is balanced in both the invariant units—erg, cm, sec—and the two connecting unit pairs—(abvolt, abvoltq) and (abwever, abweberq)—for the transition to the emuq system. Multiplying the right-hand side of Eq. (3.37b) by

$$\text{erg} \cdot \frac{\text{cm}}{\text{sec}} \cdot \frac{1}{\text{abvolt}} \cdot \frac{1}{\text{abweber}}$$

$$= \left(\frac{\text{gm} \cdot \text{cm}^2}{\text{sec}^2}\right)\left(\frac{\text{cm}}{\text{sec}}\right)\left(\frac{\text{sec}^2}{\text{gm}^{1/2} \cdot \text{cm}^{3/2}}\right)\left(\frac{\text{sec}}{\text{gm}^{1/2} \cdot \text{cm}^{3/2}}\right) = 1,$$

which is just 1 in emu units, gives

$$\vec{S}_{\text{cgs}}\frac{\text{ergs}}{\text{cm}^2\text{sec}}$$

$$= \frac{1}{4\pi}\left(\vec{E}_{\text{emu}}\frac{\text{abvolt}}{\text{cm}}\right) \times \left(\vec{B}_{\text{emu}}\frac{\text{abweber}}{\text{cm}^2}\right) \cdot \left(\frac{\text{erg} \cdot \text{cm}}{\text{sec}} \cdot \frac{1}{\text{abvolt}} \cdot \frac{1}{\text{abweber}}\right).$$

$$\tag{3.37c}$$

Equation (3.37c) is simultaneously balanced in both the invariant and connecting units, so we can now recognize charge as a new fundamental dimension to get

Table 3.12 Numeric components of physical quantities in the esu and unrationalized mks systems.

(magnetic vector potential) $$A_{mks} = A_{esu} \cdot c_{cgs} \cdot 10^{-6}$$	(current-loop magnetic dipole density) $$(M_I)_{mks} = (M_I)_{esu} \cdot 10^3 \cdot c_{cgs}^{-1}$$
(magnetic induction) $$B_{mks} = B_{esu} \cdot c_{cgs} \cdot 10^{-4}$$	(magnetic permeability) $$\mu_{mks} = c_{cgs}^2 \cdot 10^{-7} \cdot \mu_{esu}$$
(capacitance) $$C_{mks} = C_{esu} \cdot c_{cgs}^{-2} \cdot 10^9$$	(magnetic permeability of free space) $$(\mu_0)_{mks} = c_{cgs}^2 \cdot 10^{-7} \cdot (\mu_0)_{esu}$$ $$= 1 \cdot 10^{-7}$$
(electric displacement) $$D_{mks} = D_{esu} \cdot 10^5 \cdot c_{cgs}^{-1}$$	(magnetic pole strength) $$(p_H)_{mks} = c_{cgs} \cdot 10^{-8} \cdot (p_H)_{esu}$$
(electric field) $$E_{mks} = E_{esu} \cdot c_{cgs} \cdot 10^{-6}$$	(electric dipole moment) $$p_{mks} = p_{esu} \cdot 10^{-1} \cdot c_{cgs}^{-1}$$
(dielectric constant) $$\varepsilon_{mks} = \varepsilon_{esu} \cdot c_{cgs}^{-2} \cdot 10^{11}$$	(electric dipole density) $$P_{mks} = P_{esu} \cdot 10^5 \cdot c_{cgs}^{-1}$$
(permittivity of free space) $$(\varepsilon_0)_{mks} = (\varepsilon_0)_{esu} \cdot c_{cgs}^{-2} \cdot 10^{11}$$ $$= c_{cgs}^{-2} \cdot 10^{11}$$	(permeance) $$\mathcal{P}_{mks} = \mathcal{P}_{esu} \cdot 10^{-9} \cdot c_{cgs}^2$$
(magnetomotive force) $$\mathcal{F}_{mks} = \mathcal{F}_{esu} \cdot 10 \cdot c_{cgs}^{-1}$$	(charge) $$Q_{mks} = Q_{esu} \cdot 10 \cdot c_{cgs}^{-1}$$
(magnetic flux) $$(\Phi_B)_{mks} = (\Phi_B)_{esu} \cdot c_{cgs} \cdot 10^{-8}$$	(resistance) $$R_{mks} = R_{esu} \cdot 10^{-9} \cdot c_{cgs}^2$$
(conductance) $$G_{mks} = G_{esu} \cdot 10^9 \cdot c_{cgs}^{-2}$$	(reluctance) $$\mathcal{R}_{mks} = c_{cgs}^{-2} \cdot 10^9 \cdot \mathcal{R}_{esu}$$
(magnetic field) $$H_{mks} = H_{esu} \cdot 10^3 \cdot c_{cgs}^{-1}$$	(volume charge density) $$(\rho_Q)_{mks} = (\rho_Q)_{esu} \cdot 10^7 \cdot c_{cgs}^{-1}$$
(current) $$I_{mks} = I_{esu} \cdot 10 \cdot c_{cgs}^{-1}$$	(resistivity) $$(\rho_R)_{mks} = (\rho_R)_{esu} \cdot 10^{-11} \cdot c_{cgs}^2$$
(volume current density) $$J_{mks} = J_{esu} \cdot 10^5 \cdot c_{cgs}^{-1}$$	(elastance) $$S_{mks} = S_{esu} \cdot 10^{-9} \cdot c_{cgs}^2$$
(surface current density) $$(\mathcal{J}_S)_{mks} = (\mathcal{J}_S)_{esu} \cdot 10^3 \cdot c_{cgs}^{-1}$$	(surface charge density) $$(S_Q)_{mks} = (S_Q)_{esu} \cdot 10^5 \cdot c_{cgs}^{-1}$$
(inductance) $$L_{mks} = L_{esu} \cdot 10^{-9} \cdot c_{cgs}^2$$	(conductivity) $$\sigma_{mks} = \sigma_{esu} \cdot 10^{11} \cdot c_{cgs}^{-2}$$
(permanent-magnet dipole moment) $$(m_H)_{mks} = c_{cgs} \cdot 10^{-10} \cdot (m_H)_{esu}$$	(electric potential) $$V_{mks} = V_{esu} \cdot 10^{-8} \cdot c_{cgs}$$
(current-loop magnetic dipole moment) $$(m_I)_{mks} = (m_I)_{esu} \cdot 10^{-3} \cdot c_{cgs}^{-1}$$	(magnetic scalar potential) $$(\Omega_H)_{mks} = (\Omega_H)_{esu} \cdot 10 \cdot c_{cgs}^{-1}$$
(permanent-magnet dipole density) $$(M_H)_{mks} = c_{cgs} \cdot 10^{-4} \cdot (M_H)_{esu}$$	

$$\vec{S}_{cgs}\frac{\text{ergs}}{\text{cm}^2\text{sec}}$$

$$= \frac{1}{4\pi}\left(\vec{E}_{emuq}\frac{\text{abvoltq}}{\text{cm}}\right) \tag{3.37d}$$

$$\times \left(\vec{B}_{emuq}\frac{\text{abweberq}}{\text{cm}^2}\right) \cdot \left(\frac{\text{erg}\cdot\text{cm}}{\text{sec}}\cdot\frac{1}{\text{abvoltq}}\cdot\frac{1}{\text{abweberq}}\right),$$

where we have used Eq. (2.34b) to replace $\vec{E}_{emu}, \vec{B}_{emu}$ by $\vec{E}_{emuq}, \vec{B}_{emuq}$. Again we expect the block of units on the extreme right to turn out to be either ε_0, μ_0, or some combination of ε_0 and μ_0. They are in fact equal to

$$\frac{\text{erg}\cdot\text{cm}}{\text{sec}}\cdot\frac{1}{\text{abvoltq}}\cdot\frac{1}{\text{abweberq}} = \frac{\text{cm}}{\text{sec}}\cdot\frac{\text{abcoulq}}{\text{abvoltq}\cdot\text{sec}}$$

$$= \frac{\text{cm}}{\text{sec}}\cdot\frac{\text{abampq}}{\text{abvoltq}} = \frac{\text{cm}}{\text{abhenryq}}, \tag{3.37e}$$

where the unit equalities abvoltq = erg/abcoulq, abweberq = abvoltq · sec, and abhenryq = abvoltq · sec/abampq from Section 2.8 are used to simplify the units in Eq. (3.37e). From Eq. (2.46b) we recognize cm/abhenryq as the emuq units of μ_0^{-1}, and from Eq. (2.46e) we note that $\underset{\text{emuq}}{\text{N}}(\mu_0) = (\mu_0)_{emuq} = 1$. Therefore, Eq. (3.37d) can be written as

$$\vec{S}_{cgs}\frac{\text{ergs}}{\text{cm}^2\text{sec}}$$

$$= \frac{1}{4\pi}\left(\vec{E}_{emuq}\frac{\text{abvoltq}}{\text{cm}}\right) \times \left(\vec{B}_{emuq}\frac{\text{abweberq}}{\text{cm}^2}\right) \cdot \left(\frac{1}{(\mu_0)_{emuq}\text{abhenryq}/\text{cm}}\right).$$

$$\tag{3.37f}$$

Since the esuq, emuq, and unrationalized mks systems always have the same form of electromagnetic equations, we can go directly to the final result from Eq. (3.37f), which has clearly been converted to the emuq system. Thus, in the unrationalized mks system the formula for the Poynting vector in empty space must be

$$\vec{S} = \frac{1}{4\pi\mu_0}\vec{E}\times\vec{B} = \frac{1}{4\pi}\vec{E}\times\vec{H}, \tag{3.37g}$$

where we have used in the last step that $\vec{H} = \mu_0^{-1}\vec{B}$ in a vacuum. When Table 3.11 is used to check this result by converting the units of Eq. (3.37f) to the unrationalized mks system, we find that

$$\vec{S}_{\text{cgs}} \frac{10^{-7}\text{joule}}{10^{-4}\text{m}^2\text{sec}}$$

$$= \frac{1}{4\pi}\left(\vec{E}_{\text{emuq}}\frac{10^{-8}\text{volt}}{10^{-2}\text{m}}\right)$$

$$\times \left(\vec{B}_{\text{emuq}}\frac{10^{-8}\text{weber}}{10^{-4}\text{m}^2}\right) \cdot \left(\frac{1}{(\mu_0)_{\text{emuq}}10^{-9}\text{henry}/10^{-2}\text{m}}\right).$$

Remembering that $\vec{E}_{\text{emuq}}$, $\vec{B}_{\text{emuq}}$, and $(\mu_0)_{\text{emuq}}$ represent the same collection of pure numbers as $\vec{E}_{\text{emu}}$, $\vec{B}_{\text{emu}}$, and $(\mu_0)_{\text{emu}}$, we consult Table 3.13 to get

$$\vec{S}_{\text{mks}}\frac{\text{joule}}{\text{m}^2\text{sec}} = \frac{1}{4\pi}\left(\vec{E}_{\text{mks}}\frac{\text{volt}}{\text{m}}\right) \times \left(\vec{B}_{\text{mks}}\frac{\text{weber}}{\text{m}^2}\right)\left(\frac{1}{(\mu_0)_{\text{mks}}\text{henry}/\text{m}}\right), \quad (3.37\text{h})$$

where Rule I is used to write $\vec{S}_{\text{mks}} = 10^{-3}\vec{S}_{\text{cgs}}$. As before, we find that $(\mu_0)_{\text{mks}} = 10^{-7}(\mu_0)_{\text{emu}} = 10^{-7}$ for the unrationalized mks system. Clearly, Eq. (3.37h) is the same as Eq. (3.37g) when written in terms of the physical quantities $\vec{S}$, $\vec{E}$, $\vec{B}$, and μ_0.

Going from the unrationalized mks system back to esu or emu units is really quite simple. We know the esuq, emuq, and unrationalized mks systems always have the same form of equation; so given any formula in unrationalized mks, we just start thinking of it as having esu units if we want to go the esu system and as having emuq units if we want to go to the emu system. Changing from esuq to esu or from emuq to emu is discussed in Section 2.9. All we are really doing is ceasing to recognize charge as a fundamental dimension, and doing it in such a way as to make $\varepsilon_0 = 1$, $\mu_0 = c^{-2}$ for esu units and $\varepsilon_0 = c^{-2}$, $\mu_0 = 1$ for emu units. When we no longer recognized time as a fundamental dimension in Section 1.10 it was done to make $c = 1$, and any equation could be converted to these new units simply by replacing c by 1 in that equation. If c was not explicitly present in the equation, the form of the equation did not change. A similar pattern occurs when going from esuq to esu or from emuq to emu. When going from esuq to esu we replace ε_0 by 1 whenever it is explicitly mentioned; and, if desired, μ_0 can be replaced by c^{-2} whenever it is explicitly mentioned. If ε_0 is not present, the form of the equation does not change. So, for example, the Maxwell equation in unrationalized mks for $\vec{D}$,

$$\vec{\nabla} \cdot \vec{D} = 4\pi\rho,$$

does not change form when written in esu; but if we assume $\vec{P} = 0$ so that we can make the substitution $\vec{D} = \varepsilon_0\vec{E}$ in unrationalized mks and write that

$$\vec{\nabla} \cdot \vec{E} = \frac{4\pi}{\varepsilon_0}\rho,$$

Table 3.13 Numeric components of physical quantities in the emu and unrationalized mks systems.

(magnetic vector potential) $A_{mks} = A_{emu} \cdot 10^{-6}$	(volume current density) $J_{mks} = J_{emu} \cdot 10^5$	(permeance) $\mathcal{P}_{mks} = \mathcal{P}_{emu} \cdot 10^{-9}$
(magnetic induction) $B_{mks} = B_{emu} \cdot 10^{-4}$	(surface current density) $(\mathcal{J}_S)_{mks} = (\mathcal{J}_S)_{emu} \cdot 10^3$	(charge) $Q_{mks} = Q_{emu} \cdot 10$
(capacitance) $C_{mks} = C_{emu} \cdot 10^9$	(inductance) $L_{mks} = L_{emu} \cdot 10^{-9}$	(resistance) $R_{mks} = R_{emu} \cdot 10^{-9}$
(electric displacement) $D_{mks} = D_{emu} \cdot 10^5$	(permanent-magnet dipole moment) $(m_H)_{mks} = 10^{-10} \cdot (m_H)_{emu}$	(reluctance) $\mathcal{R}_{mks} = 10^9 \cdot \mathcal{R}_{emu}$
(electric field) $E_{mks} = E_{emu} \cdot 10^{-6}$	(current-loop magnetic dipole moment) $(m_I)_{mks} = (m_I)_{emu} \cdot 10^{-3}$	(volume charge density) $(\rho_Q)_{mks} = (\rho_Q)_{emu} \cdot 10^7$
(dielectric constant) $\varepsilon_{mks} = \varepsilon_{emu} \cdot 10^{11}$	(permanent-magnet dipole density) $(M_H)_{mks} = 10^{-4} \cdot (M_H)_{emu}$	(resistivity) $(\rho_R)_{mks} = (\rho_R)_{emu} \cdot 10^{-11}$
(permittivity of free space) $(\varepsilon_0)_{mks} = (\varepsilon_0)_{emu} \cdot 10^{11} = c_{cgs}^{-2} \cdot 10^{11}$	(current-loop magnetic dipole density) $(M_I)_{mks} = (M_I)_{emu} \cdot 10^3$	(elastance) $S_{mks} = S_{emu} \cdot 10^{-9}$
(magnetomotive force) $\mathcal{F}_{mks} = \mathcal{F}_{emu} \cdot 10$	(magnetic permeability) $\mu_{mks} = 10^{-7} \cdot \mu_{emu}$	(surface charge density) $(S_Q)_{mks} = (S_Q)_{emu} \cdot 10^5$
(magnetic flux) $(\Phi_B)_{mks} = (\Phi_B)_{emu} \cdot 10^{-8}$	(magnetic permeability of free space) $(\mu_0)_{mks} = 10^{-7} \cdot (\mu_0)_{emu} = 10^{-7}$	(conductivity) $\sigma_{mks} = \sigma_{emu} \cdot 10^{11}$
(conductance) $G_{mks} = G_{emu} \cdot 10^9$	(magnetic pole strength) $(p_H)_{mks} = 10^{-8} \cdot (p_H)_{emu}$	(electric potential) $V_{mks} = V_{emu} \cdot 10^{-8}$
(magnetic field) $H_{mks} = H_{emu} \cdot 10^3$	(electric dipole moment) $p_{mks} = p_{emu} \cdot 10^{-1}$	(magnetic scalar potential) $(\Omega_H)_{mks} = (\Omega_H)_{emu} \cdot 10$
(current) $I_{mks} = I_{emu} \cdot 10$	(electric dipole density) $P_{mks} = P_{emu} \cdot 10^5$	

then this becomes

$$\vec{\nabla} \cdot \vec{E} = 4\pi\rho$$

in the esu system. The substitution itself, $\vec{D} = \varepsilon_0 \vec{E}$ in unrationalized mks, becomes $\vec{D} = \vec{E}$ in esu units. When going from emuq to emu, we replace μ_0 by 1 whenever it is explicitly present and, if desired, ε_0 can be replaced by c^{-2}. If μ_0 is not present, the form of the equation does not change. Following this rule, the Maxwell equation

$$\vec{\nabla} \times \vec{H} = 4\pi \vec{J} + \frac{\partial \vec{D}}{\partial t}$$

in unrationalized mks, emuq, or esuq systems does not change when going to emu or esu units. If, however, we make the substitutions $\vec{H} = \mu_0^{-1}\vec{B}$ and $\vec{D} = \varepsilon_0\vec{E}$ appropriate for media where $\vec{P} = 0$ and $\vec{M}_I = 0$, this equation becomes

$$\vec{\nabla} \times \vec{B} = 4\pi\mu_0\vec{J} + \mu_0\varepsilon_0\frac{\partial\vec{E}}{\partial t}$$

in the unrationalized mks, esuq, or emuq systems. Then, in the emu system we have

$$\vec{\nabla} \times \vec{B} = 4\pi\vec{J} + \varepsilon_0\frac{\partial\vec{E}}{\partial t} \quad \text{or} \quad \vec{\nabla} \times \vec{B} = 4\pi\vec{J} + \frac{1}{c^2}\frac{\partial\vec{E}}{\partial t},$$

and in the esu system we have

$$\vec{\nabla} \times \vec{B} = 4\pi\mu_0\vec{J} + \mu_0\frac{\partial\vec{E}}{\partial t} \quad \text{or} \quad \vec{\nabla} \times \vec{B} = \frac{4\pi}{c^2}\vec{J} + \frac{1}{c^2}\frac{\partial\vec{E}}{\partial t},$$

where the substitution equations for $\vec{H}$ and $\vec{D}$ become $\vec{H} = \vec{B}, \vec{D} = \varepsilon_0\vec{E} = c^{-2}\vec{E}$ in emu units and $\vec{H} = \mu_0^{-1}\vec{B} = c^2\vec{B}, \vec{D} = \vec{E}$ in esu units.

Figure 3.2 shows the connections between Gaussian cgs units, esu units, emu units, and the unrationalized mks system. Conversion from the Gaussian system to the esu and emu systems has already been discussed, as has the conversion from esu or emu to unrationalized mks. To show the whole chain at once, we convert the formula for $\mathcal{E}$, the energy density of electromagnetic fields in a vacuum, from Gaussian units to the unrationalized mks system. In Gaussian units we have

$$\mathcal{E} = \frac{1}{8\pi}\left(E^2 + B^2\right) \tag{3.38a}$$

or, broken up into units and numeric components,

$$\mathcal{E}_{cgs}\frac{erg}{cm^3} = \frac{1}{8\pi}\left[\left(E_{gs}\frac{statvolt}{cm}\right)^2 + \left(B_{gs}gauss\right)^2\right]. \tag{3.38b}$$

From Table 3.5 and $1\,erg = gm \cdot cm^2 \cdot sec^{-2}$ we get

$$\frac{statvolt^2}{cm^2} = \frac{gm}{cm \cdot sec^2} = \frac{gm \cdot cm^2 \cdot sec^{-2}}{cm^3} = \frac{erg}{cm^3}$$

and

$$gauss^2 = \frac{gm}{cm \cdot sec^2} = \frac{erg}{cm^3},$$

so clearly Eq. (3.38b) has balanced units in the Gaussian cgs system. Since 1 gauss = abweber/cm², we can write Eq. (3.38b) as

$$\mathcal{E}_{cgs}\frac{erg}{cm^3} = \frac{1}{8\pi}\left[\left(E_{esu}\frac{statvolt}{cm}\right)^2 + \left(B_{emu}\frac{abweber}{cm^2}\right)^2\right] \qquad (3.38c)$$

where we use Table 3.1 to recognize E as an electric quantity in the Gaussian system measured in esu units, with $E_{gs} = E_{esu}$, and B as a magnetic quantity measured in emu units with $B_{gs} = B_{emu}$. Converting B to esu units, we use Table 2.8 to write

$$B_{emu}\frac{abweber}{cm^2} = B_{emu}\left(\frac{cm}{sec}\right)\frac{statweber}{cm^2},$$

and Table 2.7 to get

$$B_{emu}\left(\frac{cm}{sec}\right)\frac{statweber}{cm^2} = B_{esu}c_{cgs}\left(\frac{cm}{sec}\right)\frac{statweber}{cm^2}$$

$$= \left(B_{esu}\frac{statweber}{cm^2}\right)\left(c_{cgs}\frac{cm}{sec}\right).$$

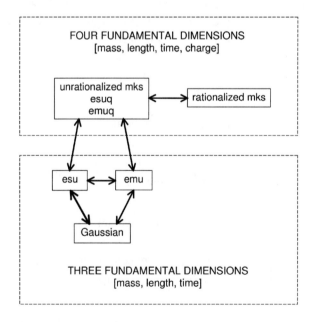

Figure 3.2 These are the recommended conversion paths between each of the five major electromagnetic systems.

Hence, Eq. (3.38c) becomes

$$
\mathcal{E}_{cgs}\frac{erg}{cm^3}
= \frac{1}{8\pi}\left[\left(E_{esu}\frac{statvolt}{cm}\right)^2 + \left(B_{esu}\frac{statweber}{cm^2}\right)^2\left(c_{cgs}\frac{cm}{sec}\right)^2\right]. \tag{3.38d}
$$

We have now taken the formula for $\mathcal{E}$ from Gaussian to esu units, moving up one level in Fig. 3.2. To climb one more level to the esuq system, we must recognize charge as a separate dimension. Rule VIII states that both the E_{esu}^2 and B_{esu}^2 terms have to match $\mathcal{E}$ in both the invariant and connecting units before the new fundamental dimension of charge can be recognized. We multiply the E_{esu}^2 term by $cm^{-1} \cdot erg/statvolt^2$. To show that this is 1 in esu units, we consult Table 2.3 and use $1\,erg = gm \cdot cm^2 \cdot sec^{-2}$ to get

$$
\frac{1}{cm} \cdot \frac{erg}{statvolt^2} = \frac{erg}{cm} \cdot \frac{sec^2}{gm \cdot cm} = \frac{gm \cdot cm}{sec^2} \cdot \frac{sec^2}{gm \cdot cm} = 1.
$$

The B_{esu}^2 term is multiplied by $(erg/cm) \cdot sec^2 \cdot statweber^{-2}$ which is also 1 in esu units. From Table 2.3 we have

$$
\frac{erg}{cm} \cdot sec^2 \cdot \frac{1}{statweber^2} = \left(\frac{gm \cdot cm}{sec^2}\right) \cdot sec^2 \cdot \frac{1}{gm \cdot cm} = 1.
$$

Equation (3.38d) now becomes

$$
\mathcal{E}_{cgs}\frac{erg}{cm^3} = \frac{1}{8\pi}\left(E_{esu}\frac{statvolt}{cm}\right)^2\left[\frac{erg}{cm \cdot statvolt^2}\right]
$$
$$
+ \frac{1}{8\pi}\left(B_{esu}\frac{statweber}{cm^2}\right)^2\left(c_{cgs}\frac{cm}{sec}\right)^2\left[\frac{erg \cdot sec^2}{cm} \cdot \frac{1}{statweber^2}\right],
$$
$$\tag{3.38e}$$

which is balanced in both the three invariant units erg, cm, sec and the two connecting units (statvolt, statvoltq) and (statweber, statweberq). Now we can transfer to the esuq system of units by recognizing charge as a fourth fundamental dimension, writing

$$
\mathcal{E}_{cgs}\frac{erg}{cm^3} = \frac{1}{8\pi}\left(E_{esu}\frac{statvoltq}{cm}\right)^2\left[\frac{erg}{cm \cdot statvoltq^2}\right]
$$
$$
+ \frac{1}{8\pi}\left(B_{esu}\frac{statweberq}{cm^2}\right)^2\left[c_{cgs}^2\frac{erg \cdot cm}{statweberq^2}\right]. \tag{3.38f}
$$

We note that, using 1 statvoltq = erg/statcoulq and 1 erg = dyne·cm [see Table 1.1 and Eq. (2.39b)],

$$\frac{\text{erg}}{\text{cm} \cdot \text{statvoltq}^2} = \frac{\text{erg} \cdot \text{statcoulq}^2}{\text{cm} \cdot \text{erg}^2} = \frac{\text{statcoulq}^2}{\text{cm}^2 \cdot \text{dyne}}.$$

From Eq. (2.37a) we recognize this as the units of ε_0 in the esuq system, and from Eq. (2.37b) we know that $\underset{\text{esuq}}{N} (\varepsilon_0) = 1$, giving

$$\frac{\text{erg}}{\text{cm} \cdot \text{statvoltq}^2} = (\varepsilon_0)_{\text{esuq}} \frac{\text{statcoulq}^2}{\text{cm}^2 \cdot \text{dyne}}. \tag{3.38g}$$

Similarly we note, using the definitions in Section 2.8 [see Eqs. (2.39b), (2.45a, b), and (2.49)], that

$$\text{statweberq} = \text{statvoltq} \cdot \text{sec} = \text{erg} \cdot \text{sec}/\text{statcoulq} = \text{erg}/\text{statampq}$$

and

$$\text{stathenryq} = \text{statvoltq} \cdot \text{sec}/\text{statampq} = \text{statweberq}/\text{statampq},$$

which makes

$$c_{\text{cgs}}^2 \frac{\text{erg} \cdot \text{cm}}{\text{statweberq}^2} = c_{\text{cgs}}^2 \frac{\text{erg} \cdot \text{cm}}{\text{statweberq}} \cdot \frac{\text{statampq}}{\text{erg}} = c_{\text{cgs}}^2 \frac{\text{cm}}{\text{stathenryq}}. \tag{3.38h}$$

From Eq. (2.45c) we recognize cm/stathenryq as the units of μ_0^{-1} in the esuq system, and from Eq. (2.45f) we see that $\underset{\text{esuq}}{N} (\mu_0) = (\mu_0)_{\text{esuq}} = c_{\text{cgs}}^{-2}$. Hence, Eq. (3.38f) becomes, using Eqs. (3.38g, h),

$$\begin{aligned}
\mathcal{E}_{\text{cgs}} \frac{\text{erg}}{\text{cm}^3} &= \frac{1}{8\pi} \left(E_{\text{esu}} \frac{\text{statvoltq}}{\text{cm}} \right)^2 \left[(\varepsilon_0)_{\text{esuq}} \frac{\text{statcoulq}^2}{\text{cm}^2 \cdot \text{dyne}} \right] \\
&+ \frac{1}{8\pi} \left(B_{\text{esu}} \frac{\text{statweberq}}{\text{cm}^2} \right)^2 \left[\frac{1}{(\mu_0)_{\text{esuq}} \text{stathenry}/\text{cm}} \right].
\end{aligned} \tag{3.38i}$$

Written as physical quantities in the esuq system, the formula for $\mathcal{E}$ becomes (remember that $E_{\text{esuq}} = E_{\text{esu}}$ and $B_{\text{esuq}} = B_{\text{esu}}$)

$$\mathcal{E} = \frac{1}{8\pi} \left(\varepsilon_0 E^2 + \frac{1}{\mu_0} B^2 \right) \tag{3.38j}$$

or

$$\mathcal{E} = \frac{1}{8\pi} (\vec{E} \cdot \vec{D} + \vec{B} \cdot \vec{H}), \tag{3.38k}$$

where in Eq. (3.38k) we make the substitutions $\vec{D} = \varepsilon_0 \vec{E}$ and $\vec{H} = \mu_0^{-1} \vec{B}$ for the $\vec{D}$ and $\vec{H}$ fields in empty space. Equations (3.38j, k) are what we set out to find, since electromagnetic equations written in physical variables must have the same form in both the esuq and unrationalized mks system. Tables 1.3, 3.10, and 3.12 show how to convert Eq. (3.38i) to unrationalized mks.

$$
\begin{aligned}
\mathcal{E}_{cgs} \frac{10^{-7} \text{joule}}{10^{-6} \text{m}^3} &= \frac{1}{8\pi} \left(E_{esu} \frac{c_{cgs}}{10^8} \cdot \frac{\text{volt}}{10^{-2}\text{m}} \right)^2 \\
&\quad \times \left[(\varepsilon_0)_{esuq} \frac{10^2}{c_{cgs}^2} \cdot \frac{\text{coul}^2}{10^{-4}\text{m}^2 \cdot 10^{-5}\text{newton}} \right] \\
&\quad + \frac{1}{8\pi} \left(B_{esu} \frac{c_{cgs}}{10^8} \cdot \frac{\text{weber}}{10^{-4}\text{m}^2} \right)^2 \left[(\mu_0)_{esuq} c_{cgs}^2 10^{-9} \frac{\text{henry}}{10^{-2}\text{m}} \right]^{-1} \\
&= \frac{1}{8\pi} \left(E_{mks} \frac{\text{volt}}{\text{m}} \right)^2 \left[(\varepsilon_0)_{esuq} \frac{10^{11}}{c_{cgs}^2} \cdot \frac{\text{coul}^2}{\text{m}^2 \cdot \text{newton}} \right] \\
&\quad + \frac{1}{8\pi} \left(B_{mks} \frac{\text{weber}}{\text{m}^2} \right)^2 \left[(\mu_0)_{esuq} c_{cgs}^2 10^{-7} \frac{\text{henry}}{\text{m}} \right]^{-1}.
\end{aligned}
$$

We know that $(\varepsilon_0)_{esuq} = (\varepsilon_0)_{esu} = 1$, so by Rule I this last step gives the numeric component of ε_0 in the unrationalized mks system,

$$
\underset{\text{mks}}{\text{N}} (\varepsilon_0) = (\varepsilon_0)_{mks} = (\varepsilon_0)_{esuq} \frac{10^{11}}{c_{cgs}^2} = \frac{10^{11}}{c_{cgs}^2}, \tag{3.39a}
$$

and also confirms the already-known value for the numeric component of μ_0 in the unrationalized mks system,

$$
\underset{\text{mks}}{\text{N}} (\mu_0) = (\mu_0)_{mks} = (\mu_0)_{esuq} \cdot c_{cgs}^2 10^{-7} = c_{cgs}^{-2} \cdot c_{cgs}^2 10^{-7} = 10^{-7}. \tag{3.39b}
$$

Rule I also shows that $\mathcal{E}_{mks} = \mathcal{E}_{cgs} \cdot 10^{-1}$, so we end up with

$$
\begin{aligned}
\mathcal{E}_{mks} \frac{\text{joule}}{\text{m}^3} &= \frac{1}{8\pi} \left(E_{mks} \frac{\text{volt}}{\text{m}} \right)^2 \left[(\varepsilon_0)_{mks} \frac{\text{farad}}{\text{m}} \right] \\
&\quad + \frac{1}{8\pi} \left(B_{mks} \frac{\text{weber}}{\text{m}^2} \right)^2 \left[(\mu_0)_{mks} \frac{\text{henry}}{\text{m}} \right]^{-1},
\end{aligned} \tag{3.39c}
$$

where we have used the definition of the mks farad in Table 3.8 and newton $\cdot$ m $=$ joule to get

$$
\frac{\text{coul}^2}{\text{newton} \cdot \text{m}^2} = \frac{\text{coul}^2}{\text{joule} \cdot \text{m}} = \frac{\text{farad}}{\text{m}},
$$

a more conventional way of writing the units of ε_0. As expected, Eq. (3.39c) when written using the physical quantities $\mathcal{E}$, E, B, ε_0, and μ_0 takes on the same form as (3.38j), the formula for $\mathcal{E}$ in the esuq system:

$$\mathcal{E} = \frac{1}{8\pi}\left(\varepsilon_0 E^2 + \frac{1}{\mu_0}B^2\right).$$

3.8 CONVERSION OF EQUATIONS TO AND FROM THE RATIONALIZED MKS SYSTEM

Having discussed the unrationalized mks system, we now move on to the predominant system of units in use today, the rationalized mks system (often called SI units). Table 3.14 shows the rationalization scheme used to rescale the electromagnetic physical quantities in the rationalized mks system. The physical quantities labeled with a prefix "f" are rescaled from their historic definitions established during the nineteenth century; those not labeled with an "f" prefix retain their historical meaning.[*,2] Unlike the Heaviside-Lorentz rationalization scheme shown in Table 3.6, which can be applied to any electromagnetic system of equations, Fessenden rationalization can only be applied to electromagnetic equations that explicitly use both the ε_0 and μ_0 constants required by the esuq, emuq, and mks systems of electromagnetic units. We note in passing that any system of units recognizing charge as a fourth fundamental dimension must contain the physical quantities ε_0 and μ_0 in its system of electromagnetic equations. As has already been mentioned above, from our point of view Fessenden rationalization, like Heaviside-Lorentz rationalization, is something that is done to the physical quantities themselves rather than their units.[†] This is why we are careful to relabel the rescaled physical quantities with an "f." Unfortunately, textbooks often refer to "rationalized mks units," implying that the units rather than the physical quantities themselves have been changed; and it is very unusual to see the rescaled physical quantities labeled as such, with an "f" or anything else.

Comparing Table 3.14 to Table 3.6, we note that Fessenden rationalization rescales very few physical quantities compared to Heaviside-Lorentz rationalization, and in particular it does not change the physical quantities used in circuit theory. This explains its appeal to electrical engineers, who do not have to recalibrate their voltmeters, ammeters, etc., or relabel the resistances and voltages of their circuits and batteries, in order to use Fessenden rationalization. This is in sharp contrast to Heaviside-Lorentz rationalization, which would force them to recalibrate and relabel almost every instrument and circuit in their laboratories.

The Fessenden rationalization scheme is designed to affect common electromagnetic formulas the same way Heaviside-Lorentz rationalization affects them.

---

[*] The "f" is for R. A. Fessenden who first proposed this rationalization method in 1900.

[†] See discussion at beginning of Section 3.3 above.

Table 3.14 Fessenden rationalization for electromagnetic physical quantities.

(magnetic vector potential, unchanged) A	(magnetic permeability, rationalized) $_f\mu = 4\pi\mu$
(magnetic induction, unchanged) B	(relative magnetic permeability, unchanged) μ_r
(capacitance, unchanged) C	(magnetic permeability of free space, rationalized) $_f\mu_0 = 4\pi\mu_0$
(electric displacement, rationalized) $_fD = D/(4\pi)$	(magnetic pole strength, rationalized) $_fp_H = 4\pi p_H$
(electric field, unchanged) E	(electric dipole moment, unchanged) p
(dielectric constant, rationalized) $_f\varepsilon = \varepsilon/(4\pi)$	(electric dipole density, unchanged) P
(relative dielectric constant, unchanged) ε_r	(permeance, rationalized) $_f\mathcal{P} = 4\pi\mathcal{P}$
(permittivity of free space, rationalized) $_f\varepsilon_0 = \varepsilon_0/(4\pi)$	(charge, unchanged) Q
(magnetomotive force, rationalized) $_f\mathcal{F} = \mathcal{F}/(4\pi)$	(resistance, unchanged) R
(magnetic flux, unchanged) Φ_B	(reluctance, rationalized) $_f\mathcal{R} = \mathcal{R}/(4\pi)$
(conductance, unchanged) G	(volume charge density, unchanged) ρ_Q
(magnetic field, rationalized) $_fH = H/(4\pi)$	(resistivity, unchanged) ρ_R
(current, unchanged) I	(elastance, unchanged) S
(volume current density, unchanged) J	(surface charge density, unchanged) S_Q
(surface current density, unchanged) $\mathcal{J}_S$	(conductivity, unchanged) σ
(inductance, unchanged) L	(electric potential, unchanged) V
(permanent-magnet dipole moment, rationalized) $_fm_H = 4\pi m_H$	(magnetic scalar potential, rationalized) $_f\Omega_H = \Omega_H/(4\pi)$
(current-loop magnetic dipole moment, unchanged) m_I	(electric susceptibility, rationalized) $_f\chi_e = 4\pi\chi_e$
(permanent-magnet dipole density, rationalized) $_fM_H = 4\pi M_H$	(magnetic susceptibility, rationalized) $_f\chi_m = 4\pi\chi_m$
(current-loop magnetic dipole density, unchanged) M_I	

Equation (3.11a), the formula for the capacitance of a parallel-plate capacitor, looks like

$$C_{\mathrm{esu}}\mathrm{statfarad} = \frac{A_{\mathrm{cgs}}\mathrm{cm}^2}{4\pi \; d_{\mathrm{cgs}}\mathrm{cm}} \tag{3.40a}$$

in the esu or—since it involves only electric quantities—the Gaussian system of units. Table 2.3 shows that 1 statfarad = cm, so (3.40a) has balanced units; this formula becomes, using the Heaviside-Lorentz rationalization shown in Table 3.6,

$$(_hC)_{\text{esu}}\text{statfarad} = \frac{A_{\text{cgs}}\text{cm}^2}{d_{\text{cgs}}\text{cm}}. \tag{3.40b}$$

This is the same equation, split up into units and numeric components, given above in Eq. (3.13). Following the path shown in Fig. 3.2, we take Eq. (3.40a) into the rationalized mks system by way of the unrationalized mks system. Equation (3.40a), although balanced in esu units, has to be balanced in both the connecting unit (statfaraf, statfaradq) as well as the invariant unit cm before we can recognize charge as a new fundamental dimension. Multiplying the right-hand side by statfarad/cm, which is clearly just 1 in the esu system because a statfarad is the same thing as a cm, gives

$$C_{\text{esu}}\text{statfarad} = \frac{A_{\text{cgs}}\text{cm}^2}{4\pi d_{\text{cgs}}\text{cm}} \cdot \frac{\text{statfarad}}{\text{cm}}.$$

Recognizing charge as a new fundamental dimension, we move to esuq units to get

$$C_{\text{esuq}}\text{statfaradq} = \frac{A_{\text{cgs}}\text{cm}^2}{4\pi d_{\text{cgs}}\text{cm}} \cdot \frac{\text{statfaradq}}{\text{cm}}, \tag{3.40c}$$

where we know $C_{\text{esuq}} = C_{\text{esu}}$ because by definition all physical quantities have the same numeric components in the esu and esuq systems. From Table 1.1 and Eqs. (2.39b) and (2.47a),

$$\text{statfaradq} = \frac{\text{statcoulq}}{\text{statvoltq}} = \frac{\text{statcoulq}^2}{\text{erg}} = \frac{\text{statcoulq}^2}{\text{dyne} \cdot \text{cm}},$$

so

$$\frac{\text{statfaradq}}{\text{cm}} = \frac{\text{statcoulq}^2}{\text{dyne} \cdot \text{cm}^2}.$$

Hence, from Eq. (2.38) we have

$$1\,\frac{\text{statfaradq}}{\text{cm}} = \varepsilon_0;$$

and Eq. (3.40c) written in terms of the physical quantities C, A, d, and ε_0 is

$$C = \frac{A\varepsilon_0}{4\pi d} \tag{3.40d}$$

in the esuq system. Equation (3.40d) is the same in the unrationalized mks system because it is written using variables representing physical quantities. Table 3.14

takes us from the unrationalized to the rationalized mks system in Fig. 3.2; making the substitution

$$f\varepsilon_0 = \frac{\varepsilon_0}{4\pi}$$

gives

$$C = \frac{A \cdot (f\varepsilon_0)}{d}. \tag{3.40e}$$

Both Eq. (3.40e) in the rationalized mks system and Eqs. (3.13) and (3.40b) in the Heaviside-Lorentz system have lost their 4π's from the same place in the formulas, showing how Fessenden rationalization has the same effect as Heaviside-Lorentz rationalization. The appearance of the extra factor $f\varepsilon_0$ may make Eq. (3.40e) look less appealing as a formula than Eqs. (3.13) and (3.40b), but in fact most parallel-plate capacitors are built with a dielectric substance rather than empty space between the plates. This means engineers would in practice almost always be calculating parallel-plate capacitances using the dimensionless relative dielectric constant ε_r in the Heaviside-Lorentz system *

$$_hC = \frac{A\varepsilon_r}{d},$$

with different dielectrics having different values of ε_r. In the rationalized mks system the same capacitor has the capacitance

$$C = \frac{A \cdot (f\varepsilon)}{d},$$

with $f\varepsilon = \varepsilon_r \cdot f\varepsilon_0$. To use one formula the engineer looks up the dimensionless relative dielectric constant from a table of relative dielectric constants; and to use the other formula the engineer looks up the dimensional dielectric constant $f\varepsilon$ from a table of dimensional dielectric constants. From this engineer's point of view, the two systems look identical, except for the special case of empty space between the plates. Greater complexity for this little-encountered special case is a small price to pay compared to the greater price of relabeling all capacitors as having capacitances $_hC$ instead of C when switching to the Heaviside-Lorentz system.

We see the same congruence between Fessenden and Heaviside-Lorentz rationalization in Coulomb's law for electric charge and magnetic poles. From the work done in Chapter 2 [see Eqs. (2.1a, b) and (2.35b)], we know that Coulomb's law in esuq units is

$$F = \frac{Q_1 Q_2}{\varepsilon_0 r^2}.$$

* The meaning of ε_r is discussed in Section 2.12.

All of the variables in this formula are physical quantities, so it has the same form
in unrationalized mks; and Table 3.14 shows that only $\varepsilon_0 = 4\pi\,_f\varepsilon_0$ is changed by
Fessenden rationalization when switching to the rationalized mks system:

$$F = \frac{Q_1 Q_2}{4\pi\,_f\varepsilon_0 r^2}. \tag{3.41a}$$

We see that the rationalized mks system puts a 4π into the denominator of
Coulombs law for electric charge, the same place it appears in the Heaviside-
Lorentz version of Coulomb's law [Eq. (3.16a)]. Coulomb's law for magnetic
poles undergoes a similar change. From (2.2a, b) and the discussion following
Eq. (2.48b), we know that Coulomb's law for magnetic poles in the esuq system—
and so also for the unrationalized mks system—has the form

$$F = \frac{(p_H)_1 (p_H)_2}{\mu_0 r^2}.$$

Making the substitutions $(p_H)_1 = (_f p_H)_1/(4\pi), (p_H)_2 = (_f p_H)_2/(4\pi)$, and
$\mu_0 = _f\mu_0/(4\pi)$ required by Table 3.14 gives

$$F = \frac{(_f p_H)_1 (_f p_H)_2}{4\pi\,_f\mu_0 r^2}, \tag{3.41b}$$

putting another 4π into the denominator, the same place it appears in Coulomb's
law for magnetic poles in the Heaviside-Lorentz system [Eq. (3.16b)].

Equation (3.41b) requires the pole strength p_H to be replaced by the rescaled
quantity $_f p_H$, so if scientists and engineers have instruments to measure magnetic
pole strength, they have to recalibrate them when switching to the rationalized mks
system. Fortunately for Fessenden rationalization, the idea of a magnetic pole as
a "real" physical quantity was rapidly falling out of favor with most physicists
by the beginning of the twentieth century, along with the idea of H rather than
B as the fundamental magnetic field. So, not only was there no strong objection
to redefining p_H as $_f p_H/(4\pi)$, there was also no strong objection to redefining
the auxiliary field H as $_f H = 4\pi H$. Although both H and p_H are rescaled, their
product stays constant:

$$p_H H = \left(\frac{_f p_H}{4\pi}\right)(4\pi\,_f H) = (_f p_H)(_f H).$$

This ensures that the force between already-characterized permanent magnets stays
the same, keeping the interface with the mechanical systems of units unchanged.
In fact, both the Heaviside-Lorentz and Fessenden rationalization schemes make
a point of preserving the interface with mechanical units by maintaining the values of all products or ratios—such as $p_H H, QE, \mu_0\varepsilon_0, I^2 R, QV$, etc.—that produce physical quantities having only mechanical dimensions (see Tables 3.6 and

3.14). Table 3.14 shows that Fessenden rationalization rescales the D field to $_fD = D/(4\pi)$; this was not as objectionable at the beginning of the twentieth century as it might have been earlier, since D was being treated ever more often as a helpful mathematical construct rather than as a "real" physical field. James Maxwell himself, when setting down his famous equations, defined his electric displacement as $_fD$ rather than D; and although none of his other electromagnetic physical quantities were rationalized and many subsequent authors did not follow his lead, it could still be argued that there was a strong precedent for using $_fD$ rather than D to represent the electric displacement.

A close look at Table 3.14 shows that the only group of technologists whose laboratories were seriously disturbed by adopting Fessenden rationalization were magnetic specialists. The magnetomotive force $\mathcal{F}$, magnetic field intensity H, permanent-magnet dipole moment and dipole density m_H and M_H, permeability of free space μ_0, magnetic pole strength p_H, permeance $\mathcal{P}$, reluctance $\mathcal{R}$, scalar magnetic potential Ω_H, and magnetic susceptibilty χ_m are all rescaled using Fessenden rationalization. Inevitably, magnetic specialists were faced with recalibrating and relabeling more than a few of their favorite instruments and magnets, making them less than enthusiastic about the rationalized mks system. To this day, magnetic fields are often measured using gauss, the Gaussian cgs unit of the B field, rather than using the mks unit tesla = weber/m^2.* By contrast, electric specialists are treated quite gently; only D—using Maxwell's preference as a precedent—ε_0 and the electric susceptibility χ_e are rescaled. Table 3.14 shows that the capacitance C, electric field E, conductance G, electric current and volume current density I and J, inductance L, electric dipole moment and dipole density p and P, electric charge and charge density Q and ρ_Q, resistance and resistivity R and ρ_R, elastance S, conductivity σ, and electric potential V are all unchanged. Everything important to circuit theory and electrical engineers is left untouched; even the magnetic quantities that escape rescaling—B, Φ_B, m_I, and M_I—are arguably those of greatest importance to electrical engineers. Fessenden rationalization offered the vast majority of practicing electromagnetic scientists and engineers the attractive choice of simplifying their most commonly used formulas without having to recalibrate equipment or relabel circuitry; this contributed greatly to its eventual acceptance.

Table 3.14 can be used to put all of the formulas that we converted to the unrationalized mks system into the rationalized mks system. In the first one, Eq. (3.35f) for the attractive force per unit length between two wires each carrying current I, we can substitute $\mu_0 = (4\pi)^{-1}{}_f\mu_0$ to get

$$\phi = \frac{I^2{}_f\mu_0}{2\pi r} \qquad (3.42a)$$

in the rationalized mks system. Note that this is another example of an equation that, like Coulomb's law, becomes more rather than less complicated in rationalized mks. An advocate for rationalization can justify this by pointing to the inherent cylindrical symmetry of the two wires, saying that we ought to expect a factor of

---

* The tesla was named in honor of Nikola Tesla (1856–1943).

2π and that the unrationalized equations mislead us by leaving it out. The second formula, Eq. (3.37g) for the Poynting vector, becomes

$$\vec{S} = \frac{1}{_f\mu_0}\vec{E} \times \vec{B} = \vec{E} \times {_f}\vec{H} \qquad (3.42\text{b})$$

when converted to the rationalized mks system using $_f\mu_0 = 4\pi\mu_0$ and $_f\vec{H} = (4\pi)^{-1}\vec{H}$. The third equation converted to unrationalized mks, the formula for the energy density of electromagnetic fields in a vacuum, is specified by Eqs. (3.38j, k). In the rationalized mks system, Eq. (3.38j) becomes, using $_f\mu_0 = 4\pi\mu_0$ and $_f\varepsilon_0 = \varepsilon_0/(4\pi)$,

$$\mathcal{E} = \frac{1}{2}\left[(_f\varepsilon_0)E^2 + \left(\frac{1}{_f\mu_0}\right)B^2\right]; \qquad (3.42\text{c})$$

and Eq. (3.38k) becomes, using $_f\vec{D} = (4\pi)^{-1}\vec{D}$ and $_f\vec{H} = (4\pi)^{-1}\vec{H}$,

$$\mathcal{E} = \frac{1}{2}(\vec{E} \cdot {_f}\vec{D} + \vec{B} \cdot {_f}\vec{H}). \qquad (3.42\text{d})$$

We now decide to find Maxwell's equations in the rationalized mks system by transforming Eqs. (3.5a–f) from Gaussian cgs units. From Fig. 3.2 they must be put into the form of the esuq, emuq, or unrationalized mks systems before they can be converted to rationalized mks.

Looking at Eq. (3.5a), we see from Table 3.1 that its physical quantities $\vec{D}$ and ρ_Q are electric; here, esu and Gaussian units are identical and Table 2.4 can be used to write

$$\left(\text{cm}^{-1}\vec{\nabla}_{\text{cgs}}\right) \cdot \left[\vec{D}_{\text{esu}}\frac{\text{statcoul}}{\text{cm}^2}\right] = 4\pi\left[(\rho_Q)_{\text{esu}}\frac{\text{statcoul}}{\text{cm}^3}\right].$$

This is already balanced in both the invariant unit cm and the connecting unit (statcoul, statcoulq), so by Rule VIII we can promote this to esuq units at once by recognizing charge as a fundamental dimension:

$$\left(\text{cm}^{-1}\vec{\nabla}_{\text{cgs}}\right) \cdot \left[\vec{D}_{\text{esuq}}\frac{\text{statcoulq}}{\text{cm}^2}\right] = 4\pi\left[(\rho_Q)_{\text{esuq}}\frac{\text{statcoulq}}{\text{cm}^3}\right]. \qquad (3.43\text{a})$$

We have also used our knowledge of how the esuq system is defined in Section 2.8 [see Eq. (2.34a)] to replace $\vec{D}_{\text{esu}}$ and $(\rho_Q)_{\text{esu}}$ with $\vec{D}_{\text{esuq}}$ and $(\rho_Q)_{\text{esuq}}$, respectively. Equation (3.43a) is in esuq units—so when written using only physical quantities, it must have the same form as in the unrationalized mks system.

$$\vec{\nabla} \cdot \vec{D} = 4\pi\rho_Q.$$

This is clearly one equation which does not change form going from Gaussian units to the unrationalized mks system. To go to the rationalized mks system we consult Table 3.14, replacing $\vec{D}$ by $4\pi_f\vec{D}$ to get

$$\vec{\nabla} \cdot {}_f\vec{D} = \rho_Q. \tag{3.43b}$$

Equation (3.5b) is even easier to convert. It only contains magnetic quantities, so we can immediately interpret it as being in the emu system. We can take the right-hand side to have any units we please because it is zero, automatically making the equation balanced in both the invariant and connecting units. Therefore, we can at once recognize charge as a fundamental dimension, putting it into the emuq system. Relationships between physical quantities in the emuq system have the same form as in the unrationalized mks system, so Eq. (3.5b) can now be interpreted as unrationalized mks. The final step is to consult Table 3.14 and note that Fessenden rationalization does not change $\vec{B}$. We conclude that Eq. (3.5b) has the same form in Gaussian, emu, emuq, unrationalized mks, and, finally, rationalized mks:

$$\vec{\nabla} \cdot \vec{B} = 0. \tag{3.43c}$$

This is altogether a most obliging equation.

Equation (3.5c) contains mixed magnetic and electric quantities; i.e., $\vec{B}$ and $\vec{E}$, respectively. Splitting it up into units and numeric components gives

$$\left(\text{cm}^{-1}\vec{\nabla}_{\text{cgs}}\right) \times \left(\vec{E}_{\text{gs}}\frac{\text{statvolt}}{\text{cm}}\right) = \frac{-1}{\left(c_{\text{cgs}}\dfrac{\text{cm}}{\text{sec}}\right)}\left(\text{sec}^{-1}\frac{\partial}{\partial t_{\text{cgs}}}\right)\left(\vec{B}_{\text{gs}}\text{gauss}\right). \tag{3.44a}$$

Following Fig. 3.2, we must convert to either esu or emu units. Choosing emu units, we see from Tables 3.1 and 3.4 that $\vec{B}$ is already in emu units with $\vec{B}_{\text{gs}} = \vec{B}_{\text{emu}}$; but the electric field $\vec{E}$ is in esu units with $\vec{E}_{\text{gs}} = \vec{E}_{\text{esu}}$ and needs to be changed. Converting (3.44a) to emu gives, using 1 gauss = abweber/cm^2 and Tables 2.7 and 2.8,

$$\left(\text{cm}^{-1}\vec{\nabla}_{\text{cgs}}\right) \times \left(c_{\text{cgs}}^{-1}\vec{E}_{\text{emu}}\left(\frac{\text{sec}}{\text{cm}}\right)\frac{\text{abvolt}}{\text{cm}}\right)$$

$$= \frac{-1}{\left(c_{\text{cgs}}\dfrac{\text{cm}}{\text{sec}}\right)}\left(\text{sec}^{-1}\frac{\partial}{\partial t_{\text{cgs}}}\right)\left(\vec{B}_{\text{emu}}\frac{\text{abweber}}{\text{cm}^2}\right)$$

or

$$\left(\text{cm}^{-1}\vec{\nabla}_{\text{cgs}}\right) \times \left(\vec{E}_{\text{emu}}\frac{\text{abvolt}}{\text{cm}}\right) + \left(\text{sec}^{-1}\frac{\partial}{\partial t_{\text{cgs}}}\right)\left(\vec{B}_{\text{emu}}\frac{\text{abweber}}{\text{cm}^2}\right) = 0.$$

Because 1 abvolt = $\text{gm}^{1/2} \cdot \text{cm}^{3/2}/\text{sec}^2$ and 1 abweber = $\text{gm}^{1/2} \cdot \text{cm}^{3/2}/\text{sec}$ (see Table 2.3), this equation is balanced in emu units but does not yet look as if it satisfies Rule VIII by being separately balanced in both the invariant units of cm, sec and the connecting units of (abvolt, abvoltq) and (abweber, abweberq). We note, however, that 1 abweber = abvolt · sec which gives us

$$
\begin{aligned}
\left(\text{cm}^{-1}\vec{\nabla}_{\text{cgs}}\right) &\times \left(\vec{E}_{\text{emu}}\frac{\text{abvolt}}{\text{cm}}\right) \\
&= -\left(\text{sec}^{-1}\frac{\partial}{\partial t_{\text{cgs}}}\right)\left(\vec{B}_{\text{emu}}\frac{\text{abvolt} \cdot \text{sec}}{\text{cm}^2}\right).
\end{aligned}
\tag{3.44b}
$$

This is balanced in both the invariant and connecting units, allowing us to recognize charge as a fundamental dimension. Because $\vec{E}_{\text{emu}} = \vec{E}_{\text{emuq}}$ and $\vec{B}_{\text{emu}} = \vec{B}_{\text{emuq}}$ [see Eq. (2.34b)], we can write this as

$$
\begin{aligned}
\left(\text{cm}^{-1}\vec{\nabla}_{\text{cgs}}\right) &\times \left(\vec{E}_{\text{emuq}}\frac{\text{abvoltq}}{\text{cm}}\right) \\
&= -\left(\text{sec}^{-1}\frac{\partial}{\partial t_{\text{cgs}}}\right)\left(\vec{B}_{\text{emuq}}\frac{\text{abvoltq} \cdot \text{sec}}{\text{cm}^2}\right)
\end{aligned}
\tag{3.44c}
$$

or, using physical variables,

$$
\vec{\nabla} \times \vec{E} + \frac{\partial \vec{B}}{\partial t} = 0.
\tag{3.44d}
$$

We know this formula must have the same form in unrationalized mks, so all that is left is to apply Fessenden rationalization, transforming the equation to the rationalized mks system. Table 3.14 shows that neither $\vec{E}$ nor $\vec{B}$ are rescaled, so Eq. (3.44d) does not change form going from unrationalized to rationalized mks.

Equation (3.5d) also contains mixed magnetic and electric quantities. Because only $\vec{H}$ is magnetic, measured in emu units in the Gaussian system, and both $\vec{J}$ and $\vec{D}$ are electric, measured in esu units in the Gaussian system, it is easier to follow the esu branch of Fig. 3.2 when going from Gaussian to rationalized mks. Breaking Eq. (3.5d) into units and numeric components gives

$$
\begin{aligned}
\left(\text{cm}^{-1}\vec{\nabla}_{\text{cgs}}\right) &\times \left(\vec{H}_{\text{gs}}\frac{\text{abamp}}{\text{cm}}\right) \\
&= \frac{4\pi}{\left(c_{\text{cgs}}\dfrac{\text{cm}}{\text{sec}}\right)}\left(\vec{J}_{\text{gs}}\frac{\text{statamp}}{\text{cm}^2}\right) \\
&\quad + \frac{1}{\left(c_{\text{cgs}}\dfrac{\text{cm}}{\text{sec}}\right)}\left(\text{sec}^{-1}\frac{\partial}{\partial t_{\text{cgs}}}\right)\left(\vec{D}_{\text{gs}}\frac{\text{statcoul}}{\text{cm}^2}\right).
\end{aligned}
\tag{3.45a}
$$

Consulting Tables 3.2 and 3.3, we note that $\vec{J}_{gs} = \vec{J}_{esu}$, $\vec{D}_{gs} = \vec{D}_{esu}$, $\vec{H}_{gs} = c_{cgs}^{-1}\vec{H}_{esu}$, and 1 abamp = (sec/cm) · statamp. Substitution of this into Eq. (3.45a) gives

$$\left(cm^{-1}\vec{\nabla}_{cgs}\right) \times \left(\vec{H}_{esu}\frac{statamp}{cm}\right)$$
$$= 4\pi\left(\vec{J}_{esu}\frac{statamp}{cm^2}\right) + \left(sec^{-1}\frac{\partial}{\partial t_{cgs}}\right)\left(\vec{D}_{esu}\frac{statcoul}{cm^2}\right). \tag{3.45b}$$

Because 1 statcoul = $gm^{1/2} \cdot cm^{3/2}/sec$ and 1 statamp = $gm^{1/2} \cdot cm^{3/2}/sec^2$ (see Table 2.3), we can use statamp = statcoul/sec to write

$$\left(cm^{-1}\vec{\nabla}_{cgs}\right) \times \left(\vec{H}_{esu}\frac{statcoul}{cm \cdot sec}\right)$$
$$= 4\pi\left(\vec{J}_{esu}\frac{statcoul}{sec \cdot cm^2}\right) + \left(sec^{-1}\frac{\partial}{\partial t_{cgs}}\right)\left(\vec{D}_{esu}\frac{statcoul}{cm^2}\right). \tag{3.45c}$$

Equation (3.45c) is balanced in both the invariant units, cm and sec, and the connecting unit, (statcoul, statcoulq). Recognizing charge as a fundamental dimension puts Eq. (3.45c) into esuq units,

$$\left(cm^{-1}\vec{\nabla}_{cgs}\right) \times \left(\vec{H}_{esuq}\frac{statcoulq}{cm \cdot sec}\right)$$
$$= 4\pi\left(\vec{J}_{esuq}\frac{statcoulq}{sec \cdot cm^2}\right) + \left(sec^{-1}\frac{\partial}{\partial t_{cgs}}\right)\left(\vec{D}_{esuq}\frac{statcoulq}{cm^2}\right), \tag{3.45d}$$

where again the defining characteristic of the esuq system is used to replace $\vec{H}_{esu}$, $\vec{J}_{esu}$, and $\vec{D}_{esu}$ by $\vec{H}_{esuq}$, $\vec{J}_{esuq}$, and $\vec{D}_{esuq}$ [see Eq. (2.34a)]. Equation (3.45d) is in esuq units and can be written using physical quantities as

$$\vec{\nabla} \times \vec{H} = 4\pi\vec{J} + \frac{\partial\vec{D}}{\partial t}. \tag{3.45e}$$

We know Eq. (3.45e) must have the same form in unrationalized mks as in the esuq system, so now we only need to apply Fessenden rationalization from Table 3.14 to get

$$\vec{\nabla} \times \left(_f\vec{H}\right) = \vec{J} + \frac{\partial(_f\vec{D})}{\partial t}, \tag{3.45f}$$

the fourth Maxwell equation in the rationalized mks system.

Equation (3.5e) has only electrical quantities $\vec{D}$, $\vec{E}$, and $\vec{P}$, so by the rules of the Gaussian system everything in this equation can immediately be interpreted as having esu units:

$$\vec{D}_{esu}\frac{statcoul}{cm^2} = \vec{E}_{esu}\frac{statvolt}{cm} + 4\pi\left(\vec{P}_{esu}\frac{statcoul}{cm^2}\right). \qquad (3.46a)$$

Because statcoul $= gm^{1/2} \cdot cm^{3/2}/sec$ and statvolt $= gm^{1/2} \cdot cm^{1/2}/sec$ [see Table 2.3], Eq. (3.46a) is balanced in the esu system but cannot yet be transferred to the esuq system because it is not balanced in both the invariant unit cm and the two connecting units (statcoul, statcoulq) and (statvolt, statvoltq). Multiplying $\vec{E}_{esu}$ by

$$\frac{statcoul}{statvolt \cdot cm} = \frac{gm^{1/2} \cdot cm^{3/2}}{sec} \cdot \frac{sec}{gm^{1/2} \cdot cm^{1/2}} \cdot \frac{1}{cm} = 1$$

gives

$$\vec{D}_{esu}\frac{statcoul}{cm^2} = \vec{E}_{esu}\frac{statvolt}{cm}\left(\frac{statcoul}{statvolt \cdot cm}\right) \\ + 4\pi\left(\vec{P}_{esu}\frac{statcoul}{cm^2}\right). \qquad (3.46b)$$

This equation is balanced in both the invariant and connecting units, letting us transfer to the esuq system by replacing $\vec{D}_{esu}$, $\vec{E}_{esu}$, $\vec{P}_{esu}$ with $\vec{D}_{esuq}$, $\vec{E}_{esuq}$, $\vec{P}_{esuq}$ [see Eq. (2.34a)] and recognizing charge as a fundamental dimension.

$$\vec{D}_{esuq}\frac{statcoulq}{cm^2} = \vec{E}_{esuq}\frac{statvoltq}{cm}\left(\frac{statcoulq}{statvoltq \cdot cm}\right) \\ + 4\pi\left(\vec{P}_{esuq}\frac{statcoulq}{cm^2}\right). \qquad (3.46c)$$

Since statvoltq $= erg/statcoulq = dyne \cdot cm/statcoulq$, we can write

$$\frac{statcoulq}{statvoltq \cdot cm} = \frac{statcoulq^2}{dyne \cdot cm^2} = \varepsilon_0$$

from Eq. (2.38). Hence, Eq. (3.46c) can be written in the esuq system, using physical quantities, as

$$\vec{D} = \varepsilon_0\vec{E} + 4\pi\,\vec{P}. \qquad (3.46d)$$

This equation must have the same form in unrationalized mks, so we can apply Fessenden rationalization from Table 3.14 to get

$$_f\vec{D} = (_f\varepsilon_0)\vec{E} + \vec{P}, \qquad (3.46e)$$

the definition of the rationalized electric displacement $_f\vec{D}$ in the rationalized mks system.

The auxiliary field $\vec{H}$ is defined two different (but equivalent) ways in Eq. (3.5f). Both definitions involve only magnetic quantities $\vec{H}$, $\vec{B}$, $\vec{M}_H$, and $\vec{M}_I$, so by the rules of the Gaussian system they can be immediately interpreted as being in emu units. Writing the $\vec{B}$ unit as gauss = abweber/cm^2 gives

$$\vec{H}_{emu}\frac{abamp}{cm} = \vec{B}_{emu}\frac{abweber}{cm^2} - 4\pi\left[\left(\vec{M}_H\right)_{emu}\frac{abweber}{cm^2}\right] \qquad (3.47a)$$

and

$$\vec{H}_{emu}\frac{abamp}{cm} = \vec{B}_{emu}\frac{abweber}{cm^2} - 4\pi\left[\left(\vec{M}_I\right)_{emu}\frac{abamp}{cm}\right]. \qquad (3.47b)$$

Since abweber = gm$^{1/2}\cdot$cm$^{3/2}$/sec and abamp = gm$^{1/2}\cdot$cm$^{1/2}$/sec (see Table 2.3), we know abweber/cm = abamp so that both Eqs. (3.47a, b) are balanced in emu units. We might just replace abweber/cm by abamp everywhere, getting

$$\vec{H}_{emu}\frac{abamp}{cm} = \vec{B}_{emu}\frac{abamp}{cm} - 4\pi\left[\left(\vec{M}_H\right)_{emu}\frac{abamp}{cm}\right]$$

and

$$\vec{H}_{emu}\frac{abamp}{cm} = \vec{B}_{emu}\frac{abamp}{cm} - 4\pi\left[\left(\vec{M}_I\right)_{emu}\frac{abamp}{cm}\right].$$

These two equations are indeed balanced in both the invariant unit cm and the connecting unit (abamp, abampq), but they fail another requirement of Rule VIII—giving the magnetic induction $\vec{B}$ the same units as $\vec{H}$ violates the physical meaning of a $\vec{B}$ field in the emuq system. In fact, neither $\vec{B}$ nor $\vec{M}_H$ can have units of abamp/cm in the emuq system. So, before charge can be recognized as a fundamental dimension, both the $\vec{B}$ and $\vec{M}_H$ fields must have units of abweber/cm^2; we want only the $\vec{H}$ and $\vec{M}_I$ fields to have units of abamp/cm. Hence, we multiply the $\vec{B}_{emu}$ and $\left(\vec{M}_H\right)_{emu}$ terms in Eqs. (3.47a, b) by abamp $\cdot$ cm/abweber = 1 to get

$$\vec{H}_{emu}\frac{abamp}{cm}$$
$$= \left\{\vec{B}_{emu}\frac{abweber}{cm^2} - 4\pi\left[\left(\vec{M}_H\right)_{emu}\frac{abweber}{cm^2}\right]\right\}\left(\frac{abamp\cdot cm}{abweber}\right) \qquad (3.47c)$$

and

$$\vec{H}_{emu}\frac{abamp}{cm}$$
$$= \left(\vec{B}_{emu}\frac{abweber}{cm^2}\right)\left(\frac{abamp\cdot cm}{abweber}\right) - 4\pi\left[\left(\vec{M}_I\right)_{emu}\frac{abamp}{cm}\right]. \qquad (3.47d)$$

Now these equations match all the requirements of Rule VIII and can be put into emuq units by recognizing charge as a fundamental dimension and replacing $\vec{H}_{emu}$, $\vec{B}_{emu}$, $(\vec{M}_H)_{emu}$, $(\vec{M}_I)_{emu}$ by $\vec{H}_{emuq}$, $\vec{B}_{emuq}$, $(\vec{M}_H)_{emuq}$, $(\vec{M}_I)_{emuq}$ respectively [see Eq. (2.34b)]:

$$\vec{H}_{emuq}\frac{abampq}{cm}$$
$$= \left\{\vec{B}_{emuq}\frac{abweberq}{cm^2} - 4\pi\left[(\vec{M}_H)_{emuq}\frac{abweberq}{cm^2}\right]\right\}\left(\frac{abampq \cdot cm}{abweberq}\right) \tag{3.47e}$$

$$\vec{H}_{emuq}\frac{abampq}{cm}$$
$$= \left(\vec{B}_{emuq}\frac{abweberq}{cm^2}\right)\left(\frac{abampq \cdot cm}{abweberq}\right) - 4\pi\left[(\vec{M}_I)_{emuq}\frac{abampq}{cm}\right]. \tag{3.47f}$$

We note from Eqs. (2.46b, e) that

$$1\frac{abampq \cdot cm}{abweberq} = 1\frac{abampq \cdot cm}{abvoltq \cdot sec} = 1\frac{cm}{abhenryq} = (\mu_0)^{-1},$$

so Eqs. (3.47e, f) can be written as

$$\vec{H} = \frac{1}{\mu_0}\left(\vec{B} - 4\pi\vec{M}_H\right) \tag{3.47g}$$

and

$$\vec{H} = \frac{\vec{B}}{\mu_0} - 4\pi\vec{M}_I, \tag{3.47h}$$

using physical quantities in the emuq system. Since all equations using only physical quantities have the same form in unrationalized mks as they do in emuq, we can now apply Fessenden rationalization (see Table 3.14) to convert Eqs. (3.47g, h) to the rationalized mks system. This gives

$$_f\vec{H} = \frac{1}{_f\mu_0}\left(\vec{B} - {_f\vec{M}_H}\right) \tag{3.47i}$$

and

$$_f\vec{H} = \frac{\vec{B}}{_f\mu_0} - \vec{M}_I \tag{3.47j}$$

for the two equivalent definitions of $_f\vec{H}$. Almost all textbooks written during the last half of the twentieth century have preferred Eq. (3.47j) to Eq. (3.47i) as the definition of $_f\vec{H}$ in Maxwell's equations.

Gathering together Eqs. (3.43b), (3.43c), (3.44d), (3.45f), (3.46e), and (3.47i, j) gives Maxwell's equations written in the rationalized mks system:

$$\vec{\nabla} \cdot {}_f \vec{D} = \rho_Q, \tag{3.48a}$$

$$\vec{\nabla} \cdot \vec{B} = 0, \tag{3.48b}$$

$$\vec{\nabla} \times \vec{E} + \frac{\partial \vec{B}}{\partial t} = 0, \tag{3.48c}$$

$$\vec{\nabla} \times ({}_f \vec{H}) = \vec{J} + \frac{\partial ({}_f \vec{D})}{\partial t}, \tag{3.48d}$$

where

$$_f \vec{D} = ({}_f \varepsilon_0) \vec{E} + \vec{P} \tag{3.48e}$$

and

$$_f \vec{H} = \frac{1}{{}_f \mu_0} (\vec{B} - {}_f \vec{M}_H) = \frac{\vec{B}}{{}_f \mu_0} - \vec{M}_I. \tag{3.48f}$$

In empty space where $_f \vec{M}_H = 0$, $\vec{M}_I = 0$, $\vec{P} = 0$, these equations reduce to

$$\vec{\nabla} \cdot \vec{E} = ({}_f \varepsilon_0)^{-1} \rho_Q, \tag{3.49a}$$

$$\vec{\nabla} \cdot \vec{B} = 0, \tag{3.49b}$$

$$\vec{\nabla} \times \vec{E} + \frac{\partial \vec{B}}{\partial t} = 0, \tag{3.49c}$$

$$\vec{\nabla} \times \vec{B} = ({}_f \mu_0) \vec{J} + ({}_f \varepsilon_0)({}_f \mu_0) \frac{\partial \vec{E}}{\partial t}. \tag{3.49d}$$

From Table 3.14 and Eq. (3.2), we note that

$$({}_f \varepsilon_0)({}_f \mu_0) = \varepsilon_0 \mu_0 = c^{-2}, \tag{3.49e}$$

which means Eq. (3.49d) can also be written as

$$\vec{\nabla} \times \vec{B} = ({}_f \mu_0) \vec{J} + \frac{1}{c^2} \frac{\partial \vec{E}}{\partial t}. \tag{3.49f}$$

Not one of the physical fields in Eqs. (3.49a–f) has undergone Fessenden rationalization; unlike $_f \vec{H}$ and $_f \vec{D}$, the $\vec{E}$, $\vec{B}$, $\vec{J}$, and ρ_Q physical quantities still have the same physical meaning as they did in Maxwell's day.

Tables 3.15 through 3.17 can be helpful in converting equations from the rationalized mks system to the Gaussian system. The equation for the magnetic field

Table 3.15 Numeric components of physical quantities in the esu system and the *Systeme International* or rationalized mks system. Physical quantities without the "f" prefix are left unchanged by Fessenden rationalization and have the same numeric components in both the rationalized and unrationalized mks systems.

(magnetic vector potential) $$A_{\mathrm{mks}} = A_{\mathrm{esu}} \cdot c_{\mathrm{cgs}} \cdot 10^{-6}$$	(current-loop magnetic dipole density) $$(M_I)_{\mathrm{mks}} = (M_I)_{\mathrm{esu}} \cdot 10^3 \cdot c_{\mathrm{cgs}}^{-1}$$
(magnetic induction) $$B_{\mathrm{mks}} = B_{\mathrm{esu}} \cdot c_{\mathrm{cgs}} \cdot 10^{-4}$$	(magnetic permeability) $$f\mu_{\mathrm{mks}} = 4\pi c_{\mathrm{cgs}}^2 \cdot 10^{-7} \cdot \mu_{\mathrm{esu}}$$
(capacitance) $$C_{\mathrm{mks}} = C_{\mathrm{esu}} \cdot c_{\mathrm{cgs}}^{-2} \cdot 10^9$$	(magnetic permeability of free space) $$(f\mu_0)_{\mathrm{mks}} = 4\pi c_{\mathrm{cgs}}^2 \cdot 10^{-7} \cdot (\mu_0)_{\mathrm{esu}}$$ $$= 4\pi \cdot 10^{-7}$$
(electric displacement) $$f D_{\mathrm{mks}} = D_{\mathrm{esu}} \cdot 10^5 \cdot (4\pi c_{\mathrm{cgs}})^{-1}$$	(magnetic pole strength) $$(f p_H)_{\mathrm{mks}} = 4\pi c_{\mathrm{cgs}} \cdot 10^{-8} \cdot (p_H)_{\mathrm{esu}}$$
(electric field) $$E_{\mathrm{mks}} = E_{\mathrm{esu}} \cdot c_{\mathrm{cgs}} \cdot 10^{-6}$$	(electric dipole moment) $$p_{\mathrm{mks}} = p_{\mathrm{esu}} \cdot 10^{-1} \cdot c_{\mathrm{cgs}}^{-1}$$
(dielectric constant) $$f\varepsilon_{\mathrm{mks}} = \varepsilon_{\mathrm{esu}} \cdot (4\pi c_{\mathrm{cgs}}^2)^{-1} \cdot 10^{11}$$	(electric dipole density) $$P_{\mathrm{mks}} = P_{\mathrm{esu}} \cdot 10^5 \cdot c_{\mathrm{cgs}}^{-1}$$
(permittivity of free space) $$(f\varepsilon_0)_{\mathrm{mks}} = (\varepsilon_0)_{\mathrm{esu}} \cdot (4\pi c_{\mathrm{cgs}}^2)^{-1} \cdot 10^{11}$$ $$= \left(4\pi \, c_{\mathrm{cgs}}^2\right)^{-1} \cdot 10^{11}$$	(permeance) $$f\mathcal{P}_{\mathrm{mks}} = \mathcal{P}_{\mathrm{esu}} \cdot 10^{-9} \cdot 4\pi c_{\mathrm{cgs}}^2$$
(magnetomotive force) $$f\mathcal{F}_{\mathrm{mks}} = \mathcal{F}_{\mathrm{esu}} \cdot 10 \cdot (4\pi c_{\mathrm{cgs}})^{-1}$$	(charge) $$Q_{\mathrm{mks}} = Q_{\mathrm{esu}} \cdot 10 \cdot c_{\mathrm{cgs}}^{-1}$$
(magnetic flux) $$(\Phi_B)_{\mathrm{mks}} = (\Phi_B)_{\mathrm{esu}} \cdot c_{\mathrm{cgs}} \cdot 10^{-8}$$	(resistance) $$R_{\mathrm{mks}} = R_{\mathrm{esu}} \cdot 10^{-9} \cdot c_{\mathrm{cgs}}^2$$
(conductance) $$G_{\mathrm{mks}} = G_{\mathrm{esu}} \cdot 10^9 \cdot c_{\mathrm{cgs}}^{-2}$$	(reluctance) $$f\mathcal{R}_{\mathrm{mks}} = (4\pi c_{\mathrm{cgs}}^2)^{-1} \cdot 10^9 \cdot \mathcal{R}_{\mathrm{esu}}$$
(magnetic field) $$f H_{\mathrm{mks}} = H_{\mathrm{esu}} \cdot 10^3 \cdot (4\pi c_{\mathrm{cgs}})^{-1}$$	(volume charge density) $$(\rho_Q)_{\mathrm{mks}} = (\rho_Q)_{\mathrm{esu}} \cdot 10^7 \cdot c_{\mathrm{cgs}}^{-1}$$
(current) $$I_{\mathrm{mks}} = I_{\mathrm{esu}} \cdot 10 \cdot c_{\mathrm{cgs}}^{-1}$$	(resistivity) $$(\rho_R)_{\mathrm{mks}} = (\rho_R)_{\mathrm{esu}} \cdot 10^{-11} \cdot c_{\mathrm{cgs}}^2$$
(volume current density) $$J_{\mathrm{mks}} = J_{\mathrm{esu}} \cdot 10^5 \cdot c_{\mathrm{cgs}}^{-1}$$	(elastance) $$S_{\mathrm{mks}} = S_{\mathrm{esu}} \cdot 10^{-9} \cdot c_{\mathrm{cgs}}^2$$
(surface current density) $$(\mathcal{J}_S)_{\mathrm{mks}} = (\mathcal{J}_S)_{\mathrm{esu}} \cdot 10^3 \cdot c_{\mathrm{cgs}}^{-1}$$	(surface charge density) $$(S_Q)_{\mathrm{mks}} = (S_Q)_{\mathrm{esu}} \cdot 10^5 \cdot c_{\mathrm{cgs}}^{-1}$$
(inductance) $$L_{\mathrm{mks}} = L_{\mathrm{esu}} \cdot 10^{-9} \cdot c_{\mathrm{cgs}}^2$$	(conductivity) $$\sigma_{\mathrm{mks}} = \sigma_{\mathrm{esu}} \cdot 10^{11} \cdot c_{\mathrm{cgs}}^{-2}$$

Table 3.15 (Continued).

(permanent-magnet dipole moment)	(electric potential)
$(_f m_H)_{mks} = 4\pi c_{cgs} \cdot 10^{-10} \cdot (m_H)_{esu}$	$V_{mks} = V_{esu} \cdot 10^{-8} \cdot c_{cgs}$
(current-loop magnetic dipole moment)	(magnetic scalar potential)
$(m_I)_{mks} = (m_I)_{esu} \cdot 10^{-3} \cdot c_{cgs}^{-1}$	$(_f \Omega_H)_{mks} = (\Omega_H)_{esu} \cdot 10 \cdot (4\pi c_{cgs})^{-1}$
(permanent-magnet dipole density)	
$(_f M_H)_{mks} = 4\pi c_{cgs} \cdot 10^{-4} \cdot (M_H)_{esu}$	

$_f \vec{H}$ in a vacuum at a time t for a field point which is a distance r from an oscillating electric dipole $\vec{p}$ is, in the rationalized mks system,

$$\left. _f \vec{H} \right|_{\text{at time } t} = \frac{-1}{4\pi c r} \left[\hat{e}_r \times \frac{d^2}{dt^2}(\vec{p}) \right] \Bigg|_{\substack{\text{at time} \\ t'=t-(r/c)}} , \qquad (3.50a)$$

where $\hat{e}_r = \hat{r}/r$ is the dimensionless unit vector pointing from the dipole to the field point and the physical quantity on the right-hand side is evaluated at an earlier time $t' = t - r/c$, giving the electromagnetic radiation time to reach the field point. Writing Eq. (3.50a) in terms of units and numeric components gives

$$\left[(_f \vec{H})_{mks} \frac{\text{amp}}{\text{m}} \right] \Bigg|_{\text{at time } t}$$
$$= \frac{-1}{4\pi \left(c_{mks} \dfrac{\text{m}}{\text{sec}} \right) (r_{mks}\text{m})} \left[\hat{e}_r \times \sec^{-2} \frac{d^2}{dt^2_{[\text{sec}]}} (\vec{p}_{mks} \, \text{coul} \cdot \text{m}) \right] \Bigg|_{\substack{\text{at time} \\ t'=t-(r/c)}} . \qquad (3.50b)$$

From Tables 3.17 and 3.10 we get

$$\left[\left(\frac{10^3}{4\pi} \vec{H}_{gs} \right) \frac{10^{-1} c_{cgs} \cdot \text{statampq}}{10^2 \text{cm}} \right] \Bigg|_{\text{at time } t}$$
$$= \frac{-1}{4\pi \left(c_{cgs} \dfrac{\text{cm}}{\text{sec}} \right) (r_{cgs}\text{cm})}$$
$$\cdot \left[\hat{e}_r \times \frac{1}{\sec^2} \frac{d^2}{dt^2_{[\text{sec}]}} \left(\frac{10^{-1}}{c_{cgs}} \vec{p}_{gs} \frac{c_{cgs}}{10} \, \text{statcoulq} \cdot 10^2 \text{cm} \right) \right] \Bigg|_{\substack{\text{at time} \\ t'=t-(r/c)}}$$

or, cancelling out the 4π and factors of 10,

Table 3.16 Numeric components of physical quantities in the emu system and the *Systeme International* or rationalized mks system. Physical quantities without the "f" prefix are left unchanged by Fessenden rationalization and have the same numeric components in both the rationalized and unrationalized mks systems.

(magnetic vector potential) $A_{\text{mks}} = A_{\text{emu}} \cdot 10^{-6}$	(current-loop magnetic dipole density) $(M_I)_{\text{mks}} = (M_I)_{\text{emu}} \cdot 10^3$
(magnetic induction) $B_{\text{mks}} = B_{\text{emu}} \cdot 10^{-4}$	(magnetic permeability) $f\mu_{\text{mks}} = 4\pi \cdot 10^{-7} \cdot \mu_{\text{emu}}$
(capacitance) $C_{\text{mks}} = C_{\text{emu}} \cdot 10^9$	(magnetic permeability of free space) $(f\mu_0)_{\text{mks}} = 4\pi \cdot 10^{-7} \cdot (\mu_0)_{\text{emu}}$ $= 4\pi \cdot 10^{-7}$
(electric displacement) $f D_{\text{mks}} = D_{\text{emu}} \cdot 10^5 \cdot (4\pi)^{-1}$	(magnetic pole strength) $(f p_H)_{\text{mks}} = 4\pi \cdot 10^{-8} \cdot (p_H)_{\text{emu}}$
(electric field) $E_{\text{mks}} = E_{\text{emu}} \cdot 10^{-6}$	(electric dipole moment) $p_{\text{mks}} = p_{\text{emu}} \cdot 10^{-1}$
(dielectric constant) $f\varepsilon_{\text{mks}} = \varepsilon_{\text{emu}} \cdot (4\pi)^{-1} \cdot 10^{11}$	(electric dipole density) $P_{\text{mks}} = P_{\text{emu}} \cdot 10^5$
(permittivity of free space) $(f\varepsilon_0)_{\text{mks}} = (\varepsilon_0)_{\text{emu}} \cdot (4\pi)^{-1} \cdot 10^{11}$ $= (4\pi c_{\text{cgs}}^2)^{-1} \cdot 10^{11}$	(permeance) $f\mathcal{P}_{\text{mks}} = \mathcal{P}_{\text{emu}} \cdot 10^{-9} \cdot 4\pi$
(magnetomotive force) $f\mathcal{F}_{\text{mks}} = \mathcal{F}_{\text{emu}} \cdot 10 \cdot (4\pi)^{-1}$	(charge) $Q_{\text{mks}} = Q_{\text{emu}} \cdot 10$
(magnetic flux) $(\Phi_B)_{\text{mks}} = (\Phi_B)_{\text{emu}} \cdot 10^{-8}$	(resistance) $R_{\text{mks}} = R_{\text{emu}} \cdot 10^{-9}$
(conductance) $G_{\text{mks}} = G_{\text{emu}} \cdot 10^9$	(reluctance) $f\mathcal{R}_{\text{mks}} = (4\pi)^{-1} \cdot 10^9 \cdot \mathcal{R}_{\text{emu}}$
(magnetic field) $f H_{\text{mks}} = H_{\text{emu}} \cdot 10^3 \cdot (4\pi)^{-1}$	(volume charge density) $(\rho_Q)_{\text{mks}} = (\rho_Q)_{\text{emu}} \cdot 10^7$
(current) $I_{\text{mks}} = I_{\text{emu}} \cdot 10$	(resistivity) $(\rho_R)_{\text{mks}} = (\rho_R)_{\text{emu}} \cdot 10^{-11}$
(volume current density) $J_{\text{mks}} = J_{\text{gs}} \cdot 10^5 \cdot c_{\text{cgs}}^{-1}$	(elastance) $S_{\text{mks}} = S_{\text{emu}} \cdot 10^{-9}$
(surface current density) $(\mathcal{J}_S)_{\text{mks}} = (\mathcal{J}_S)_{\text{gs}} \cdot 10^3 \cdot c_{\text{cgs}}^{-1}$	(surface charge density) $(S_Q)_{\text{mks}} = (S_Q)_{\text{emu}} \cdot 10^5$
(inductance) $L_{\text{mks}} = L_{\text{gs}} \cdot 10^{-9} \cdot c_{\text{cgs}}^2$	(conductivity) $\sigma_{\text{mks}} = \sigma_{\text{emu}} \cdot 10^{11}$

Table 3.16 (Continued).

(permanent-magnet dipole moment)	(electric potential)
$({}_f m_H)_{\text{mks}} = 4\pi \cdot 10^{-10} \cdot (m_H)_{\text{gs}}$	$V_{\text{mks}} = V_{\text{emu}} \cdot 10^{-8}$
(current-loop magnetic dipole moment)	(magnetic scalar potential)
$(m_I)_{\text{mks}} = (m_I)_{\text{gs}} \cdot 10^{-3}$	$({}_f \Omega_H)_{\text{mks}} = (\Omega_H)_{\text{emu}} \cdot 10 \cdot (4\pi)^{-1}$
(permanent-magnet dipole density)	
$({}_f M_H)_{\text{mks}} = 4\pi \cdot 10^{-4} \cdot (M_H)_{\text{emu}}$	

$$\left[(\vec{H}_{\text{gs}}) c_{\text{cgs}} \frac{\text{statampq}}{\text{cm}} \right]\Bigg|_{\text{at time } t}$$

$$= \left[\frac{-1}{\left(c_{\text{cgs}} \dfrac{\text{cm}}{\text{sec}} \right)(r_{\text{cgs}}\text{cm})} \right] \cdot \left[\hat{e}_r \times \frac{1}{\text{sec}^2} \frac{d^2}{dt_{[\text{sec}]}^2} \left(\vec{p}_{\text{gs}}\text{statcoulq} \cdot \text{cm} \right) \right]\Bigg|_{\substack{\text{at time} \\ t'=t-(r/c)}} .$$

Since statampq = statcoulq/sec [see Eq. (2.45a)], we have

$$\left[(\vec{H}_{\text{gs}}) c_{\text{cgs}} \frac{\text{statcoulq}}{\text{cm} \cdot \text{sec}} \right]\Bigg|_{\text{at time } t}$$

$$= \frac{-1}{\left(c_{\text{cgs}} \dfrac{\text{cm}}{\text{sec}} \right)\left(r_{\text{cgs}}\text{cm} \right)} \left[\hat{e}_r \times \frac{1}{\text{sec}^2} \frac{d^2}{dt_{[\text{sec}]}^2} \left(\vec{p}_{\text{gs}}\text{statcoulq} \cdot \text{cm} \right) \right]\Bigg|_{\substack{\text{at time} \\ t'=t-(r/c)}} .$$

This is balanced in the invariant units cm, sec and the connecting unit (statcoul, statcoulq), so we can stop recognizing charge as a fundamental dimension by dropping the q suffix:

$$\left[(\vec{H}_{\text{gs}}) c_{\text{cgs}} \frac{\text{statcoul}}{\text{cm} \cdot \text{sec}} \right]\Bigg|_{\text{at time } t}$$

$$= \frac{-1}{\left(c_{\text{cgs}} \dfrac{\text{cm}}{\text{sec}} \right)\left(r_{\text{cgs}}\text{cm} \right)} \left[\hat{e}_r \times \frac{1}{\text{sec}^2} \frac{d^2}{dt_{[\text{sec}]}^2} \left(\vec{p}_{\text{gs}}\text{statcoul} \cdot \text{cm} \right) \right]\Bigg|_{\substack{\text{at time} \\ t'=t-(r/c)}} .$$

$$(3.50c)$$

The only problem with Eq. (3.50c) is that in the Gaussian system, $\vec{H}$ is a magnetic quantity (see Table 3.1) and should have units of

$$\text{abcoul/sec/cm} = \text{abamp/cm}$$

instead of

$$\text{statcoul/sec/cm} = \text{statamp/cm}.$$

Table 3.17 Num eric components of physical quantities in the Gaussian system and the *Systeme International* or rationalized mks system. Physical quantities without the "*f*" prefix are left unchanged by Fessenden rationalization and have the same numeric components in both the rationalized and unrationalized mks systems.

(magnetic vector potential)	(current-loop magnetic dipole density)
$A_{\mathrm{mks}} = A_{\mathrm{gs}} \cdot 10^{-6}$	$(M_I)_{\mathrm{mks}} = (M_I)_{\mathrm{gs}} \cdot 10^{3}$
(magnetic induction)	(magnetic permeability)
$B_{\mathrm{mks}} = B_{\mathrm{gs}} \cdot 10^{-4}$	$f\mu_{\mathrm{mks}} = 4\pi \cdot 10^{-7} \cdot \mu_{\mathrm{gs}}$ $= 4\pi \cdot 10^{-7} \cdot \mu_r$
(capacitance)	(magnetic permeability of free space)
$C_{\mathrm{mks}} = C_{\mathrm{gs}} \cdot c_{\mathrm{cgs}}^{-2} \cdot 10^{9}$	$(f\mu_0)_{\mathrm{mks}} = 4\pi \cdot 10^{-7} \cdot (\mu_0)_{\mathrm{gs}}$ $= 4\pi \cdot 10^{-7}$
(electric displacement)	(magnetic pole strength)
$f D_{\mathrm{mks}} = D_{\mathrm{gs}} \cdot 10^{5} \cdot (4\pi c_{\mathrm{cgs}})^{-1}$	$(f p_H)_{\mathrm{mks}} = 4\pi \cdot 10^{-8} \cdot (p_H)_{\mathrm{gs}}$
(electric field)	(electric dipole moment)
$E_{\mathrm{mks}} = E_{\mathrm{gs}} \cdot c_{\mathrm{cgs}} \cdot 10^{-6}$	$p_{\mathrm{mks}} = p_{\mathrm{gs}} \cdot 10^{-1} \cdot c_{\mathrm{cgs}}^{-1}$
(dielectric constant)	(electric dipole density)
$f\varepsilon_{\mathrm{mks}} = \varepsilon_{\mathrm{gs}} \cdot (4\pi c_{\mathrm{cgs}}^{2})^{-1} \cdot 10^{11}$ $= \varepsilon_r \cdot (4\pi c_{\mathrm{cgs}}^{2})^{-1} \cdot 10^{11}$	$P_{\mathrm{mks}} = P_{\mathrm{gs}} \cdot 10^{5} \cdot c_{\mathrm{cgs}}^{-1}$
(permittivity of free space)	(permeance)
$(f\varepsilon_0)_{\mathrm{mks}} = (\varepsilon_0)_{\mathrm{gs}} \cdot (4\pi c_{\mathrm{cgs}}^{2})^{-1} \cdot 10^{11}$ $= (4\pi c_{\mathrm{cgs}}^{2})^{-1} \cdot 10^{11}$	$f\mathcal{P}_{\mathrm{mks}} = \mathcal{P}_{\mathrm{gs}} \cdot 10^{-9} \cdot 4\pi$
(magnetomotive force)	(charge)
$f\mathcal{F}_{\mathrm{mks}} = \mathcal{F}_{\mathrm{gs}} \cdot 10 \cdot (4\pi)^{-1}$	$Q_{\mathrm{mks}} = Q_{\mathrm{gs}} \cdot 10 \cdot c_{\mathrm{cgs}}^{-1}$
(magnetic flux)	(resistance)
$(\Phi_B)_{\mathrm{mks}} = (\Phi_B)_{\mathrm{gs}} \cdot 10^{-8}$	$R_{\mathrm{mks}} = R_{\mathrm{gs}} \cdot 10^{-9} \cdot c_{\mathrm{cgs}}^{2}$
(conductance)	(reluctance)
$G_{\mathrm{mks}} = G_{\mathrm{gs}} \cdot 10^{9} \cdot c_{\mathrm{cgs}}^{-2}$	$f\mathcal{R}_{\mathrm{mks}} = (4\pi)^{-1} \cdot 10^{9} \cdot \mathcal{R}_{\mathrm{gs}}$
(magnetic field)	(volume charge density)
$f H_{\mathrm{mks}} = H_{\mathrm{gs}} \cdot 10^{3} \cdot (4\pi)^{-1}$	$(\rho_Q)_{\mathrm{mks}} = (\rho_Q)_{\mathrm{gs}} \cdot 10^{7} \cdot c_{\mathrm{cgs}}^{-1}$
(current)	(resistivity)
$I_{\mathrm{mks}} = I_{\mathrm{gs}} \cdot 10 \cdot c_{\mathrm{cgs}}^{-1}$	$(\rho_R)_{\mathrm{mks}} = (\rho_R)_{\mathrm{gs}} \cdot 10^{-11} \cdot c_{\mathrm{cgs}}^{2}$
(volume current density)	(elastance)
$J_{\mathrm{mks}} = J_{\mathrm{gs}} \cdot 10^{5} \cdot c_{\mathrm{cgs}}^{-1}$	$S_{\mathrm{mks}} = S_{\mathrm{gs}} \cdot 10^{-9} \cdot c_{\mathrm{cgs}}^{2}$
(surface current density)	(surface charge density)
$(\mathcal{J}_S)_{\mathrm{mks}} = (\mathcal{J}_S)_{\mathrm{gs}} \cdot 10^{3} \cdot c_{\mathrm{cgs}}^{-1}$	$(S_Q)_{\mathrm{mks}} = (S_Q)_{\mathrm{gs}} \cdot 10^{5} \cdot c_{\mathrm{cgs}}^{-1}$
(inductance)	(conductivity)
$L_{\mathrm{mks}} = L_{\mathrm{gs}} \cdot 10^{-9} \cdot c_{\mathrm{cgs}}^{2}$	$\sigma_{\mathrm{mks}} = \sigma_{\mathrm{gs}} \cdot 10^{11} \cdot c_{\mathrm{cgs}}^{-2}$

Table 3.17 (Continued).

(permanent-magnet dipole moment)	(electric potential)
$(_fm_H)_{mks} = 4\pi \cdot 10^{-10} \cdot (m_H)_{gs}$	$V_{mks} = V_{gs} \cdot 10^{-8} \cdot c_{cgs}$
(current-loop magnetic dipole moment)	(magnetic scalar potential)
$(m_I)_{mks} = (m_I)_{gs} \cdot 10^{-3}$	$(_f\Omega_H)_{mks} = (\Omega_H)_{gs} \cdot 10 \cdot (4\pi)^{-1}$
(permanent-magnet dipole density)	
$(_fM_H)_{mks} = 4\pi \cdot 10^{-4} \cdot (M_H)_{gs}$	

This equation is really in esu units rather than Gaussian units. According to Fig. 3.2 this is only to be expected, because the road from the rationalized mks system to the Gaussian system must pass through either the esu or emu systems of units. We therefore consult Table 3.3 to write (3.50c) as

$$\left[\left(c_{cgs}\frac{cm}{sec}\right)\vec{H}_{gs}\frac{abamp}{cm}\right]\Bigg|_{\text{at time } t}$$

$$= \frac{-1}{\left(c_{cgs}\dfrac{cm}{sec}\right)\left(r_{cgs}cm\right)}\left[\hat{e}_r \times \frac{1}{sec^2}\frac{d^2}{dt^2_{[sec]}}(\vec{p}_{gs}\text{statcoul} \cdot cm)\right]\Bigg|_{\substack{\text{at time} \\ t'=t-(r/c)}}.$$

(3.50d)

Now the magnetic quantity is in emu units and the electric quantity is in esu units, so Eq. (3.50d) is in Gaussian units and can be written as

$$\vec{H}\Bigg|_{\text{at time } t} = \frac{-1}{rc^2}\left[\hat{e}_r \times \frac{d^2}{dt^2}(\vec{p})\right]\Bigg|_{\substack{\text{at time} \\ t'=t-(r/c)}} \tag{3.50e}$$

or

$$\vec{B}\Bigg|_{\text{at time } t} = \frac{-1}{rc^2}\left[\hat{e}_r \times \frac{d^2}{dt^2}(\vec{p})\right]\Bigg|_{\substack{\text{at time} \\ t'=t-(r/c)}}, \tag{3.50f}$$

because in Gaussian units $\vec{B} = \vec{H}$ for empty space.

3.9 EVALUATION OF THE RATIONALIZED MKS SYSTEM

Although the contest between the rationalized and unrationalized mks systems has long since been decided by history—with overwhelming victory going to the rationalized mks system—it does no harm to revisit briefly the main point of contention: which system leads to a simpler set of equations? The only good answer seems to be that it all depends on which equations are more important. We note that Eq. (3.50f) in the unrationalized Gaussian system is simpler than Eq. (3.50a), the rationalized equation we started with. In fact, we can remove the extra power of c in the Gaussian system merely by converting Eq. (3.50a) to unrationalized mks, making the substitution $_f \vec{H} = \vec{H}/(4\pi)$ to get

$$\left. \vec{H} \right|_{\text{at time } t} = \frac{-1}{cr} \left[\hat{e}_r \times \frac{d^2}{dt^2} \left(\vec{p} \right) \right] \Bigg|_{\substack{\text{at time} \\ t'=t-(r/c)}} . \tag{3.51}$$

All that rationalization of any sort accomplishes, as was pointed out in the discussion after Eq. (3.16b), is to redistribute the factors of 4π from one group of equations to another. The price for removing the 4π's from Maxwell's equations is to have them reappear in Coulomb's law and other equations involving spherical or cylindrical symmetry, such as Eq. (3.42a) and Eq. (3.50a). At the beginning of the twentieth century, when rectilinear geometries dominated the electrical engineer's laboratory, the inconvenient factors of 4π undoubtedly caused annoyance; but by the middle of the twentieth century radio, radar, television, etc., had become important fields of engineering built around the idea of outwardly propagating spherical wavefronts. It is one of the small jokes of history that, had electrical engineers stuck with unrationalized units, they would have been pleasantly surprised to find themselves working with equations like Eq. (3.51) rather than Eq. (3.50a). Today, with the dominance of computers, the added complexity or simplicity of having or not having to account for an extra factor of 4π seems trivial, given that most engineers solve their problems by programming computers and manipulating spreadsheets. It is hard to imagine how disheartening it must have been a hundred years ago to face, once again, the prospect of multiplying or dividing by 4π to get the final answer.

REFERENCES

1. D. J. Griffiths, *Introduction to Electrodynamics*, 2nd edition, Prentice-Hall, Englewood Cliffs, New Jersey, 1989, p. 5.
2. F. B. Silsbee, "Systems of Electrical Units," *Journal of Research of the National Bureau of Standards–C. Engineering and Instrumentation*, Vol. 66C, No. 2, April-June 1962, p. 166.
3. F. B. Silsbee, "Systems of Electrical Units," *Journal of Research of the National Bureau of Standards—C. Engineering and Instrumentation*, Vol. 66C, No. 2, April-June 1962, pp. 166–167.
4. J. D. Jackson, *Classical Electrodynamics*, 3rd edition, John Wiley and Sons, Inc., New York, 1999, pp 238–239.

CHAPTER 4

TWO STANDARD SHORTCUTS USED TO TRANSFORM ELECTROMAGNETIC EQUATIONS

The last several chapters have explained how the standard rules for changing units apply to electromagnetic physical quantities. Having become familiar with these rules, we are now sure that electromagnetic equations and formulas transform in a way that makes sense when going from one system of units to another. We also know, however, that following these rules can be algebraically cumbersome, forcing us always to watch for the appearance or disappearance of constants ε_0 and μ_0 as we recognize or refuse to recognize charge as a new dimension. Engineers and physicists are no more eager than anyone else to do unnecessary work; consequently, they have come up with both the free-parameter method and substitution tables, two shortcuts that can greatly reduce the time required to convert electromagnetic equations and formulas from one system of units to another. Unfortunately, neither shortcut is perfect: substitution tables can give ambiguous answers in unusual situations, and to apply the free-parameter method we must first relate our equation or formula to one or more of a predefined list of equations and formulas. Nevertheless, these shortcuts often provide a quick and easy way of transforming electromagnetic expressions; and whenever there is any doubt about the result, the transformation can be checked using the procedures explained in the previous chapters.

4.1 THE FREE-PARAMETER METHOD

Table 4.1 lists Maxwell's equations and the Lorentz force law for the six major electromagnetic systems discussed in this book. As pointed out at the beginning of Chapter 3, any classical electromagnetic formula can be derived from Maxwell's equations and the Lorentz force law. This means we can consult Table 4.1, select the appropriate equations in the desired set of units, and from them derive the formulas we need to know. Although this process gets the job done, it usually requires a lot of work. To avoid the unpleasant prospect of deriving all of our formulas and equations from Maxwell's equations and the Lorentz force law, we construct instead a long list of basic electromagnetic equations that contains everything (including Maxwell's equations and the Lorentz force law) likely to be useful. Instead of providing six long lists—one for every electromagnetic system—we use the four free parameters $\widetilde{\varepsilon}$, $\widetilde{\mu}$, k_0, and Π shown in Table 4.2 to reduce the six lists to one.[1]

As an example of how this works, consider what happens when we disregard the "h" and "f" prefixes and write Maxwell's equations as

$$\vec{\nabla} \cdot \vec{D} = \Pi \rho_Q, \tag{4.1a}$$

$$\vec{\nabla} \cdot \vec{B} = 0, \tag{4.1b}$$

$$\vec{\nabla} \times \vec{H} = k_0 \left(\Pi \vec{J} + \frac{\partial \vec{D}}{\partial t} \right), \tag{4.1c}$$

$$\vec{\nabla} \times \vec{E} + k_0 \frac{\partial \vec{B}}{\partial t} = 0, \tag{4.1d}$$

where

$$\vec{D} = \widetilde{\varepsilon} \vec{E} + \Pi \vec{P} \tag{4.1e}$$

and

$$\vec{H} = \frac{1}{\widetilde{\mu}} \vec{B} - \frac{\Pi}{\widetilde{\mu}} \vec{M}_H$$
$$= \frac{1}{\widetilde{\mu}} \vec{B} - \Pi \vec{M}_I. \tag{4.1f}$$

As always, $\vec{E}$ and $\vec{D}$ are the electric field and electric displacement, respectively; $\vec{H}$ and $\vec{B}$ are the magnetic field and magnetic induction, respectively; ρ_Q is the volume charge density; $\vec{J}$ is the volume current density; $\vec{P}$ is the electric dipole density; $\vec{M}_H$ is the permanent-magnet dipole density; and $\vec{M}_I$ is the current-loop magnetic dipole density. Clearly, Eqs. (4.1a–f) reduce to the correct set of equations in Table 4.1 when $\widetilde{\varepsilon}$, $\widetilde{\mu}$, k_0, and Π are given the appropriate values from Table 4.2. The same thing can be done to the Lorentz force law; if it is written as

$$\vec{F} = Q\vec{E} + k_0 Q (\vec{v} \times \vec{B}), \tag{4.2}$$

then it too reduces to the correct equation in Table 4.1 when k_0 is given the appropriate value from Table 4.2.

As we have just seen, the free-parameter method works most easily and naturally when we neglect the distinction between rationalized and unrationalized electromagnetic quantities—that is, neglect the "h" and "f" prefixes —which so far we have been careful to preserve. It should be emphasized that the distinction between a change of units and a rescaling of an electromagnetic physical quantity is just as important as before; the free-parameter method just makes it inconvenient to keep track of this distinction using a single table. If we want to preserve the distinction between rationalized and unrationalized physical quantities, we can consult Tables 4.3(a) or 4.3(b) after putting an equation or formula into the rationalized mks* or Heaviside-Lorentz systems, respectively.

* As pointed out in Section 3.6 of Chapter 3, most textbooks written today using the rationalized mks system say that they are using SI units.

Table 4.1 Maxwell's equations and the Lorentz force law in the rationalized mks system (which is also called SI units), the unrationalized mks system, Gaussian cgs units, the Heaviside-Lorentz cgs system, esu units, and emu units.

rationalized mks system, also called SI units	$\vec{\nabla} \cdot {}_f\vec{D} = \rho_Q, \quad \vec{\nabla} \cdot \vec{B} = 0, \quad \vec{\nabla} \times \vec{E} + \dfrac{\partial \vec{B}}{\partial t} = 0,$ $\vec{\nabla} \times {}_f\vec{H} = \vec{J} + \dfrac{\partial {}_f\vec{D}}{\partial t}, \quad {}_f\vec{D} = {}_f\varepsilon_0\vec{E} + \vec{P},$ ${}_f\vec{H} = \dfrac{\vec{B}}{{}_f\mu_0} - \vec{M}_I, \quad \vec{F} = Q\vec{E} + Q(\vec{v} \times \vec{B})$
unrationalized mks system	$\vec{\nabla} \cdot \vec{D} = 4\pi\rho_Q, \quad \vec{\nabla} \cdot \vec{B} = 0, \quad \vec{\nabla} \times \vec{E} + \dfrac{\partial \vec{B}}{\partial t} = 0,$ $\vec{\nabla} \times \vec{H} = 4\pi\vec{J} + \dfrac{\partial \vec{D}}{\partial t}, \quad \vec{D} = \varepsilon_0\vec{E} + 4\pi\vec{P},$ $\vec{H} = \dfrac{\vec{B}}{\mu_0} - 4\pi\vec{M}_I, \quad \vec{F} = Q\vec{E} + Q(\vec{v} \times \vec{B})$
Gaussian cgs units	$\vec{\nabla} \cdot \vec{D} = 4\pi\rho_Q, \quad \vec{\nabla} \cdot \vec{B} = 0, \quad \vec{\nabla} \times \vec{E} + \dfrac{1}{c}\dfrac{\partial \vec{B}}{\partial t} = 0,$ $\vec{\nabla} \times \vec{H} = \dfrac{4\pi}{c}\vec{J} + \dfrac{1}{c}\dfrac{\partial \vec{D}}{\partial t}, \quad \vec{D} = \vec{E} + 4\pi\vec{P},$ $\vec{H} = \vec{B} - 4\pi\vec{M}_I, \quad \vec{F} = Q\vec{E} + \dfrac{Q}{c}(\vec{v} \times \vec{B})$
Heaviside-Lorentz cgs system	$\vec{\nabla} \cdot {}_h\vec{D} = {}_h\rho_Q, \quad \vec{\nabla} \cdot {}_h\vec{B} = 0, \quad \vec{\nabla} \times {}_h\vec{E} + \dfrac{1}{c}\dfrac{\partial {}_h\vec{B}}{\partial t} = 0,$ $\vec{\nabla} \times {}_h\vec{H} = \dfrac{1}{c}{}_h\vec{J} + \dfrac{1}{c}\dfrac{\partial {}_h\vec{D}}{\partial t}, \quad {}_h\vec{D} = {}_h\vec{E} + {}_h\vec{P},$ ${}_h\vec{H} = {}_h\vec{B} - {}_h\vec{M}_I, \quad \vec{F} = {}_hQ{}_h\vec{E} + \dfrac{{}_hQ}{c}(\vec{v} \times {}_h\vec{B})$
esu units	$\vec{\nabla} \cdot \vec{D} = 4\pi\rho_Q, \quad \vec{\nabla} \cdot \vec{B} = 0, \quad \vec{\nabla} \times \vec{E} + \dfrac{\partial \vec{B}}{\partial t} = 0,$ $\vec{\nabla} \times \vec{H} = 4\pi\vec{J} + \dfrac{\partial \vec{D}}{\partial t}, \quad \vec{D} = \vec{E} + 4\pi\vec{P},$ $\vec{H} = c^2\vec{B} - 4\pi\vec{M}_I, \quad \vec{F} = Q\vec{E} + Q(\vec{v} \times \vec{B})$
emu units	$\vec{\nabla} \cdot \vec{D} = 4\pi\rho_Q, \quad \vec{\nabla} \cdot \vec{B} = 0, \quad \vec{\nabla} \times \vec{E} + \dfrac{\partial \vec{B}}{\partial t} = 0,$ $\vec{\nabla} \times \vec{H} = 4\pi\vec{J} + \dfrac{\partial \vec{D}}{\partial t}, \quad \vec{D} = \dfrac{1}{c^2}\vec{E} + 4\pi\vec{P},$ $\vec{H} = \vec{B} - 4\pi\vec{M}_I, \quad \vec{F} = Q\vec{E} + Q(\vec{v} \times \vec{B})$

To show how the free parameter method, with or without prefixes, works, we apply the first row of Table 4.2 to Eq. (4.1e), reducing it to the rationalized mks system:

$$\vec{D} = \left(\frac{10^{11}}{4\pi c_{\text{cgs}}^2} \frac{\text{farad}}{\text{m}} \right) \vec{E} + \vec{P}. \tag{4.3a}$$

Table 4.2 Free-parameter values for the rationalized mks system which is also called SI units, the unrationalized mks system, Gaussian cgs units, the Heaviside-Lorentz cgs system, esu units, and emu units.

	Rationalization free parameter	Permittivity free parameter	Permeability free parameter	Light-Speed free parameter	Is this a rationalized system?
rationalized mks system, also called SI units	$\Pi = 1$	$\tilde{\varepsilon} = \dfrac{10^{11}}{4\pi\, c_{cgs}^2}\dfrac{farad}{m}$	$\tilde{\mu} = 4\pi \cdot 10^{-7}\dfrac{henry}{m}$	$k_0 = 1$	Yes, use $\tilde{\varepsilon} = f\varepsilon_0$, $\tilde{\mu} = f\mu_0$, and consult Table 4.3(a) to see where the other prefixes go.
unrationalized mks system	$\Pi = 4\pi$	$\tilde{\varepsilon} = \dfrac{10^{11}}{c_{cgs}^2}\dfrac{farad}{m}$	$\tilde{\mu} = 10^{-7}\dfrac{henry}{m}$	$k_0 = 1$	No, use $\tilde{\varepsilon} = \varepsilon_0$, $\tilde{\mu} = \mu_0$, and there are no prefixes.
Gaussian cgs units	$\Pi = 4\pi$	$\tilde{\varepsilon} = 1$	$\tilde{\mu} = 1$	$k_0 = \dfrac{1}{c}$	No
Heaviside-Lorentz cgs system	$\Pi = 1$	$\tilde{\varepsilon} = 1$	$\tilde{\mu} = 1$	$k_0 = \dfrac{1}{c}$	Yes, consult Table 4.3b to see where the prefixes go.
esu units	$\Pi = 4\pi$	$\tilde{\varepsilon} = 1$	$\tilde{\mu} = \dfrac{1}{c^2}$	$k_0 = 1$	No
emu units	$\Pi = 4\pi$	$\tilde{\varepsilon} = \dfrac{1}{c^2}$	$\tilde{\mu} = 1$	$k_0 = 1$	No

Table 4.3(a) Rationalized and unrationalized physical quantities in the rationalized mks system, which is also referred to as SI units.

magnetic vector potential	volume current density	permeance
Unrationalized, A	*Unrationalized*, J	*Rationalized*, $_f\mathcal{P}$
magnetic induction	surface current density	charge
Unrationalized, B	*Unrationalized*, $\mathcal{J}_S$	*Unrationalized*, Q
capacitance	inductance	resistance
Unrationalized, C	*Unrationalized*, L	*Unrationalized*, R
electric displacement	permanent-magnet dipole moment reluctance	reluctance
Rationalized, $_fD$	*Rationalized*, $_fm_H$	*Rationalized*, $_f\mathcal{R}$
electric field	current-loop magnetic dipole moment	volume charge density
Unrationalized, E	*Unrationalized*, m_I	*Unrationalized*, ρ_Q
dielectric constant	permanent-magnet dipole density	resistivity
Rationalized, $_f\varepsilon$	*Rationalized*, $_fM_H$	*Unrationalized*, ρ_R
relative dielectric constant	current-loop magnetic dipole density	elastance
Unrationalized, ε_r	*Unrationalized*, M_I	*Unrationalized*, S
permittivity of free space	magnetic permeability	surface charge density
Rationalized, $_f\varepsilon_0$	*Rationalized*, $_f\mu$	*Unrationalized*, S_Q
magnetomotive force	relative magnetic permeability	conductivity
Rationalized, $_f\mathcal{F}$	*Unrationalized*, μ_r	*Unrationalized*, σ
magnetic flux	magnetic permeability of free space	electric potential
Unrationalized, Φ_B	*Rationalized*, $_f\mu_0$	*Unrationalized*, V
conductance	magnetic pole strength	magnetic scalar potential
Unrationalized, G	*Rationalized*, $_fp_H$	*Rationalized*, $_f\Omega_H$
magnetic field	electric dipole moment	
Rationalized, $_fH$	*Unrationalized*, p	
current	electric dipole density	
Unrationalized, I	*Unrationalized*, P	

From the fifth entry of row one, we note that

$$\widetilde{\varepsilon} \rightarrow \frac{10^{11}}{4\pi c_{\mathrm{cgs}}^2}\frac{\mathrm{farad}}{\mathrm{m}} = {}_f\varepsilon_0;$$

and from Table 4.3(a) we see that

$$\vec{D} \rightarrow {}_f\vec{D},$$

Table 4.3(b) Rationalized and unrationalized physical quantities in the Heaviside-Lorentz cgs system.

magnetic vector potential	volume current density	permeance
Rationalized, $_h A$	*Rationalized*, $_h J$	*Unrationalized*, $\mathcal{P}$
magnetic induction	surface current density	charge
Rationalized, $_h B$	*Rationalized*, $_h \mathcal{J}_S$	*Rationalized*, $_h Q$
capacitance	inductance	resistance
Rationalized, $_h C$	*Rationalized*, $_h L$	*Rationalized*, $_h R$
electric displacement	permanent-magnet dipole moment	reluctance
Rationalized, $_h D$	*Rationalized*, $_h m_H$	*Unrationalized*, $\mathcal{R}$
electric field	current-loop magnetic dipole moment	volume charge density
Rationalized, $_h E$	*Rationalized*, $_h m_I$	*Rationalized*, $_h \rho_Q$
relative dielectric constant	permanent-magnet dipole density	resistivity
Unrationalized, ε_r	*Rationalized*, $_h M_H$	*Rationalized*, $_h \rho_R$
magnetomotive force	current-loop magnetic dipole density	elastance
Rationalized, $_h \mathcal{F}$	*Rationalized*, $_h M_I$	*Rationalized*, $_h S$
magnetic flux	relative magnetic permeability	surface charge density
Rationalized, $_h \Phi_B$	*Unrationalized*, μ_r	*Rationalized*, $_h S_Q$
conductance	magnetic pole strength	conductivity
Rationalized, $_h G$	*Rationalized*, $_h p_H$	*Rationalized*, $_h \sigma$
magnetic field	electric dipole moment	electric potential
Rationalized, $_h H$	*Rationalized*, $_h p$	*Rationalized*, $_h V$
current	electric dipole density	magnetic scalar potential
Rationalized, $_h I$	*Rationalized*, $_h P$	*Rationalized*, $_h \Omega_H$

showing that Eq. (4.3a) should be written as

$$_f \vec{D} = {}_f \varepsilon_0 \vec{E} + \vec{P} \tag{4.3b}$$

to match the notation of the previous three chapters of this book. Nothing stops us, however, from following the notation of most modern textbooks by dropping the prefix "f" from ε_0 and $\vec{D}$ to write Eq. (4.3a) as

$$\vec{D} = \varepsilon_0 \vec{E} + \vec{P}. \tag{4.3c}$$

When putting Eq. (4.1e) into the other rationalized system, the Heaviside-Lorentz system, we apply the fourth row of Table 4.2 to get

$$\vec{D} = \vec{E} + \vec{P}. \tag{4.4a}$$

This result, which lacks prefixes to show which electromagnetic quantities have been rationalized, is already in the form used by most modern authors working with the Heaviside-Lorentz system. Nothing stops us, however, from consulting Table 4.3(b), which shows that all three quantities $\vec{D}$, $\vec{E}$, and $\vec{P}$ are rationalized in the Heaviside-Lorentz system. We can show this by following the convention of the previous chapters and adding an "h" to get

$$_h\vec{D} = {_h}\vec{E} + {_h}\vec{P}. \tag{4.4b}$$

The formula for the radiant electric field generated by an oscillating current-loop magnetic dipole gives us a somewhat more complicated example of how to use Tables 4.2 and 4.3(a)–(b). The free-parameter formula for the $\vec{E}$ field at a field point that is a distance r from an oscillating current-loop magnetic dipole $\vec{m}_I$ is

$$\vec{E}\bigg|_{\substack{\text{evaluated} \\ \text{at time } t}} = \frac{k_0 \tilde{\mu} \Pi}{4\pi c r} \left\{ \hat{e}_r \times \frac{d^2 \vec{m}_I}{dt^2} \right\}\bigg|_{\substack{\text{evaluated at} \\ \text{the retarded} \\ \text{time } t'=t-r/c}} . \tag{4.5}$$

Here, r is large enough to put the field point in the dipole's far-field region, the dimensionless unit vector pointing from the current-loop magnetic dipole to the field point is $\hat{e}_r$, and c is the speed of light.

To put Eq. (4.5) into the rationalized mks system, we consult the first row of Table 4.2 and write

$$\vec{E}\bigg|_{\substack{\text{evaluated} \\ \text{at time } t}} = \left(4\pi \cdot 10^{-7} \frac{\text{henry}}{\text{m}} \right) \cdot \frac{1}{4\pi c r} \cdot \left\{ \hat{e}_r \times \frac{d^2 \vec{m}_I}{dt^2} \right\}\bigg|_{\substack{\text{evaluated at} \\ \text{the retarded} \\ \text{time } t'=t-r/c}} . \tag{4.6a}$$

The fifth entry of row one suggests that this result be written as

$$\vec{E}\bigg|_{\substack{\text{evaluated} \\ \text{at time } t}} = \frac{f\mu_0}{4\pi c r} \cdot \left\{ \hat{e}_r \times \frac{d^2 \vec{m}_I}{dt^2} \right\}\bigg|_{\substack{\text{evaluated at} \\ \text{the retarded} \\ \text{time } t'=t-r/c}} , \tag{4.6b}$$

but of course nothing stops us from dropping the "f" prefix and writing this as

$$\vec{E}\bigg|_{\substack{\text{evaluated} \\ \text{at time } t}} = \frac{\mu_0}{4\pi c r} \cdot \left\{ \hat{e}_r \times \frac{d^2 \vec{m}_I}{dt^2} \right\}\bigg|_{\substack{\text{evaluated at} \\ \text{the retarded} \\ \text{time } t'=t-r/c}} , \tag{4.6c}$$

if we want to follow the convention of most authors using the rationalized mks system. Table 4.3(a) assures us that neither $\vec{m}_I$ nor $\vec{E}$ are rationalized quantities, which means that neither one needs to be written with an "f" prefix to match the convention of the previous chapters.

To put Eq. (4.5) into the unrationalized mks system, we consult the second row of Table 4.2 to get

$$\Pi \to 4\pi, \quad \widetilde{\mu} \to 10^{-7} \text{henry/m} = \mu_0, \quad \text{and} \quad k_0 \to 1.$$

This gives us

$$\vec{E}\bigg|_{\substack{\text{evaluated} \\ \text{at time } t}} = \frac{\mu_0}{cr} \cdot \left\{ \hat{e}_r \times \frac{d^2\vec{m}_I}{dt^2} \right\}\bigg|_{\substack{\text{evaluated at} \\ \text{the retarded} \\ \text{time } t'=t-r/c}}, \tag{4.7}$$

When we write the free parameters $\widetilde{\varepsilon}$ and $\widetilde{\mu}$ for the unrationalized mks system using the symbols "ε_0" and "μ_0" respectively, we know from the discussion in Section 3.6 that the equation must have the same form in the esuq and emuq systems. Consequently, Eq. (4.7) can also be regarded as being written in esuq or emuq units. In fact, all we need to do to put any of the free-parameter expressions in Sections 4.1 or 4.2 into esuq units is to set

$$\Pi \to 4\pi, \quad \widetilde{\varepsilon} \to 1 \frac{\text{statfaradq}}{\text{cm}} = \varepsilon_0, \quad \widetilde{\mu} \to c_{\text{cgs}}^{-2} \frac{\text{stathenryq}}{\text{cm}} = \mu_0, \quad k_0 \to 1;$$

and all we need to do to put any of the free-parameter expressions in Sections 4.1 or 4.2 into emuq units is to set

$$\Pi \to 4\pi, \quad \widetilde{\varepsilon} \to c_{\text{cgs}}^{-2} \frac{\text{abfaradq}}{\text{cm}} = \varepsilon_0, \quad \widetilde{\mu} \to 1 \frac{\text{abhenryq}}{\text{cm}} = \mu_0, \quad k_0 \to 1.$$

To put Eq. (4.5) into Gaussian cgs units, we consult row three of Table 4.2 to get

$$\Pi \to 4\pi, \quad \widetilde{\mu} \to 1, \quad \text{and} \quad k_0 \to c^{-1},$$

which, when substituted into Eq. (4.5), leads to

$$\vec{E}\bigg|_{\substack{\text{evaluated} \\ \text{at time } t}} = \frac{1}{c^2 r} \cdot \left\{ \hat{e}_r \times \frac{d^2\vec{m}_I}{dt^2} \right\}\bigg|_{\substack{\text{evaluated at} \\ \text{the retarded} \\ \text{time } t'=t-r/c}}. \tag{4.8a}$$

Glancing back at the discussion following Eq. (2.68d), we remember that the distinction made between $\vec{m}_I$, $\vec{m}_H$ and $\vec{M}_I$, $\vec{M}_H$ is irrelevant in systems of units in which the magnetic permeability of free space is 1. According to the third entry of row three, the permeability of free space is 1 in the Gaussian cgs system, which means that in this set of units $\vec{m}_I = \vec{m}_H$, allowing us to drop the "I" subscript and write Eq. (4.8a) as

$$\vec{E}\Big|_{\substack{\text{evaluated} \\ \text{at time } t}} = \frac{1}{c^2 r} \cdot \left\{ \hat{e}_r \times \frac{d^2 \vec{m}}{dt^2} \right\}\Big|_{\substack{\text{evaluated at} \\ \text{the retarded} \\ \text{time } t'=t-r/c}} . \qquad (4.8b)$$

In Gaussian cgs units, we are always allowed to drop the "I" and "H" subscripts from the physical quantities $\vec{m}_I$, $\vec{m}_H$ and $\vec{M}_I$, $\vec{M}_H$.

To write Eq. (4.5) in the Heaviside-Lorentz system, we see from row four of Table 4.2 that

$$\Pi \to 1, \quad \widetilde{\mu} \to 1, \quad \text{and} \quad k_0 \to c^{-1}.$$

Substitution of these values into Eq. (4.5) gives

$$\vec{E}\Big|_{\substack{\text{evaluated} \\ \text{at time } t}} = \frac{1}{4\pi c^2 r} \cdot \left\{ \hat{e}_r \times \frac{d^2 \vec{m}_I}{dt^2} \right\}\Big|_{\substack{\text{evaluated at} \\ \text{the retarded} \\ \text{time } t'=t-r/c}} . \qquad (4.9a)$$

Following the advice of the fifth entry of row four, we consult Table 4.3(b) and note that both $\vec{E}$ and $\vec{m}_I$ are rationalized under the Heaviside-Lorentz system, suggesting that Eq. (4.9a) should be written as

$$_h\vec{E}\Big|_{\substack{\text{evaluated} \\ \text{at time } t}} = \frac{1}{4\pi c^2 r} \cdot \left\{ \hat{e}_r \times \frac{d^2{}_h\vec{m}_I}{dt^2} \right\}\Big|_{\substack{\text{evaluated at} \\ \text{the retarded} \\ \text{time } t'=t-r/c}} \qquad (4.9b)$$

to match the usage of the previous chapters. If we want to match the usage of most scientists using the Heaviside-Lorentz system, we omit the "h" prefixes and—since the Heaviside-Lorentz system also has the free parameter $\widetilde{\mu}$ for the permeability of free space set to 1—drop the "I" subscript to get

$$\vec{E}\Big|_{\substack{\text{evaluated} \\ \text{at time } t}} = \frac{1}{4\pi c^2 r} \cdot \left\{ \hat{e}_r \times \frac{d^2 \vec{m}}{dt^2} \right\}\Big|_{\substack{\text{evaluated at} \\ \text{the retarded} \\ \text{time } t'=t-r/c}} , \qquad (4.9c)$$

Just like in Gaussian cgs units, in the standard Heaviside-Lorentz system we are always allowed to drop the "I" and "H" subscripts from the physical quantities $\vec{m}_I$, $\vec{m}_H$ and $\vec{M}_I$, $\vec{M}_H$.

The fifth and sixth rows of Table 4.2 show us how to put Eq. (4.5) into the esu and emu systems of units, respectively. To put the formula into esu units we specify

$$\Pi \to 4\pi, \quad \widetilde{\mu} \to c^{-2}, \quad \text{and} \quad k_0 \to 1$$

to get

$$\vec{E}\Big|_{\substack{\text{evaluated} \\ \text{at time } t}} = \frac{1}{c^3 r} \cdot \left\{ \hat{e}_r \times \frac{d^2 \vec{m}_I}{dt^2} \right\}\Big|_{\substack{\text{evaluated at} \\ \text{the retarded} \\ \text{time } t'=t-r/c}} \quad ; \tag{4.10a}$$

and to put the formula into emu units we specify

$$\Pi \to 4\pi, \quad \tilde{\mu} \to 1, \quad \text{and} \quad k_0 \to 1$$

to get

$$\vec{E}\Big|_{\substack{\text{evaluated} \\ \text{at time } t}} = \frac{1}{cr} \cdot \left\{ \hat{e}_r \times \frac{d^2 \vec{m}_I}{dt^2} \right\}\Big|_{\substack{\text{evaluated at} \\ \text{the retarded} \\ \text{time } t'=t-r/c}} \quad . \tag{4.10b}$$

Note that we can drop the "I" and "H" subscripts from the physical quantities $\vec{m}_I$, $\vec{m}_H$ and $\vec{M}_I$, $\vec{M}_H$ when working in emu units where the free parameter $\tilde{\mu}$ specifying the magnetic permeability of free space is 1, but we cannot do so when working in esu units where it is not. Hence, Eq. (4.10b) can be written as

$$\vec{E}\Big|_{\substack{\text{evaluated} \\ \text{at time } t}} = \frac{1}{cr} \cdot \left\{ \hat{e}_r \times \frac{d^2 \vec{m}}{dt^2} \right\}\Big|_{\substack{\text{evaluated at} \\ \text{the retarded} \\ \text{time } t'=t-r/c}} \quad .$$

Comparing this result to its counterparts in the other five electromagnetic systems—Eqs. (4.6c), (4.7), (4.8b), (4.9c), and (4.10a)—we see that this one is the simplest.

An equation using all four free parameters simultaneously is

$$\nabla^2 \vec{B} - k_0^2 \sigma (\mu_r \tilde{\mu}) \Pi \frac{\partial \vec{B}}{\partial t} - k_0^2 (\mu_r \tilde{\mu})(\varepsilon_r \tilde{\varepsilon}) \frac{\partial^2 \vec{B}}{\partial t^2} = 0. \tag{4.11a}$$

This is the wave equation for a magnetic induction field $\vec{B}$ propagating in a homogeneous medium characterized by a dielectric constant ($\varepsilon_r \tilde{\varepsilon}$), a magnetic permeability ($\mu_r \tilde{\mu}$), and a conductivity σ. The Laplacian operator ∇^2 is explained at the end of Appendix 2B of Chapter 2. Consulting the first row of Table 4.2, we get

$$\nabla^2 \vec{B} - \sigma (\mu_{rf} \mu_0) \frac{\partial \vec{B}}{\partial t} - (\mu_{rf} \mu_0)(\varepsilon_{rf} \varepsilon_0) \frac{\partial^2 \vec{B}}{\partial t^2} = 0 \tag{4.11b}$$

in terms of the rationalized mks notation used in the previous chapters (according to Table 4.3(a), only the permittivity and the magnetic permeability of free space are rationalized quantities). Again, we take c to be the speed of light and use Eq. (3.49e) to write

$$_f\mu_0{}_f\varepsilon_0 = c^{-2}.$$

Now Eq. (4.11b) can be written as (dropping the "f" prefix to follow the usage of most modern textbooks)

$$\nabla^2\vec{B} - \mu_r\mu_0\sigma\frac{\partial\vec{B}}{\partial t} - \left(\frac{\mu_r\varepsilon_r}{c^2}\right)\frac{\partial^2\vec{B}}{\partial t^2} = 0. \tag{4.11c}$$

Consulting the second row of Table 4.2, we see that in the unrationalized mks system the wave equation for the $\vec{B}$ field becomes

$$\nabla^2\vec{B} - 4\pi\sigma\mu_r\mu_0\frac{\partial\vec{B}}{\partial t} - (\mu_r\mu_0)(\varepsilon_r\varepsilon_0)\frac{\partial^2\vec{B}}{\partial t^2} = 0.$$

According to Eq. (3.49e), the product of μ_0 and ε_0 as well as the product of $_f\mu_0$ and $_f\varepsilon_0$ is equal to c^{-2}, which means this equation can be written as

$$\nabla^2\vec{B} - 4\pi\mu_r\mu_0\sigma\frac{\partial\vec{B}}{\partial t} - \left(\frac{\mu_r\varepsilon_r}{c^2}\right)\frac{\partial^2\vec{B}}{\partial t^2} = 0. \tag{4.11d}$$

The form of the wave equation for $\vec{B}$ in Gaussian cgs units comes from the third row of Table 4.2,

$$\nabla^2\vec{B} - \left(\frac{4\pi\sigma\mu_r}{c^2}\right)\frac{\partial\vec{B}}{\partial t} - \left(\frac{\mu_r\varepsilon_r}{c^2}\right)\frac{\partial^2\vec{B}}{\partial t^2} = 0. \tag{4.11e}$$

The fourth row of Table 4.2, combined with Table 4.3(b), shows that in the standard Heaviside-Lorentz system, the wave equation for the $\vec{B}$ field is

$$\nabla^2{}_h\vec{B} - \left(\frac{_h\sigma\mu_r}{c^2}\right)\frac{\partial_h\vec{B}}{\partial t} - \left(\frac{\mu_r\varepsilon_r}{c^2}\right)\frac{\partial^2{}_h\vec{B}}{\partial t^2} = 0. \tag{4.11f}$$

We can always drop the "h" prefixes to put it into the form preferred by most scientists working in the Heaviside-Lorentz system,

$$\nabla^2\vec{B} - \left(\frac{\sigma\mu_r}{c^2}\right)\frac{\partial\vec{B}}{\partial t} - \left(\frac{\mu_r\varepsilon_r}{c^2}\right)\frac{\partial^2\vec{B}}{\partial t^2} = 0. \tag{4.11g}$$

The fifth and sixth rows of Table 4.2 show how to put this wave equation into esu and emu units, respectively. The wave equation for the $\vec{B}$ field in esu units is

$$\nabla^2\vec{B} - \left(\frac{4\pi\sigma\mu_r}{c^2}\right)\frac{\partial\vec{B}}{\partial t} - \left(\frac{\mu_r\varepsilon_r}{c^2}\right)\frac{\partial^2\vec{B}}{\partial t^2} = 0, \tag{4.11h}$$

and the wave equation for the $\vec{B}$ field in emu units is

$$\nabla^2 \vec{B} - 4\pi\sigma\mu_r \frac{\partial \vec{B}}{\partial t} - \left(\frac{\mu_r \varepsilon_r}{c^2}\right) \frac{\partial^2 \vec{B}}{\partial t^2} = 0. \qquad (4.11\text{i})$$

When $\sigma = 0$ and we neglect the "h" and "f" prefixes, all of the different versions of the wave equation reduce to the same formula:

$$\nabla^2 \vec{B} - \left(\frac{\mu_r \varepsilon_r}{c^2}\right) \frac{\partial^2 \vec{B}}{\partial t^2} = 0. \qquad (4.11\text{j})$$

When $\sigma \neq 0$ it is hard to say, looking at all the different coefficients of $\partial \vec{B}/\partial t$, which system of electromagnetic physical quantities of units gives the simplest equation. Depending on whether $(\mu_r \mu_0 \sigma)$, $(\mu_r \sigma / c^2)$, or $(4\pi\mu_r \sigma)$ seems most simple, we can work with Eqs. (4.11c), (4.11g), or (4.11i), respectively.

One final point worth making is that the free parameters in Table 4.2 can be either dimensionless numbers or dimensional physical quantities. The free parameter k_0, for example, is the dimensional physical quantity c^{-1} in the Heaviside-Lorentz system and the dimensionless numeric 1 in the rationalized mks system. Consequently, we cannot use the techniques explained in Chapters 1 through 3 to do a dimensional analysis of equations like Eqs. (4.1c–f), (4.2), (4.5), or (4.11a) because they contain free parameters $\widetilde{\varepsilon}$, $\widetilde{\mu}$, and k_0 which, depending on the electromagnetic system, are dimensionless numerics or dimensional physical quantities. We can, of course, do dimensional analysis on equations and formulas like Eq. (4.1a) that only contain the free parameter Π because Π is a dimensionless numeric on all six rows of Table 4.2.

4.2 BASIC EQUATIONS USING THE FREE PARAMETERS k_0, $\widetilde{\mu}$, $\widetilde{\varepsilon}$, AND Π

Although we could begin this list with Maxwell's equations and the Lorentz force law, using them to derive all the other basic formulas, such a procedure would take up space and bury the most useful results in unnecessary detail. We start instead with the simplest equations, the ones learned first in introductory courses, and then move on—using the absolute minimum number of words and diagrams—to more sophisticated formulations of electromagnetic theory. Consequently, the reader should expect to find elementary equations near the beginning of the list, advanced equations near the end, and intermediate equations somewhere in the middle. Electromagnetic physical quantities and formulas that are separate in some electromagnetic systems can merge into the same physical quantity or formula in other electromagnetic systems, and we try to note when this occurs. We have included in the list even those basic equations and formulas that do not contain any of the free parameters $\widetilde{\varepsilon}$, $\widetilde{\mu}$, k_0, and Π, because it may be important to know that an algebraic expression has the same form in all the electromagnetic systems of Table 4.2.

4.2.1 Coulomb's Law for Electric Charge

The force $\vec{F}_{12}$ exerted by point charge Q_1 on point charge Q_2, with both charges located in a vacuum and no other material objects nearby, is

$$\vec{F}_{12} = \frac{\Pi Q_1 Q_2}{4\pi \widetilde{\varepsilon} r^2} \hat{r}_{12},$$

where r is the distance between the two charges and $\hat{r}_{12}$ is the dimensionless unit vector pointing from Q_1 to Q_2. The values of Π and $\widetilde{\varepsilon}$ come from Table 4.2. Table 4.3(a) shows that in the rationalized mks system $\widetilde{\varepsilon} = \varepsilon_0$ should be given the prefix "f" to match the rationalized mks notation used in previous chapters, and Table 4.3(b) shows that in the Heaviside-Lorentz system Q_1 and Q_2 should be given the prefix "h" to match the Heaviside-Lorentz notation used in previous chapters.

4.2.2 Coulomb's Law for Magnetic Poles

The force $\vec{F}_{12}$ exerted by point magnetic pole $(p_H)_1$ on point magnetic pole $(p_H)_2$, with both poles located in a vacuum and no other material objects nearby, is

$$\vec{F}_{12} = \frac{\Pi (p_H)_1 (p_H)_2}{4\pi \widetilde{\mu} r^2} \hat{r}_{12},$$

where r is the distance between the two poles and $\hat{r}_{12}$ is the dimensionless unit vector pointing from $(p_H)_1$ to $(p_H)_2$. The values of Π and $\widetilde{\mu}$ come from Table 4.2. Table 4.3(a) shows that in the rationalized mks system, $(p_H)_{1,2}$ and $\widetilde{\mu} = \mu_0$ should be given the prefix "f" to match the rationalized mks notation used in the previous chapters; and Table 4.3(b) shows that in the Heaviside-Lorentz system $(p_H)_{1,2}$ should be given the prefix "h" to match the Heaviside-Lorentz notation used in the previous chapters. Magnetic poles always occur in pairs as described in Section 2.1. In Section 2.1 most of the electromagnetic systems listed in Table 4.2 had not yet been introduced, which means that we were always implicitly working in either esu or emu units, depending on the values chosen for ε_0 and μ_0.

4.2.3 The Electric Field of a Point Charge

The electric field $\vec{E}$ of a point charge Q located in a vacuum is

$$\vec{E} = \frac{\Pi Q}{4\pi \widetilde{\varepsilon} r^2} \hat{e}_r,$$

where r is the distance between the point charge Q and the field point where the electric field is being measured, and $\hat{e}_r$ is the dimensionless unit vector pointing from Q to the field point where the electric field is being measured. The values of Π and $\widetilde{\varepsilon}$ come from Table 4.2. Table 4.3(a) shows that in the rationalized mks system, $\widetilde{\varepsilon} = \varepsilon_0$ should be given the prefix "f" to match the rationalized mks notation used in previous chapters; and Table 4.3(b) shows that in the Heaviside-Lorentz system, $\vec{E}$ and Q should be given the prefix "h" to match the Heaviside-Lorentz notation used in previous chapters.

4.2.4 THE MAGNETIC FIELD OF A POINT MAGNETIC POLE

The magnetic field $\vec{H}$ of a point magnetic pole p_H located in a vacuum is

$$\vec{H} = \frac{\Pi p_H}{4\pi \widetilde{\mu} r^2} \hat{e}_r,$$

where r is the distance between the point magnetic pole p_H and the field point where the magnetic field is being measured, and $\hat{e}_r$ is the dimensionless unit vector pointing from p_H to the field point where the magnetic field is being measured. The values of Π and $\widetilde{\mu}$ come from Table 4.2. Table 4.3(a) shows that in the rationalized mks system, $\vec{H}$, p_H, and $\widetilde{\mu} = \mu_0$ should be given the prefix "f" to match the rationalized mks notation used in previous chapters; and Table 4.3(b) shows that in the Heaviside-Lorentz system, $\vec{H}$ and p_H should be given the prefix "h" to match the Heaviside-Lorentz notation used in previous chapters. Magnetic poles always occur in pairs as described in Section 2.1. In Section 2.1, most of the electromagnetic systems listed in Table 4.2 had not yet been introduced, which means that we were always implicitly working in either esu or emu units, depending on the values chosen for ε_0 and μ_0.

4.2.5 THE ELECTRIC POTENTIAL OF A POINT CHARGE

The electric potential V of a point charge Q located in a vacuum is

$$V = \frac{\Pi Q}{4\pi \widetilde{\varepsilon} r},$$

where r is the distance between the point charge Q and the field point where the electric potential is being measured. The potential is taken to be zero at field points infinitely distant from Q, and the values of Π and $\widetilde{\varepsilon}$ come from Table 4.2. Table 4.3(a) shows that in the rationalized mks system, $\widetilde{\varepsilon} = \varepsilon_0$ should be given the prefix "f" to match the rationalized mks notation used in previous chapters; and Table 4.3(b) shows that in the Heaviside-Lorentz system, Q and V should be given the prefix "h" to match the Heaviside-Lorentz notation used in previous chapters.

4.2.6 The Magnetic Potential of a Point Magnetic Pole

The magnetic potential Ω_H of a point magnetic pole p_H located in a vacuum is

$$\Omega_H = \frac{\Pi p_H}{4\pi \widetilde{\mu} r},$$

where r is the distance between p_H and the field point where the magnetic potential Ω_H is being measured. The potential is taken to be zero at field points infinitely distant from p_H, and the values of Π and $\widetilde{\mu}$ come from Table 4.2. Table 4.3(a) shows that in the rationalized mks system, Ω_H, p_H, and $\widetilde{\mu} = \mu_0$ should be given the prefix "f" to match the rationalized mks notation used in previous chapters; and Table 4.3(b) shows that in the Heaviside-Lorentz system, Ω_H and p_H should be given the prefix "h" to match the Heaviside-Lorentz notation used in previous chapters. Magnetic poles always occur in pairs as described in Section 2.1. In Section 2.1, most of the electromagnetic systems listed in Table 4.2 had not yet been introduced, which means that we were always implicitly working in either esu or emu units, depending on the values chosen for ε_0 and μ_0.

4.2.7 The Force on a Charge in an Electric Field

An electric field $\vec{E}$ exerts a force $\vec{F}$ on a charge Q, which is given by the formula

$$\vec{F} = Q\vec{E}.$$

Note that this formula does not contain any of the free parameters $\widetilde{\varepsilon}$, $\widetilde{\mu}$, k_0, or Π, so it has the same form for all the electromagnetic systems listed in Table 4.2. Table 4.3(b) shows that in the Heaviside-Lorentz system, Q and $\vec{E}$ should be given the prefix "h" to match the Heaviside-Lorentz notation used in previous chapters; and Table 4.3(a) shows that no prefixes are to be expected in the rationalized mks system.

4.2.8 The Force on a Magnetic Pole in a Magnetic Field

A magnetic field $\vec{H}$ exerts a force $\vec{F}$ on a magnetic pole p_H, which is given by the formula

$$\vec{F} = p_H \vec{H}.$$

When using this formula it should be remembered that, as pointed out in the discussion following Eq. (2.5), isolated magnetic poles do not exist. Because the formula does not contain any of the free parameters $\widetilde{\varepsilon}$, $\widetilde{\mu}$, k_0, or Π, it has the same form for all the electromagnetic systems listed in Table 4.2. Table 4.3(a) shows that in the rationalized mks system, p_H and $\vec{H}$ should be given the prefix "f" to match the rationalized mks notation used in previous chapters; and Table 4.3(b) shows that in the Heaviside-Lorentz system, p_H and $\vec{H}$ should be given the prefix "h" to match the Heaviside-Lorentz notation used in previous chapters.

4.2.9 THE POTENTIAL ENERGY OF A CHARGE IN AN ELECTRIC POTENTIAL FIELD

At a point where there is an electric potential V, a point charge Q has associated with the electric potential a potential energy U, given by the formula

$$U = QV.$$

This formula does not contain any of the free parameters $\tilde{\varepsilon}$, $\tilde{\mu}$, k_0, or Π, so it has the same form in all the electromagnetic systems listed in Table 4.2. Table 4.3(b) shows that in the Heaviside-Lorentz system, both Q and V should be given the prefix "h" to match the Heaviside-Lorentz notation used in previous chapters; and Table 4.3(a) shows that no prefixes are to be expected in the rationalized mks system.

4.2.10 THE POTENTIAL ENERGY OF A MAGNETIC POLE IN A MAGNETIC POTENTIAL FIELD

At a point where there is an magnetic potential Ω_H, a point magnetic pole p_H has associated with the magnetic potential a potential energy U, given by the formula

$$U = p_H \Omega_H.$$

When using this formula it should be remembered that, as pointed out in the discussion following Eq. (2.5), isolated magnetic poles do not exist. The formula does not contain any of the free parameters $\tilde{\varepsilon}$, $\tilde{\mu}$, k_0, or Π, so it has the same form in all the electromagnetic systems listed in Table 4.2. Table 4.3(a) shows that in the rationalized mks system, both p_H and Ω_H should be given the prefix "f" to match the rationalized mks notation used in previous chapters; and Table '4.3(b) shows that in the Heaviside-Lorentz system, both p_H and Ω_H should be given the prefix "h" to match the Heaviside-Lorentz notation used in previous chapters.

4.2.11 THE FORMULA FOR AN ELECTRIC DIPOLE

The formula for an electric dipole $\vec{p}$ is

$$\vec{p} = |Q| L \hat{e},$$

where L is a small distance separating two point charges that have charges of equal and opposite magnitude, $+|Q|$ and $-|Q|$, and $\hat{e}$ is a dimensionless unit vector pointing from the $-|Q|$ point charge to the $+|Q|$ point charge. We can construct a point electric dipole by taking the limit as $|Q| \to \infty$ and $L \to 0$ in such a way as to keep the $(|Q| \cdot L)$ product constant. This formula does not contain any of the free parameters $\widetilde{\varepsilon}$, $\widetilde{\mu}$, k_0, or Π, which means it has the same form in all the electromagnetic systems listed in Table 4.2. Table 4.3(b) shows that in the Heaviside-Lorentz system both $\vec{p}$ and Q should be given the prefix "h" to match the Heaviside-Lorentz notation used in previous chapters; and Table 4.3(a) shows that no prefixes are to be expected in the rationalized mks system.

4.2.12 THE FORMULA FOR A PERMANENT-MAGNET DIPOLE

The formula for a permanent-magnet dipole $\vec{m}_H$ is

$$\vec{m}_H = |p_H|L\hat{e},$$

where L is a small distance separating two point magnetic poles that have pole strengths of equal and opposite magnitude, $+|p_H|$ and $-|p_H|$, and $\hat{e}$ is a dimensionless unit vector pointing from the $-|p_H|$ point magnetic pole to the $+|p_H|$ point magnetic pole. We can construct a point permanent-magnet dipole by taking the limit as $|p_H| \to \infty$ and $L \to 0$ in such a way as to keep the $(|p_H| \cdot L)$ product constant. The formula for $\vec{m}_H$ does not contain any of the free parameters $\widetilde{\varepsilon}$, $\widetilde{\mu}$, k_0, or Π, so it has the same form in all of the electromagnetic systems listed in Table 4.2. Table 4.3(a) shows that in the rationalized mks system, both $\vec{m}_H$ and p_H should be given the prefix "f" to match the rationalized mks notation used in previous chapters; and Table 4.3(b) shows that in the Heaviside-Lorentz system, $\vec{m}_H$ and p_H should be given the prefix "h" to match the Heaviside-Lorentz notation used in previous chapters. The formula for the current-loop magnetic dipole given in Section 4.2.13 below shows that for those electromagnetic systems in Table 4.2 where $\widetilde{\mu} = 1$, a permanent-magnet dipole is the same thing as a current-loop magnetic dipole.

4.2.13 THE RELATIONSHIP BETWEEN THE CURRENT-LOOP MAGNETIC DIPOLE MOMENT AND THE PERMANENT-MAGNET DIPOLE MOMENT

The current-loop magnetic dipole moment $\vec{m}_I$ and the permanent-magnet dipole moment $\vec{m}_H$ of the same system are always proportional to each other, with

$$\vec{m}_I = \vec{m}_H/\widetilde{\mu} \quad \text{or} \quad \vec{m}_H = \widetilde{\mu}\vec{m}_I.$$

The value of $\widetilde{\mu}$ comes from Table 4.2. Note that for the electromagnetic systems of Table 4.2 where $\widetilde{\mu} = 1$, a current-loop magnetic dipole moment is the same thing as a permanent-magnet dipole moment. Table 4.3(a) shows that in the rationalized mks system, both $\vec{m}_H$ and $\tilde{\mu} = \mu_0$ should be given the prefix "f" to match the rationalized mks notation used in previous chapters; and Table 4.3(b) shows that in the Heaviside-Lorentz system, $\vec{m}_I$ and $\vec{m}_H$ should be given the prefix "h" to match the Heaviside-Lorentz notation used in previous chapters.

4.2.14 THE RELATIONSHIP BETWEEN THE ELECTRIC-DIPOLE DENSITY AND THE ELECTRIC DIPOLE MOMENT

The electric dipole moment $\vec{p}$ at a point in space where there is an electric dipole density $\vec{P}$ is

$$\vec{p} = \vec{P} \cdot dV,$$

where dV is an infinitesimal volume surrounding the position of the electric dipole moment. The electric dipole moment can be thought of as a point electric dipole $\vec{p}$ at the position of the infinitesimal volume dV. This formula does not contain any of the free parameters $\tilde{\varepsilon}$, $\tilde{\mu}$, k_0, or Π, so it has the same form in all the electromagnetic systems listed in Table 4.2. Table 4.3(b) shows that in the Heaviside-Lorentz system, both $\vec{p}$ and $\vec{P}$ should be given the prefix "h" to match the Heaviside-Lorentz notation used in previous chapters; and Table 4.3(a) shows that no prefixes are to be expected in the rationalized mks system.

4.2.15 THE RELATIONSHIP BETWEEN THE PERMANENT-MAGNET DIPOLE DENSITY AND THE PERMANENT-MAGNET DIPOLE MOMENT

The permanent-magnet dipole moment $\vec{m}_H$ at a point in space where there is a permanent-magnet dipole density $\vec{M}_H$ is

$$\vec{m}_H = \vec{M}_H \cdot dV,$$

where dV is an infinitesimal volume surrounding the position of the permanent-magnet dipole moment. The permanent-magnet dipole moment can be thought of as a point permanent-magnet dipole $\vec{m}_H$ at the position of the infinitesimal volume dV. This formula does not contain any of the free parameters $\tilde{\varepsilon}$, $\tilde{\mu}$, k_0, or Π, which means it has the same form in all of the electromagnetic systems listed in Table 4.2. The formulas given in Sections 4.2.13 and 4.2.17 show that for those electromagnetic systems in Table 4.2 where $\tilde{\mu} = 1$, there is no difference between a permanent-magnet and current-loop magnetic dipole and no difference between a permanent-magnet and current-loop magnetic dipole density. Table 4.3(a) shows that in the rationalized mks system, both $\vec{m}_H$ and $\vec{M}_H$ should be given an "f" prefix to match the rationalized mks notation used in previous chapters; and Table 4.3(b) shows that in the Heaviside-Lorentz system, $\vec{m}_H$ and $\vec{M}_H$ should be given an "h" prefix to match the Heaviside-Lorentz notation used in previous chapters.

4.2.16 THE RELATIONSHIP BETWEEN THE CURRENT-LOOP MAGNETIC DIPOLE DENSITY AND THE CURRENT-LOOP MAGNETIC DIPOLE MOMENT

The current-loop magnetic dipole moment $\vec{m}_I$ at a point in space where there is a current-loop magnetic dipole density $\vec{M}_I$ is

$$\vec{m}_I = \vec{M}_I \cdot dV,$$

where dV is an infinitesimal volume surrounding the position of the current-loop magnetic dipole moment. The current-loop magnetic dipole moment can be thought of as a point current-loop magnetic dipole $\vec{m}_I$ at the position of the infinitesimal volume dV. This formula does not contain any of the free parameters $\widetilde{\varepsilon}$, $\widetilde{\mu}$, k_0, or Π, so it has the same form in all of the electromagnetic systems listed in Table 4.2. Table 4.3(b) shows that in the Heaviside-Lorentz system, $\vec{m}_I$ and $\vec{M}_I$ should be given an "h" prefix to match the Heaviside-Lorentz notation used in the previous chapters; and Table 4.3(a) shows that no prefixes are to be expected in the rationalized mks system.

4.2.17 THE RELATIONSHIP BETWEEN THE CURRENT-LOOP MAGNETIC DIPOLE DENSITY AND THE PERMANENT-MAGNET DIPOLE DENSITY

At every point in space, the current-loop magnetic dipole density $\vec{M}_I$ and the permanent-magnet dipole density $\vec{M}_H$ are proportional to each other, obeying the relationship

$$\vec{M}_I = \vec{M}_H / \widetilde{\mu} \quad \text{or} \quad \vec{M}_H = \widetilde{\mu} \vec{M}_I.$$

Note that for all of the electromagnetic systems of Table 4.2 where $\widetilde{\mu} = 1$, the two types of magnetic dipole density are the same. Table 4.3(a) shows that in the rationalized mks system, both $\vec{M}_H$ and $\widetilde{\mu} = \mu_0$ should be given an "f" prefix to match the rationalized mks notation used in previous chapters; and Table 4.3(b) shows that in the Heaviside-Lorentz system both $\vec{M}_I$ and $\vec{M}_H$ should be given an "h" prefix to match the Heaviside-Lorentz notation used in previous chapters.

4.2.18 THE TORQUE ON AN ELECTRIC DIPOLE IN AN ELECTRIC FIELD

The torque $\vec{T}$ experienced by a small electric dipole $\vec{p}$ located at a point where there is an external electric field $\vec{E}$ is

$$\vec{T} = \vec{p} \times \vec{E}.$$

This formula does not contain the free parameters $\widetilde{\varepsilon}$, $\widetilde{\mu}$, k_0, or Π, so it has the same form in all the electromagnetic systems of Table 4.2. When this formula is used in the Heaviside-Lorentz system, Table 4.3(b) shows that both $\vec{p}$ and $\vec{E}$ should be given the prefix "h" to match the Heaviside-Lorentz notation used in previous chapters; and Table 4.3(a) shows that no prefixes are to be expected in the rationalized mks system.

4.2.19 THE TORQUE ON A MAGNETIC DIPOLE IN A MAGNETIC FIELD

The torque $\vec{T}$ experienced by a small current-loop magnetic dipole $\vec{m}_I$ located at a point where there is an external magnetic field $\vec{H}$ is

$$\vec{T} = (\widetilde{\mu}\vec{m}_I) \times \vec{H}.$$

The torque $\vec{T}$ experienced by a small permanent-magnet dipole $\vec{m}_H$ located at a point where there is an external magnetic field $\vec{H}$ is

$$\vec{T} = \vec{m}_H \times \vec{H}.$$

The value of $\widetilde{\mu}$ in the first formula comes from Table 4.2; and since the second formula contains none of the free parameters $\widetilde{\varepsilon}, \widetilde{\mu}, k_0$, or Π, it has the same form in all of the electromagnetic systems of Table 4.2. The two formulas become identical in electromagnetic systems where $\widetilde{\mu} = 1$ because, according to Section 4.2.13, $\vec{m}_I = \vec{m}_H$ when $\widetilde{\mu} = 1$. When these formulas are used in the rationalized mks system, Table 4.3(a) shows that $\vec{H}, \vec{m}_H$, and $\widetilde{\mu} = \mu_0$ should all be given the prefix "f" to match the rationalized mks notation used in the previous chapters; and when these formulas are used in the Heaviside-Lorentz system, Table 4.3(b) shows that $\vec{H}, \vec{m}_I$, and $\vec{m}_H$ should all be given the prefix "h" to match the Heaviside-Lorentz notation used in the previous chapters.

4.2.20 THE POTENTIAL ENERGY OF AN ELECTRIC DIPOLE IN AN ELECTRIC FIELD

The potential energy U associated with the orientation of a small electric dipole $\vec{p}$ located at a point where there is an external electric field $\vec{E}$ is

$$U = -\vec{p} \cdot \vec{E}.$$

This formula does not contain any of the free parameters $\widetilde{\varepsilon}, \widetilde{\mu}, k_0$, or Π, which means it has the same form in all the electromagnetic systems listed in Table 4.2. Table 4.3(b) shows that in the Heaviside-Lorentz system, $\vec{p}$ and $\vec{E}$ should be given an "h" prefix to match the Heaviside-Lorentz notation used in the previous chapters; and Table 4.3(a) shows that no prefixes are to be expected in the rationalized mks system.

4.2.21 THE POTENTIAL ENERGY OF A MAGNETIC DIPOLE IN A MAGNETIC FIELD

The potential energy U associated with the orientation of a small current-loop magnetic dipole $\vec{m}_I$ located at a point where there is an external magnetic field $\vec{H}$ is

$$U = -(\widetilde{\mu}\vec{m}_I) \cdot \vec{H}.$$

The potential energy associated with the orientation of a small permanent-magnet dipole $\vec{m}_H$ located at a point where there is an external magnetic field $\vec{H}$ is

$$U = -\vec{m}_H \cdot \vec{H}.$$

The value of $\widetilde{\mu}$ in the first formula comes from Table 4.2; and the second formula has the same form in all the electromagnetic systems of Table 4.2 because it contains none of the free parameters $\widetilde{\varepsilon}$, $\widetilde{\mu}$, k_0, or Π. The two formulas become identical in electromagnetic systems where $\widetilde{\mu} = 1$, because $\vec{m}_I = \vec{m}_H$ when $\widetilde{\mu} = 1$ according to Section 4.2.13. When the formulas are used in the rationalized mks system, Table 4.3(a) shows that $\vec{H}, \vec{m}_H$, and $\widetilde{\mu} = \mu_0$ should be given the prefix "f" to match the rationalized mks notation used in the previous chapters; and when the formulas are used in the Heaviside-Lorentz system, Table 4.3(b) shows that $\vec{H}, \vec{m}_I$, and $\vec{m}_H$ should be given the prefix "h" to match the Heaviside-Lorentz notation used in the previous chapters.

4.2.22 THE ELECTRIC POTENTIAL FIELD OF AN ELECTRIC DIPOLE

The electric potential V generated by a small electric dipole $\vec{p}$ located in a vacuum is

$$V = \left(\frac{\Pi}{4\pi\widetilde{\varepsilon}}\right) \cdot \left(\frac{\vec{p} \cdot \hat{e}_r}{r^2}\right),$$

where r is the distance between the electric dipole and the field point at which V is being measured, and $\hat{e}_r$ is the dimensionless unit vector pointing from the dipole to the field point. The values of Π and $\widetilde{\varepsilon}$ come from Table 4.2. When this formula is used in the rationalized mks system, Table 4.3(a) shows that $\widetilde{\varepsilon} = \varepsilon_0$ should be given the prefix "f" to match the rationalized mks notation used in previous chapters; and when it is used in the Heaviside-Lorentz system, Table 4.3(b) shows that $\vec{p}$ and V should be given the prefix "h" to match the Heaviside-Lorentz notation used in the previous chapters.

4.2.23 THE MAGNETIC POTENTIAL FIELD OF A MAGNETIC DIPOLE

The magnetic potential Ω_H generated by a small current-loop magnetic dipole $\vec{m}_I$ located in a vacuum is

$$\Omega_H = \left(\frac{\Pi}{4\pi}\right) \cdot \left(\frac{\vec{m}_I \cdot \hat{e}_r}{r^2}\right),$$

where r is the distance between the dipole and the field point at which Ω_H is being evaluated, and $\hat{e}_r$ is the dimensionless unit vector pointing from the dipole to the field point. The magnetic potential generated by a small permanent-magnet dipole $\vec{m}_H$ located in a vacuum is

$$\Omega_H = \left(\frac{\Pi}{4\pi\widetilde{\mu}}\right) \cdot \left(\frac{\vec{m}_H \cdot \hat{e}_r}{r^2}\right),$$

where again r is the distance between the dipole and the field point, and $\hat{e}_r$ is the dimensionless unit vector pointing from the dipole to the field point. The values of Π and $\widetilde{\mu}$ come from Table 4.2. These two formulas become identical for those electromagnetic systems with $\widetilde{\mu} = 1$, because $\vec{m}_H = \vec{m}_I$ when $\widetilde{\mu} = 1$ according to Section 4.2.13. When these two formulas are used in the rationalized mks system, Table 4.3(a) shows that $\vec{m}_H$, Ω_H, and $\widetilde{\mu} = \mu_0$ should be given the prefix "f" to match the rationalized mks notation used in previous chapters; and when they are used in the Heaviside-Lorentz system, Table 4.3(b) shows that $\vec{m}_I$, $\vec{m}_H$, and Ω_H should be given the prefix "h" to match the Heaviside-Lorentz notation used in previous chapters.

4.2.24 THE ELECTRIC FIELD OF AN ELECTRIC DIPOLE

The electric field $\vec{E}$ generated by a small electric dipole $\vec{p}$ located in a vacuum is

$$\vec{E} = \left(\frac{\Pi}{4\pi\widetilde{\varepsilon}r^3}\right) \cdot \left[3\hat{e}_r\left(\hat{e}_r \cdot \vec{p}\right) - \vec{p}\right],$$

where r is the distance between the electric dipole and the position of the field point at which the electric field is measured, and $\hat{e}_r$ is the dimensionless unit vector pointing from the dipole to the field point. The values of Π and $\widetilde{\varepsilon}$ come from Table 4.2. When this equation is used in the rationalized mks system, Table 4.3(a) shows that $\widetilde{\varepsilon} = \varepsilon_0$ should have a prefix "f" to match the rationalized mks notation used in previous chapters; and when the equation is used in the Heaviside-Lorentz system, Table 4.3(b) shows that $\vec{p}$ and $\vec{E}$ should have a prefix "h" to match the Heaviside-Lorentz notation used in previous chapters.

4.2.25 THE MAGNETIC FIELD OF A MAGNETIC DIPOLE

The magnetic field $\vec{H}$ generated by a small current-loop magnetic dipole $\vec{m}_I$ located in a vacuum is

$$\vec{H} = \left(\frac{\Pi}{4\pi r^3}\right) \cdot \left[3\hat{e}_r\left(\hat{e}_r \cdot \vec{m}_I\right) - \vec{m}_I\right],$$

where r is the distance between the dipole and the position of the field point at which the magnetic field is evaluated, and $\hat{e}_r$ is the dimensionless unit vector pointing from the dipole to the field point. The magnetic field $\vec{H}$ generated by a small permanent-magnet dipole $\vec{m}_H$ is

$$\vec{H} = \left(\frac{\Pi}{4\pi \tilde{\mu} r^3}\right) \cdot \left[3\hat{e}_r\left(\hat{e}_r \cdot \vec{m}_H\right) - \vec{m}_H\right],$$

where again r is the distance between the dipole and the position of the field point and $\hat{e}_r$ is the dimensionless unit vector pointing from the dipole to the field point. The values of Π and $\tilde{\mu}$ come from Table 4.2. The two formulas are identical in those electromagnetic systems where $\tilde{\mu} = 1$ because, according to Section 4.2.13, $\vec{m}_H = \vec{m}_I$ when $\tilde{\mu} = 1$. When these equations are used in the rationalized mks system, Table 4.3(a) shows that $\vec{m}_H$, $\vec{H}$, and $\tilde{\mu} = \mu_0$ should be given the prefix "f" to match the rationalized mks notation used in the previous chapters; and when they are used in the Heaviside-Lorentz system, Table 4.3(b) shows that $\vec{m}_I$, $\vec{m}_H$, and $\vec{H}$ should be given the prefix "h" to match the Heaviside-Lorentz notation used in the previous chapters.

4.2.26 THE RELATIONSHIP BETWEEN THE ELECTRIC DISPLACEMENT, ELECTRIC FIELD, AND THE ELECTRIC DIPOLE DENSITY

The electric displacement field $\vec{D}$ is

$$\vec{D} = \tilde{\varepsilon}\vec{E} + \Pi\vec{P},$$

where $\vec{E}$ is the electric field and $\vec{P}$ is the electric dipole density. The values of $\tilde{\varepsilon}$ and Π come from Table 4.2. Note that when $\vec{P} = 0$, such as in a vacuum, the electric displacement $\vec{D}$ and the electric field $\vec{E}$ are the same field in those electromagnetic systems where $\tilde{\varepsilon} = 1$. When this formula is used in the rationalized mks system, Table 4.3(a) shows that $\vec{D}$ and $\tilde{\varepsilon} = \varepsilon_0$ should be given the prefix "f" to match the rationalized mks notation used in previous chapters; and when it is used in the Heaviside-Lorentz system, Table 4.3(b) shows that $\vec{E}$, $\vec{P}$, and $\vec{D}$ should be given the prefix "h" to match the Heaviside-Lorentz notation used in previous chapters.

4.2.27 THE RELATIONSHIP BETWEEN THE MAGNETIC INDUCTION,
        MAGNETIC FIELD, AND PERMANENT-MAGNET DIPOLE DENSITY

The magnetic induction field $\vec{B}$ satisfies the relationship

$$\vec{B} = \widetilde{\mu}\vec{H} + \Pi\vec{M}_H \quad \text{or} \quad \vec{H} = \frac{1}{\widetilde{\mu}}(\vec{B} - \Pi\vec{M}_H),$$

where $\vec{H}$ is the magnetic field and $\vec{M}_H$ is the permanent-magnet dipole density.
The values of Π and $\widetilde{\mu}$ come from Table 4.2. When these equations are used in the
rationalized mks system, Table 4.3(a) shows that $\vec{H}$, $\vec{M}_H$, and $\widetilde{\mu} = \mu_0$ should have
a prefix "f" to match the rationalized mks notation used in the previous chapters;
and when these equations are used in the Heaviside-Lorentz system, Table 4.3(b)
shows that $\vec{B}$, $\vec{M}_H$, and $\vec{H}$ should have a prefix "h" to match the Heaviside-Lorentz
notation used in the previous chapters. Note that when $\vec{M}_H = 0$, such as in a vacuum, the $\vec{B}$ and $\vec{H}$ fields are identical in those electromagnetic systems that have
$\widetilde{\mu} = 1$. We also note that $\vec{M}_H$ and $\vec{M}_I$ become equal in Section 4.2.17 when $\widetilde{\mu}$ is 1,
which means the equations in Sections 4.2.27 and 4.2.28 become the same in the
electromagnetic systems of Table 4.2 where $\widetilde{\mu} = 1$.

4.2.28 THE RELATIONSHIP BETWEEN THE MAGNETIC INDUCTION,
        MAGNETIC FIELD, AND THE CURRENT-LOOP MAGNETIC DIPOLE
        DENSITY

The magnetic induction field $\vec{B}$ satisfies the relationship

$$\vec{B} = \widetilde{\mu}(\vec{H} + \Pi\vec{M}_I) \quad \text{or} \quad \vec{H} = \frac{1}{\widetilde{\mu}}\vec{B} - \Pi\vec{M}_I,$$

where $\vec{H}$ is the magnetic field and $\vec{M}_I$ is the current-loop magnetic dipole density.
The values of Π and $\widetilde{\mu}$ come from Table 4.2. When these equations are used in
the rationalized mks system, Table 4.3(a) shows that $\vec{H}$, and $\widetilde{\mu} = \mu_0$ should have
a prefix "f" to match the rationalized mks notation used in the previous chapters;
and when these equations are used in the Heaviside-Lorentz system, Table 4.3(b)
shows that $\vec{H}$, $\vec{M}_I$, and $\vec{B}$ should have a prefix "h" to match the Heaviside-Lorentz
notation used in the previous chapters. Note that when $\vec{M}_I = 0$, such as in a vacuum, the $\vec{B}$ and $\vec{H}$ fields are identical in those electromagnetic systems where
$\widetilde{\mu} = 1$. We also note that $\vec{M}_H$ and $\vec{M}_I$ become equal in Section 4.2.17 when $\widetilde{\mu}$ is 1,
which means the equations in Sections 4.2.27 and 4.2.28 become the same in the
electromagnetic systems of Table 4.2 where $\widetilde{\mu} = 1$.

4.2.29 THE LORENTZ FORCE LAW

The force $\vec{F}$ experienced by a point charge Q moving at a velocity $\vec{v}$ at a location where there is an electric field $\vec{E}$ and a magnetic induction field $\vec{B}$ is

$$\vec{F} = Q\vec{E} + k_0 Q(\vec{v} \times \vec{B}).$$

The value of k_0 comes from Table 4.2. When this equation is written in the Heaviside-Lorentz system, Table 4.3(b) shows that Q, $\vec{E}$, and $\vec{B}$ should be given the prefix "h" to match the Heaviside-Lorentz notation used in the previous chapters; and Table 4.3(a) shows that no prefixes are to be expected when the equation is written in the rationalized mks system.

4.2.30 THE SURFACE CONDITIONS ON THE ELECTRIC FIELD AND THE ELECTRIC DISPLACEMENT ACROSS A BOUNDARY

Figure 4.1 shows an electric field $\vec{E}$ and an electric displacement field $\vec{D}$ crossing a boundary between medium 1 and medium 2. The dimensionless unit normal vector $\hat{n}$ of the boundary surface points from medium 2 to medium 1. The electric field and electric displacement field in medium 1 are called $\vec{E}_1$ and $\vec{D}_1$, respectively; and the electric field and electric displacement field in medium 2 are called $\vec{E}_2$ and $\vec{D}_2$, respectively. The tangential component of the electric field is always continuous across the boundary:

$$(\vec{E}_2 - \vec{E}_1) \times \hat{n}\bigg|_{\substack{\text{at any point} \\ \text{on the boundary}}} = 0;$$

and if there is a surface charge density S_Q at the boundary, then the normal component of the electric displacement field is discontinuous across the boundary;

$$(\vec{D}_1 - \vec{D}_2) \cdot \hat{n}\bigg|_{\substack{\text{at any point} \\ \text{on the boundary}}} = \Pi S_Q.$$

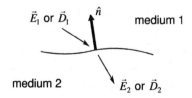

Figure 4.1 The tangential component of the E field is always continuous across the surface between medium 1 and medium 2 and the normal component of the D field may be discontinuous.

The value of Π in the second equation comes from Table 4.2; and since the first equation contains none of the free parameters $\widetilde{\varepsilon}$, $\widetilde{\mu}$, k_0, or Π, it has the same form in all the electromagnetic systems of Table 4.2. When the first equation is written in the rationalized mks system, there is no reason to expect a prefix "f," but when the second equation is written in rationalized mks, Table 4.3(a) shows that $\vec{D}_1$ and $\vec{D}_2$ should be given the prefix "f" to match the rationalized mks notation used in previous chapters. When either the first or second equations are written in the Heaviside-Lorentz system, Table 4.3(b) shows that $\vec{E}_1$, $\vec{E}_2$, $\vec{D}_1$, $\vec{D}_2$, and S_Q should be given a prefix "h" to match the Heaviside-Lorentz notation used in the previous chapters.

4.2.31 THE SURFACE CONDITIONS ON THE MAGNETIC FIELD AND THE MAGNETIC INDUCTION ACROSS A BOUNDARY

Figure 4.2 shows a magnetic field $\vec{H}$ and a magnetic induction field $\vec{B}$ crossing a boundary between medium 1 and medium 2. The dimensionless unit normal vector $\hat{n}$ of the boundary surface points from medium 2 to medium 1. The magnetic field and magnetic induction field in medium 1 are called $\vec{H}_1$ and $\vec{B}_1$, respectively; and the magnetic field and magnetic induction field in medium 2 are called $\vec{H}_2$ and $\vec{B}_2$, respectively. The tangential component of the magnetic field across the boundary is discontinuous if there is a surface current density $\vec{\mathcal{J}}_S$:

$$(\vec{H}_2 - \vec{H}_1) \times \hat{n} \Big|_{\substack{\text{at any point} \\ \text{on the boundary}}} = k_0 \Pi \vec{\mathcal{J}}_S;$$

and the normal component of the magnetic induction is always continuous across the boundary:

$$(\vec{B}_1 - \vec{B}_2) \cdot \hat{n} \Big|_{\substack{\text{at any point} \\ \text{on the boundary}}} = 0.$$

The surface current density $\vec{\mathcal{J}}_S$ points out of the page in Fig. 4.2. The values of k_0 and Π in the first equation come from Table 4.2; and since the second equation contains none of the free parameters $\widetilde{\varepsilon}$, $\widetilde{\mu}$, k_0, or Π, it has the same form in all the electromagnetic systems of Table 4.2. When the first equation is written in the rationalized mks system, Table 4.3(a) shows that $\vec{H}$ should be given the prefix "f" to match the rationalized mks notation used in the previous chapters; but when the second equation is written in rationalized mks, there is no reason to expect prefixes. When the first and second equations are written in the Heaviside-Lorentz system, Table 4.3(b) shows that $\vec{H}_1$, $\vec{H}_2$, $\vec{B}_1$, $\vec{B}_2$, and $\vec{\mathcal{J}}_S$ should be given the prefix "h" to match the Heaviside-Lorentz notation used in the previous chapters.

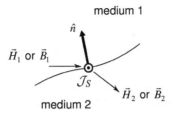

Figure 4.2 The normal component of the B field is always continuous across the surface between medium 1 and medium 2 and the tangential component of the H field may be discontinuous. The surface current $\vec{\mathcal{J}}_S$ points out of the page.

4.2.32 The Torque on a Magnetic Dipole in a Magnetic Induction Field

The torque $\vec{T}$ experienced by a small current-loop magnetic dipole $\vec{m}_I$ located at a point where there is an external magnetic induction field $\vec{B}$ is

$$\vec{T} = \vec{m}_I \times \vec{B},$$

and the torque experienced by a small permanent-magnet dipole $\vec{m}_H$ located at a point where there is an external magnetic induction field $\vec{B}$ is

$$\vec{T} = \left(\frac{1}{\widetilde{\mu}} \vec{m}_H \right) \times \vec{B}.$$

The value of $\widetilde{\mu}$ in the second formula comes from Table 4.2; the first formula does not contain any of the free parameters $\widetilde{\varepsilon}$, $\widetilde{\mu}$, k_0, or Π, which means it has the same form in all of the electromagnetic systems listed in Table 4.2. These two equations become identical when written in the electromagnetic systems of Table 4.2 for which $\widetilde{\mu} = 1$, because in these systems $\vec{m}_H = \vec{m}_I$ (see Section 4.2.13 above). When these equations are written in the Heaviside-Lorentz system, Table 4.3(b) shows that $\vec{m}_I$, $\vec{m}_H$, and $\vec{B}$ should be given the prefix "h" to match the Heaviside-Lorentz notation used in the previous chapters. When the second equation is written in the rationalized mks system, Table 4.3(a) shows that $\vec{m}_H$ and $\widetilde{\mu} = \mu_0$ should both be given the prefix "f" to match the rationalized mks notation used in the previous chapters. [Table 4.3(a) also shows that no prefixes are to be expected when the first equation is put into rationalized mks.]

4.2.33 The Potential Energy of a Magnetic Dipole in a Magnetic Induction Field

The potential energy U associated with the orientation of a small current-loop magnetic dipole $\vec{m}_I$ located at a point where there is an external magnetic induction field $\vec{B}$ is

$$U = -\vec{m}_I \cdot \vec{B},$$

and the potential energy U associated with the orientation of a small permanent-magnet dipole $\vec{m}_H$ located at a point where there is an external magnetic induction field $\vec{B}$ is

$$U = -\left(\frac{1}{\tilde{\mu}}\vec{m}_H\right) \cdot \vec{B}.$$

The value of $\tilde{\mu}$ in the second formula comes from Table 4.2; the first formula does not contain any of the free parameters $\tilde{\varepsilon}$, $\tilde{\mu}$, k_0, or Π, so it has the same form in all the electromagnetic systems listed in Table 4.2. The two equations become identical in the electromagnetic systems of Table 4.2 for which $\tilde{\mu} = 1$, because in these systems $\vec{m}_H = \vec{m}_I$ (see Section 4.2.13 above). When these equations are put into the Heaviside-Lorentz system, Table 4.3(b) shows that $\vec{m}_I$, $\vec{m}_H$, and $\vec{B}$ should be given the prefix "h" to match the Heaviside-Lorentz notation used in the previous chapters. When the second equation is put into the rationalized mks system, Table 4.3(a) shows that $\vec{m}_H$ and $\tilde{\mu} = \mu_0$ should both be given the prefix "f" to match the rationalized mks notation used in the previous chapters. [Table 4.3(a) also shows that no prefixes are to be expected when the first equation is put into rationalized mks.]

4.2.34 THE MAGNETIC INDUCTION FIELD OF A MAGNETIC DIPOLE

The magnetic induction field $\vec{B}$ generated by a small current-loop magnetic dipole located in a vacuum is

$$\vec{B} = \left(\frac{\Pi\tilde{\mu}}{4\pi r^3}\right) \cdot \left[3\hat{e}_r\left(\hat{e}_r \cdot \vec{m}_I\right) - \vec{m}_I\right],$$

and the magnetic induction field $\vec{B}$ generated by a small permanent-magnet dipole located in a vacuum is

$$\vec{B} = \left(\frac{\Pi}{4\pi r^3}\right) \cdot \left[3\hat{e}_r\left(\hat{e}_r \cdot \vec{m}_H\right) - \vec{m}_H\right].$$

In both equations, r is the distance between the dipole and the field point where $\vec{B}$ is evaluated, and $\hat{e}_r$ is the dimensionless unit vector pointing from the dipole to the field point. The values of $\tilde{\mu}$ and Π come from Table 4.2. These two equations become identical when written in the electromagnetic systems of Table 4.2 which have $\tilde{\mu} = 1$, because in these systems $\vec{m}_H = \vec{m}_I$ (see Section 4.2.13 above). When these two equations are written in the rationalized mks system, Table 4.3(a) shows that $\vec{m}_H$ and $\widetilde{\mu} = \mu_0$ should be given the prefix "f" to match the rationalized mks notation used in the previous chapters. When they are written in the Heaviside-Lorentz system, Table 4.3(b) shows that $\vec{m}_I$, $\vec{m}_H$, and $\vec{B}$ should be given the prefix "h" to match the Heaviside-Lorentz notation used in the previous chapters.

4.2.35 THE RELATIONSHIP BETWEEN THE MAGNETIC INDUCTION AND THE MAGNETIC VECTOR POTENTIAL

The magnetic induction $\vec{B}$ can always be written as

$$\vec{B} = \vec{\nabla} \times \vec{A}$$

for a vector field $\vec{A}$ called the magnetic vector potential. This formula does not contain any of the free parameters $\widetilde{\varepsilon}$, $\widetilde{\mu}$, k_0, or Π, which means it has the same form in all the electromagnetic systems listed in Table 4.2. When this formula is used in the rationalized mks system, Table 4.3(a) shows that no prefixes are to be expected; and when it is used in the Heaviside-Lorentz system, Table 4.3(b) shows that $\vec{A}$ and $\vec{B}$ should be given the prefix "h" to match the Heaviside-Lorentz notation used in previous chapters.

4.2.36 THE RELATIONSHIP BETWEEN THE ELECTRIC FIELD, ELECTRIC POTENTIAL, AND THE MAGNETIC VECTOR POTENTIAL

The electric field $\vec{E}$ can always be written as

$$\vec{E} = -\vec{\nabla}V - k_0 \frac{\partial \vec{A}}{\partial t},$$

where V is the electric potential and $\vec{A}$ is the magnetic vector potential. The value of k_0 comes from Table 4.2. When this formula is used in the rationalized mks system, Table 4.3(a) shows that no prefixes are to be expected; and when it is used in the Heaviside-Lorentz system, Table 4.3(b) shows that $\vec{A}$, V, and $\vec{E}$ should be given the prefix "h" to match the Heaviside-Lorentz notation used in previous chapters.

4.2.37 THE MAGNETIC VECTOR POTENTIAL OF A MAGNETIC DIPOLE

The magnetic vector potential $\vec{A}$ of a current-loop magnetic dipole $\vec{m}_I$ located in a vacuum is

$$\vec{A} = \left(\frac{\Pi \widetilde{\mu}}{4\pi r^3} \right) \cdot \left[\vec{m}_I \times (r\hat{e}_r) \right]$$

and the magnetic vector potential $\vec{A}$ of a permanent-magnet dipole $\vec{m}_H$ located in a vacuum is

$$\vec{A} = \left(\frac{\Pi}{4\pi r^3} \right) \cdot [\vec{m}_H \times (r\hat{e}_r)].$$

In both formulas, r is the distance between the permanent-magnet dipole and the field point where $\vec{A}$ is evaluated, and $\hat{e}_r$ is the dimensionless unit vector pointing from the dipole to the field point. The values of $\tilde{\mu}$ and Π come from Table 4.2. These two formulas become identical when written in the electromagnetic systems of Table 4.2 that have $\tilde{\mu} = 1$, because in these systems, $\vec{m}_H = \vec{m}_I$ (see Section 4.2.13). When the two formulas are used in the rationalized mks system, Table 4.3(a) shows that $\vec{m}_H$ and $\tilde{\mu} = \mu_0$ should be given the prefix "f" to match the rationalized mks notation used in the previous chapters. When they are used in the Heaviside-Lorentz system, Table 4.3(b) shows that $\vec{m}_I, \vec{m}_H$, and $\vec{A}$ should be given the prefix "h" to match the Heaviside-Lorentz notation used in the previous chapters.

4.2.38 THE ELECTRIC-CIRCUIT EQUATIONS

Ideal capacitors in electric circuits obey the equation

$$Q = C \cdot V,$$

where Q is the charge on the capacitor, V is the capacitor's electric potential difference (often called the voltage across the capacitor), and C is its capacitance. This equation can also be written as

$$V = S \cdot Q,$$

where S, which is clearly equal to C^{-1}, is called the elastance of the capacitor. Resistors in electric circuits obey the equation

$$V = I \cdot R,$$

where V is the electric potential or voltage across the resistor, I is the current through the resistor, and the resistance of the resistor is R. This equation is also written as

$$I = G \cdot V,$$

where G, which is clearly equal to R^{-1}, is called the conductance of the resistor. The relationship between charge Q and current I in a circuit is

$$I = \frac{dQ}{dt},$$

with dQ/dt being the derivative of charge Q with respect to time t, showing the rate at which charge is moving through the circuit. Ideal inductors in electric circuits obey the equation

$$V = L \cdot \frac{dI}{dt},$$

where V is the electric potential or voltage across the inductor, I is the current through the inductor, dI/dt is the derivative of the current with respect to time, showing the rate at which the current flowing through the inductor is changing, and L is the inductance of the inductor. There are no free parameters $\widetilde{\varepsilon}, \widetilde{\mu}, k_0$, or Π in these electric-circuit equations, which means they have the same form in all the electromagnetic systems of Table 4.2. Table 4.3(a) shows that when these equations are written in the rationalized mks system, no prefixes are to be expected; and Table 4.3(b) shows that when these equations are written in the Heaviside-Lorentz system, Q, C, S, V, I, R, G, and L should all be given the prefix "h" to match the Heaviside-Lorentz notation used in the previous chapters.

4.2.39 THE MAGNETIC-CIRCUIT EQUATIONS

For a magnetic circuit like the one shown in Fig. 4.3, the magnetomotive force $\mathcal{F}$ obeys the equation

$$\mathcal{R} \cdot \Phi_B = \mathcal{F}.$$

Here $\mathcal{R}$ is the reluctance of the magnetic circuit and Φ_B is the magnetic flux through the cross-sectional surface a_B in Fig. 4.3. The formula for Φ_B is

$$\Phi_B = \int_{a_B} (\vec{B} \cdot \hat{n}) da,$$

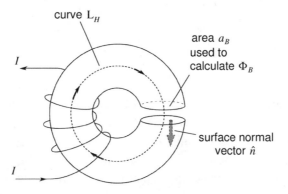

Figure 4.3 Curve L_H is used to define the magnetomotive force $\mathcal{F}$.

where $\hat{n}$ is a dimensionless unit vector normal to surface a_B in Fig. 4.3 and $\vec{B}$ is the magnetic induction field over the surface. Figure 4.3 also specifies the closed curve L_H, which is used to define the magnetomotive force $\mathcal{F}$ as a line integral over the magnetic field $\vec{H}$:

$$\mathcal{F} = \int_{L_H} \vec{H} \cdot d\vec{l}.$$

Curve L_H must go through all of the current-carrying loops around the metal core. The relationship between $\mathcal{F}$ and Φ_B can also be written as

$$\Phi_B = \mathcal{P} \cdot \mathcal{F},$$

where $\mathcal{P}$, which clearly equals $\mathcal{R}^{-1}$, is called the permeance. There are no free parameters $\tilde{\varepsilon}, \tilde{\mu}, k_0$, or Π in these magnetic-circuit equations, which means they have the same form in all of the electromagnetic systems of Table 4.2. When these magnetic-circuit equations are written in the rationalized mks system, Table 4.3(a) shows that $\mathcal{F}, \mathcal{R}, \mathcal{P}$, and $\vec{H}$ should be given the prefix "f" to match the rationalized mks notation used in the previous chapters; and when they are written in the Heaviside-Lorentz system, Table 4.3(b) shows that $\Phi_B, \mathcal{F}, \vec{H}$, and $\vec{B}$ should be given the prefix "h" to match the Heaviside-Lorentz notation used in previous chapters.

4.2.40 THE RELATIONSHIP BETWEEN MAGNETOMOTIVE FORCE AND CURRENT

The magnetic circuit shown in Fig. 4.3 has a magnetomotive force $\mathcal{F}$, which is given by

$$\mathcal{F} = (k_0 \Pi) N I,$$

where I is the current flowing in the circuit and N is the total number of turns of wire about the metal core. The values of k_0 and Π come from Table 4.2. When this equation is written in the rationalized mks system, Table 4.3(a) shows that $\mathcal{F}$ should be given the prefix "f" to match the rationalized mks notation used in previous chapters; and when this equation is written in the Heaviside-Lorentz system, Table 4.3(b) shows that I and F should be given the prefix "h" to match the Heaviside-Lorentz notation used in previous chapters.

4.2.41 Conductivity, Resistivity, Volume Current Density, and the Electric Field

The conductivity σ is the proportionality constant between the electric field $\vec{E}$ and the volume current density $\vec{J}$:

$$\vec{J} = \sigma \cdot \vec{E}.$$

Of course in empty charge-free space, $\sigma = 0$ for all electromagnetic systems and units. The equation for the conservation of charge is

$$\frac{\partial \rho_Q}{\partial t} + \vec{\nabla} \cdot \vec{J} = 0,$$

where ρ_Q is the volume charge density. The resistivity ρ_R is

$$\rho_R = \frac{1}{\sigma}.$$

The resistivity of a homogeneous wire whose resistance is R and that has a length ℓ and cross-sectional area a is

$$\rho_R = \frac{aR}{\ell}.$$

The current I in the wire can be written as

$$I = \int_a (\vec{J} \cdot \hat{n}) da,$$

where $\hat{n}$ is the dimensionless unit normal vector of the wire's cross-sectional surface a and $\vec{J}$ is the volume current density inside the wire. There are no free parameters $\tilde{\varepsilon}, \tilde{\mu}, k_0$, or Π in these five equations, which means they have the same form in all the electromagnetic systems of Table 4.2. Table 4.3(a) shows there is no prefix expected on any of the variables of these equations when they are written in the rationalized mks system. Table 4.3(b) shows that when these equations are written in the Heaviside-Lorentz system, $\vec{E}, \sigma, \vec{J}, \rho_Q, \rho_R, R$, and I should all be given the prefix "h" to match the Heaviside-Lorentz notation used in the previous chapters.

4.2.42 THE DEFINING EQUATIONS FOR THE ELECTRIC SUSCEPTIBILITY AND THE RELATIVE DIELECTRIC CONSTANT

Inside a substance or medium where the electric dipole density $\vec{P}$ is parallel to the electric field $\vec{E}$,

$$\vec{P} = \chi_e \cdot \widetilde{\varepsilon} \cdot \vec{E},$$

where χ_e is called the electric susceptibility of the substance or medium. The relative dielectric constant ε_r of the substance or medium is

$$\varepsilon_r = 1 + \Pi \chi_e,$$

and the electric displacement $\vec{D}$ inside the substance or medium is

$$\vec{D} = \varepsilon_r \cdot \widetilde{\varepsilon} \cdot \vec{E}.$$

In electromagnetic systems where $\widetilde{\varepsilon} \neq 1$, this last formula is often written as

$$\vec{D} = \varepsilon \cdot \vec{E}, \quad \text{with } \varepsilon = \varepsilon_r \cdot \widetilde{\varepsilon}$$

called the dielectric constant of the substance or medium. The values $\widetilde{\varepsilon}$ and Π come from Table 4.2. When these equations are used in the rationalized mks system, Table 4.3(a) shows that $\vec{D}$, χ_e, ε, and $\widetilde{\varepsilon} = \varepsilon_0$ should be given the prefix "f" to match the rationalized mks notation of previous chapters; and when they are used in the Heaviside-Lorentz system, Table 4.3(b) shows that χ_e, $\vec{E}$, $\vec{P}$, and $\vec{D}$ should be given the prefix "h" to match the Heaviside-Lorentz notation use in previous chapters. Of course, in empty space $\chi_e = 0$ and $\varepsilon_r = 1$ for all electromagnetic systems and units. Consequently in empty space the $\vec{E}$ and $\vec{D}$ fields are always the same in those electromagnetic systems where $\widetilde{\varepsilon} = 1$. (Note that this matches the analysis of the $\vec{E}$ and $\vec{D}$ fields in Section 4.2.26.) In electromagnetic systems where $\widetilde{\varepsilon} = 1$, the relative dielectric constant ε_r is sometimes called just the dielectric constant and the "r" subscript is dropped, giving $\vec{D} = \varepsilon_r \vec{E} = \varepsilon \vec{E}$.

4.2.43 THE DEFINING EQUATIONS FOR THE MAGNETIC SUSCEPTIBILITY AND THE RELATIVE MAGNETIC PERMEABILITY

Inside a substance or medium where the current-loop magnetic dipole density $\vec{M}_I$ and the permanent-magnet dipole density $\vec{M}_H$ are parallel to the magnetic field $\vec{H}$, the magnetic susceptibility χ_m satisfies the equations

$$\vec{M}_H = \chi_m \cdot \widetilde{\mu} \cdot \vec{H} \quad \text{and} \quad \vec{M}_I = \chi_m \cdot \vec{H}.$$

The relative magnetic permeability of the substance or medium is

$$\mu_r = 1 + \Pi \chi_m,$$

and the magnetic induction field $\vec{B}$ inside the substance or medium satisfies the equation

$$\vec{B} = \mu_r \cdot \tilde{\mu} \cdot \vec{H} \quad \text{or} \quad \vec{H} = (\mu_r \tilde{\mu})^{-1} \cdot \vec{B}.$$

In electromagnetic systems where $\tilde{\mu} \neq 1$, these last two formulas are often written as

$$\vec{B} = \mu \cdot \vec{H} \quad \text{and} \quad \vec{H} = \mu^{-1} \cdot \vec{B}, \quad \text{with } \mu = \mu_r \cdot \tilde{\mu}$$

called the magnetic permeability of the substance or medium. The values of $\tilde{\mu}$ and Π come from Table 4.2. When these equations are used in the rationalized mks system, Table 4.3(a) shows that $\vec{H}$, $\vec{M}_H$, χ_m, μ, and $\tilde{\mu} = \mu_0$ should be given the prefix "f" to match the rationalized mks notation used in the previous chapters; and when the equations are used in the Heaviside-Lorentz system, Table 4.3(b) shows that $\vec{H}$, $\vec{M}_H$, $\vec{M}_I$, $\vec{B}$, and χ_m should be given the prefix "h" to match the Heaviside-Lorentz notation used in the previous chapters. In empty space $\chi_m = 0$ and $\mu_r = 1$ for all electromagnetic systems and units. The above formulas then require the $\vec{B}$ and $\vec{H}$ fields to be the same in empty space in those electromagnetic systems where $\tilde{\mu} = 1$. Note that this matches what was said about the $\vec{B}$ and $\vec{H}$ fields in Sections 4.2.27 and 4.2.28. Note also that, according to Section 4.2.17, $\vec{M}_I$ and $\vec{M}_H$ become the same when $\tilde{\mu} = 1$. Consequently, the first two equations given here—the formulas for $\vec{M}_I$ and $\vec{M}_H$ in terms of the magnetic field—become the same equation when $\tilde{\mu} = 1$. In electromagnetic systems where $\tilde{\mu} = 1$, the relative magnetic permeability μ_r is sometimes called just the magnetic permeability and the "r" subscript is dropped, giving

$$\vec{B} = \mu_r \cdot \vec{H} = \mu \cdot \vec{H} \quad \text{and} \quad \vec{H} = \mu_r^{-1} \cdot \vec{B} = \mu^{-1} \cdot \vec{B}.$$

4.2.44 THE CAPACITANCE OF A PARALLEL-PLATE CAPACITOR

The capacitance C of a parallel-plate capacitor is

$$C = \frac{\varepsilon_r \tilde{\varepsilon} a}{\Pi s},$$

where ε_r is the relative dielectric constance of the substance between the plates, a is the area of either one of the two equal-sized plates, and s is the distance between the plates. (Of course $\varepsilon_r = 1$ in all electromagnetic systems and units when the two capacitor plates are separated by empty space.) The values of $\tilde{\varepsilon}$ and Π come from

Table 4.2. When this formula is used in the rationalized mks system, Table 4.3(a) shows that $\widetilde{\varepsilon} = \varepsilon_0$ should be given the prefix "f" to match the rationalized mks notation used in previous chapters; and when it is used in the Heaviside-Lorentz system, Table 4.3(b) shows that C should be given the prefix "h" to match the Heaviside-Lorentz notation used in previous chapters.

4.2.45 THE INDUCTANCE OF A SOLENOIDAL INDUCTOR

The self-inductance L of a solenoidal inductor is

$$L = \Pi k_0^2 \left(\frac{N^2}{s} \right) \mu_r \widetilde{\mu} \left(\pi r^2 \right),$$

where the inductor is N coils of wire around the walls of a tube which has a length s and a radius r. The material filling the tube has a relative magnetic permeability μ_r. (Of course, $\mu_r = 1$ in all electromagnetic systems and units when the tube contains just empty space.) The values of Π, k_0, and $\widetilde{\mu}$ come from Table 4.2. When this formula is used in the rationalized mks system, Table 4.3(a) shows that $\widetilde{\mu} = \mu_0$ should be given the prefix "f" to match the rationalized mks notation used in previous chapters; and when it is used in the Heaviside-Lorentz system, Table 4.3(b) shows that L should be given the prefix "h" to match the Heaviside-Lorentz notation used in previous chapters.

4.2.46 THE FORCE PER UNIT LENGTH BETWEEN TWO PARALLEL, CURRENT-CARRYING WIRES

The force per unit length of wire, ϕ, between two long parallel wires inside a substance or medium of relative magnetic permeability μ_r and carrying currents I_1 and I_2 is

$$\phi = \left(\frac{\mu_r \widetilde{\mu} \Pi k_0^2}{2\pi} \right) \cdot \left(\frac{I_1 I_2}{s} \right),$$

where s is the distance between the two wires, as shown in Fig. 4.4. (Of course, in empty space $\mu_r = 1$ in all electromagnetic systems and units.) The values of $\widetilde{\mu}$, Π, and k_0 come from Table 4.2. When this equation is used in the rationalized mks system, Table 4.3(a) shows that $\widetilde{\mu} = \mu_0$ should be given the prefix "f" to match the rationalized mks notation used in previous chapters; and when it is used in the Heaviside-Lorentz system, Table 4.3(b) shows that I_1 and I_2 should be given the prefix "h" to match the Heaviside-Lorentz notation used in previous chapters. The two wires attract each other when the two currents are flowing in the same direction and repel each other when the two currents are flowing in opposite directions.

Figure 4.4 The two wires separated by a distance s attract each other when their currents I_1 and I_2 flow in the same direction and repel each other when their currents I_1 and I_2 flow in opposite directions.

4.2.47 THE MAGNETIC FIELD AND MAGNETIC INDUCTION OF A LONG, CURRENT-CARRYING WIRE

The magnetic induction field $\vec{B}$ generated by a long straight wire inside a substance or medium of relative magnetic permeability μ_r and carrying a current I is

$$\vec{B} = \left(\frac{\Pi \mu_r \widetilde{\mu} k_0}{2\pi}\right) \cdot \left(\frac{I}{s}\right)\hat{e},$$

and the magnetic field $\vec{H}$ generated by the wire is

$$\vec{H} = \left(\frac{\Pi k_0}{2\pi}\right) \cdot \left(\frac{I}{s}\right)\hat{e}.$$

In these two formulas, s is the distance between the wire and the field point at which $\vec{B}$ or $\vec{H}$ is evaluated, and $\hat{e}$ is the dimensionless unit vector specifying the direction of the $\vec{B}$ and $\vec{H}$ fields (see Fig. 4.5). The dimensionless unit vector $\hat{e}$ is perpendicular both to the wire and to the line segment of length s joining the field point and the wire. When the current I is flowing directly at an observer looking along the wire, the observer sees vector $\hat{e}$ pointing in a counterclockwise direction about the wire. The values of $\widetilde{\mu}$, k_0, and Π come from Table 4.2. When these two equations are written in the rationalized mks system, Table 4.3(a) shows that $\vec{H}$ and $\widetilde{\mu} = \mu_0$ should be given the prefix "f" to match the rationalized mks notation used in the previous chapters; and when the two equations are written in the Heaviside-Lorentz system, Table 4.3(b) shows that I, $\vec{B}$, and $\vec{H}$ should be given the prefix "h" to match the Heaviside-Lorentz notation used in the previous chapters. In empty space, $\mu_r = 1$ in all electromagnetic systems and units, which means that these two formulas become identical in electromagnetic systems where $\widetilde{\mu} = 1$. This is just another example of how the $\vec{H}$ and $\vec{B}$ fields become identical in empty space when $\widetilde{\mu} = 1$ (see Sections 4.2.27 and 4.2.28).

Figure 4.5 The current *I* flowing in a long, straight wire generates *B* and *H* fields pointing in the direction specified by unit vector $\hat{e}$ at field points a distance *s* from the wire.

4.2.48 INTEGRAL FORMULA FOR THE FORCE BETWEEN TWO CURRENT LOOPS

The infinitesimal component of force $d^2\vec{F}_{12}$ on length element $d\vec{\ell}_1$ of circuit loop $\mathcal{L}_1$ in Fig. 4.6 from length element $d\vec{\ell}_2$ of circuit loop $\mathcal{L}_2$ in Fig. 4.6 is

$$d^2\vec{F}_{12} = \left(\frac{\Pi k_0^2}{4\pi}\right)\mu_r\widetilde{\mu}I_1 I_2\left[\frac{d\vec{\ell}_1 \times (d\vec{\ell}_2 \times \vec{r}_{12})}{|\vec{r}_{12}|^3}\right],$$

where I_1 is the current in the direction of $d\vec{\ell}_1$ in loop $\mathcal{L}_1$, I_2 is the current in the direction of $d\vec{\ell}_2$ in loop $\mathcal{L}_2$, $\vec{r}_{12}$ is the distance vector pointing from $d\vec{\ell}_2$ to $d\vec{\ell}_1$, and μ_r is the relative magnetic permeability of the substance or medium in which the two loops reside. (Of course, in empty space $\mu_r = 1$ for all electromagnetic systems and units.) Although, according to this formula, the force on one current element is not necessarily equal and opposite to the force on the other current element, which is in disagreement with Newton's third law, this can be fixed by integrating over both current loops in Fig. 4.6 to find that $\vec{F}_{12}$, the force on current loop $\mathcal{L}_1$ by current loop $\mathcal{L}_2$, is

$$\vec{F}_{12} = \left(\frac{\Pi k_0^2}{4\pi}\right)\mu_r\widetilde{\mu}I_1 I_2\int_{\mathcal{L}_1}\int_{\mathcal{L}_2}\frac{d\vec{\ell}_1 \times (d\vec{\ell}_2 \times \vec{r}_{12})}{|\vec{r}_{12}|^3}$$

$$= -\left(\frac{\Pi k_0^2}{4\pi}\right)\mu_r\widetilde{\mu}I_1 I_2\int_{\mathcal{L}_1}\int_{\mathcal{L}_2}\frac{(d\vec{\ell}_1 \cdot d\vec{\ell}_2)\vec{r}_{12}}{|\vec{r}_{12}|^3}.$$

Here the second double integral is explicitly symmetrical over all the pairs of length elements $d\vec{\ell}_1$ and $d\vec{\ell}_2$ (remember that for every vector $\vec{r}_{12}$ inside the double integrals over $\mathcal{L}_1$ and $\mathcal{L}_2$ there is a corresponding $\vec{r}_{21}$ vector).[2] We can therefore conclude that, even though Newton's third law is not obeyed for the physically unrealistic formula for the force between two "pieces" of a circuit, it is obeyed for the physically realistic formula for the force between two complete circuits. The values of $\widetilde{\mu}, k_0$, and Π come from Table 4.2. When these equations are written in the rationalized mks system, Table 4.3(a) shows that $\widetilde{\mu} = \mu_0$ should be given the prefix "*f*" to match the rationalized mks notation used in the previous chapters;

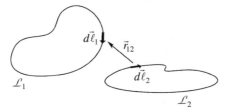

Figure 4.6 Current I_1 flows around loop $\mathcal{L}_1$ in the direction specified by $d\vec{\ell}_1$ and current I_2 flows around loop $\mathcal{L}_2$ in the direction specified by $d\vec{\ell}_2$.

and when these equations are written in the Heaviside-Lorentz system, Table 4.3(b) shows that I_1 and I_2 should be given the prefix "h" to match the Heaviside-Lorentz notation used in the previous chapters.

4.2.49 The magnetic field and magnetic induction of a current loop (Biot-Savart law)

The Biot-Savart law states that when the length element $d\vec{\ell}$ carries a current I travelling in the direction in which $d\vec{\ell}$ is pointing, there is an infinitesimal magnetic induction field $d\vec{B}$ given by

$$d\vec{B} = \left(\frac{\Pi k_0 \mu_r \tilde{\mu} I}{4\pi} \right) \frac{d\vec{\ell} \times \vec{r}}{|\vec{r}|^3}$$

and an infinitesimal magnetic field $d\vec{H}$ given by

$$d\vec{H} = \left(\frac{\Pi k_0 I}{4\pi} \right) \frac{d\vec{\ell} \times \vec{r}}{|\vec{r}|^3}.$$

Here, μ_r is the relative permeability of the substance or medium in which $d\vec{\ell}$ and the field point are embedded, and $\vec{r}$ is the distance vector pointing from the current element to the field point. Of course, in empty space μ_r is 1 for all electromagnetic systems and units, which means that in empty space the right-hand sides of these two formulas become the same when working in electromagnetic systems that have $\tilde{\mu} = 1$. This is another example of how in empty space the $\vec{H}$ and $\vec{B}$ fields become identical in electromagnetic systems where $\tilde{\mu} = 1$ (see Sections 4.2.27 and 4.2.28). Integration over all the length elements $d\vec{\ell}$ that make up a single current loop L_I carrying a current I shows that the magnetic induction $\vec{B}$ at the field point must be

$$\vec{B} = \left(\frac{\Pi k_0 \mu_r \tilde{\mu} I}{4\pi} \right) \int_{L_I} \frac{d\vec{\ell} \times \vec{r}}{|\vec{r}|^3}$$

and the magnetic field $\vec{H}$ at the field point must be

$$\vec{H} = \left(\frac{\Pi k_0 I}{4\pi}\right) \int_{L_I} \frac{d\vec{\ell} \times \vec{r}}{|\vec{r}|^3}.$$

The values of $\tilde{\mu}$, k_0, and Π come from Table 4.2. When these four equations are written in the rationalized mks system, Table 4.3(a) shows that $d\vec{H}$, $\vec{H}$, and $\tilde{\mu} = \mu_0$ should be given the prefix "f" to match the rationalized mks notation used in the previous chapters; and when the four equations are written in the Heaviside-Lorentz system, Table 4.3(b) shows that $d\vec{H}$, $\vec{H}$, I, and $\vec{B}$ should be given the prefix "h" to match the Heaviside-Lorentz notation used in the previous chapters.

4.2.50 INTEGRAL FORMULA FOR THE ELECTRIC FIELD AN ELECTRIC DISPLACEMENT OF A STATIC CHARGE-DENSITY DISTRIBUTION

A static* volume charge density ρ_Q is located inside a substance or material with a constant relative permittivity ε_r, which fills all space. The value of ρ_Q is a function of a position vector $\vec{r}'$ and, at a field point whose position vector is $\vec{r}$, generates an electric field $\vec{E}$ such that

$$\vec{E}\bigg|_{\text{at } \vec{r}} = \frac{\Pi}{4\pi \varepsilon_r \tilde{\varepsilon}} \int \frac{[\rho_Q(\vec{r}')](\vec{r} - \vec{r}')}{|\vec{r} - \vec{r}'|^3} d^3 r'.$$

The corresponding electric displacement $\vec{D}$ at that field point is

$$\vec{D}\bigg|_{\text{at } \vec{r}} = \frac{\Pi}{4\pi} \int \frac{[\rho_Q(\vec{r}')](\vec{r} - \vec{r}')}{|\vec{r} - \vec{r}'|^3} d^3 r'.$$

In both formulas, $d^3 r'$ is an infinitesimal volume element used in an integration over the entire volume of space for which $\rho_Q(\vec{r}')$, the volume charge density at position $\vec{r}'$, is not zero. For this integration to make sense, the charge density should occupy only a finite volume of space. In empty space, $\varepsilon_r = 1$ for all electromagnetic systems and units, which means that in empty space the right-hand sides of these two formulas become identical when working in electromagnetic systems where $\tilde{\varepsilon} = 1$. This is another example of how in empty space the $\vec{E}$ and $\vec{D}$ fields become identical when $\tilde{\varepsilon} = 1$ (see Section 4.2.26 above). The values of $\tilde{\varepsilon}$ and Π in these two formulas come from Table 4.2. When the two formulas are written in the rationalized mks system, Table 4.3(a) shows that $\vec{D}$ and $\tilde{\varepsilon} = \varepsilon_0$ should be given the prefix "f" to match the rationalized mks notation used in the previous chapters; and when the formulas are written in the Heaviside-Lorentz system, Table 4.3(b) shows that ρ_Q, $\vec{E}$, and $\vec{D}$ should be given the prefix "h" to match the Heaviside-Lorentz notation used in the previous chapters.

* Here "static" means that the volume charge density ρ_Q does not change with time.

4.2.51 INTEGRAL FORMULA FOR THE MAGNETIC FIELD AND MAGNETIC INDUCTION OF A STATIC CURRENT-DENSITY DISTRIBUTION

A static* volume current density $\vec{J}$ is located inside a substance or medium with a relative permeability μ_r, which fills all space. The value of $\vec{J}$ is a function of a position vector $\vec{r}'$ and, at a field point whose position vector is $\vec{r}$, generates a magnetic induction $\vec{B}$ such that

$$\vec{B}\bigg|_{\text{at } \vec{r}} = \frac{\Pi k_0 \mu_r \widetilde{\mu}}{4\pi} \int \frac{[\vec{J}(\vec{r}')] \times (\vec{r} - \vec{r}')}{|\vec{r} - \vec{r}'|^3} d^3 r'.$$

The corresponding magnetic field $\vec{H}$ at that field point is

$$\vec{H}\bigg|_{\text{at } \vec{r}} = \frac{\Pi k_0}{4\pi} \int \frac{[\vec{J}(\vec{r}')] \times (\vec{r} - \vec{r}')}{|\vec{r} - \vec{r}'|^3} d^3 r'.$$

In both formulas $d^3 r'$ is an infinitesimal volume element used in an integration over the entire volume of space for which $\vec{J}(\vec{r}')$, the volume current density at position $\vec{r}'$, is not zero. For this integration to make sense, the charge density should occupy only a finite volume of space. In empty space, $\mu_r = 1$ for all electromagnetic systems and units, which means that in empty space the right-hand sides of these two equations become identical when working in electromagnetic systems where $\widetilde{\mu} = 1$. This is another example of how in empty space the $\vec{B}$ and $\vec{H}$ fields become identical when $\widetilde{\mu} = 1$ (see Sections 4.2.27 and 4.2.28 above). The values of $\widetilde{\mu}$, k_0, and Π in these equations come from Table 4.2. When the equations are written in the rationalized mks system, Table 4.3(a) shows that $\vec{H}$ and $\widetilde{\mu} = \mu_0$ should be given the prefix "f" to match the rationalized mks notation used in the previous chapters; and when the equations are written in the Heaviside-Lorentz system, Table 4.3(b) shows that $\vec{J}, \vec{B}$, and $\vec{H}$ should be given the prefix "h" to match the Heaviside-Lorentz notation used in previous chapters.

4.2.52 THE INTEGRAL OF THE MAGNETIC INDUCTION OVER A CLOSED SURFACE

The integral of a magnetic induction field $\vec{B}$ over a closed surface $\mathbf{S}_c$ is

$$\int_{\mathbf{S}_c} (\vec{B} \cdot \hat{n}) da = 0.$$

Here, $\hat{n}$ is the dimensionless unit vector that is normal to the area da of an infinitesimal patch of surface $\mathbf{S}_c$ and points outwards from the volume enclosed by $\mathbf{S}_c$.

* Here "static" means that the volume current density J does not change with time.

This formula contains none of the free parameters $\widetilde{\varepsilon}$, $\widetilde{\mu}$, k_0, or Π, so it has the same form for all the electromagnetic systems listed in Table 4.2. When this formula is written in the Heaviside-Lorentz system, Table 4.3(b) shows that $\vec{B}$ should be given the prefix "h" to match the Heaviside-Lorentz notation used in the previous chapters. Table 4.3(a) shows that there is no reason to expect a prefix when the formula is written in the rationalized mks system.

4.2.53 THE INTEGRAL OF THE ELECTRIC FIELD AND THE ELECTRIC DISPLACEMENT OVER A CLOSED SURFACE

The integral of an electric displacement field $\vec{D}$ over a closed surface $\mathcal{S}_c$ is

$$\int_{\mathcal{S}_c} (\vec{D} \cdot \hat{n}) da = \Pi Q,$$

where Q is the total charge contained inside surface $\mathcal{S}_c$, and $\hat{n}$ is the dimensionless unit vector that is normal to the area da of an infinitesimal patch of surface $\mathcal{S}_c$ and points outwards from the volume enclosed by $\mathcal{S}_c$. If the surface is located in a substance or medium that has a constant relative dielectric constant ε_r, then the integral of the electric field $\vec{E}$ over $\mathcal{S}_c$ is

$$\int_{\mathcal{S}_c} (\vec{E} \cdot \hat{n}) da = \frac{\Pi}{\varepsilon_r \widetilde{\varepsilon}} Q,$$

with $\hat{n}$, Q, and da having the same meaning as before. In empty space, $\varepsilon_r = 1$ for all electromagetic systems and units, which means that in empty space the right-hand sides of these two equations become the same when working in electromagnetic systems where $\widetilde{\varepsilon} = 1$. This is another example of how in empty space the $\vec{E}$ and $\vec{D}$ fields become identical when $\widetilde{\varepsilon} = 1$ (see Section 4.2.26 above). The values of $\widetilde{\varepsilon}$ and Π in these two equations come from Table 4.2. When the two equations are written in the rationalized mks system, Table 4.3(a) shows that $\vec{D}$ and $\widetilde{\varepsilon} = \varepsilon_0$ should be given the prefix "f" to match the rationalized mks notation used in previous chapters; and when the two equations are written in the Heaviside-Lorentz system, Table 4.3(b) shows that $\vec{D}$, $\vec{E}$, and Q should be given the prefix "h" to match the Heaviside-Lorentz notation used in previous chapters.

4.2.54 THE INTEGRAL OF THE MAGNETIC FIELD AND THE MAGNETIC INDUCTION OVER A CLOSED LOOP

The line integral of a magnetic field $\vec{H}$ around a closed loop $\mathbf{L}_c$ is

$$\int_{\mathbf{L}_c} \vec{H} \cdot d\vec{\ell} = (k_0 \Pi) I,$$

where I is the total current flowing through loop $\mathbf{L}_c$ and $d\vec{\ell}$ is a length element of loop $\mathbf{L}_c$. From the point of view of an observer toward whom the current I is flowing, length element $d\vec{\ell}$ points counterclockwise around the loop (see Fig. 4.7). The magnetic field is one of a static collection of electromagnetic fields—that is, the electromagnetic fields are not changing with time. If the loop is located in a substance or medium having a constant magnetic permeability μ_r, then the line integral around $\mathbf{L}_c$ of the magnetic induction $\vec{B}$, which is also part of a static collection of electromagnetic fields, is

$$\int_{\mathbf{L}_c} \vec{B} \cdot d\vec{\ell} = (k_0 \Pi \mu_r \widetilde{\mu}) I,$$

with I and $d\vec{\ell}$ having the same meaning as before. In empty space, $\mu_r = 1$ in all electromagnetic systems and units, which means that in empty space the right-hand sides of these equations become identical in those electromagnetic systems where $\widetilde{\mu} = 1$. This is another example of how the $\vec{B}$ and $\vec{H}$ fields become identical in empty space when $\widetilde{\mu} = 1$ (see Sections 4.2.27 and 4.2.28 above). The values of $\widetilde{\mu}, k_0$, and Π in these two equations come from Table 4.2. When the two equations are written in the rationalized mks system, Table 4.3(a) shows that $\vec{H}$ and $\widetilde{\mu} = \mu_0$ should be given the prefix "f" to match the rationalized mks notation used in the previous chapters; and when the equations are written in the Heaviside-Lorentz system, Table 4.3(b) shows that $\vec{H}, \vec{B}$, and I should be given the prefix "h" to match the Heaviside-Lorentz notation used in the previous chapters.

4.2.55 THE RELATIONSHIP BETWEEN THE ELECTRIC POTENTIAL OR VOLTAGE AROUND A CLOSED CIRCUIT OR LOOP AND THE MAGNETIC FLUX THROUGH THE CLOSED CIRCUIT OR LOOP

The line integral of the electric field $\vec{E}$ around a closed circuit or loop $\mathbf{L}_c$ is called the voltage V around the closed circuit or loop,

$$V = \int_{\mathbf{L}_c} \vec{E} \cdot d\vec{\ell}.$$

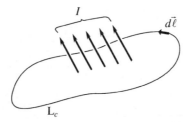

Figure 4.7 Loop L_c encloses current I. The line integral of the H or B field around the loop in the direction shown by vector $d\vec{\ell}$ is proportional to I.

If $\mathbf{S}_B$ is a surface (not a closed surface) whose edges are curve $\mathbf{L}_c$, as shown in Fig. 4.8, then the magnetic flux Φ_B of the magnetic induction field $\vec{B}$ through loop $\mathbf{L}_c$ is defined to be

$$\Phi_B = \int_{\mathbf{S}_B} (\vec{B} \cdot \hat{n}) da,$$

where da is the area of an infinitesimal patch of surface $\mathbf{S}_B$, and $\hat{n}$ is the dimensionless unit vector normal to the infinitesimal patch. Figure 4.8 specifies the relationship between the infinitesimal length elements $d\vec{\ell}$ of curve $\mathbf{L}_c$ and the dimensionless normal vectors $\hat{n}$ of surface $\mathbf{S}_B$. If an observer looking down on loop $\mathbf{L}_c$ sees $d\vec{\ell}$ pointing counterclockwise around the loop, then the $\hat{n}$ normal vectors of surface $\mathbf{S}_B$ must point generally toward the observer rather than away from the observer. The formula relating the line integral to the magnetic flux can now be written as

$$V = -k_0 \frac{d\Phi_B}{dt},$$

with the value of the free parameter k_0 coming from Table 4.2. Note that the definitions of Φ_B and V do not contain of the free parameters $\widetilde{\varepsilon}, \widetilde{\mu}, k_0$, or Π, which means they have the same form in all of the electromagnetic systems listed in Table 4.2. When these three equations are written in the Heaviside-Lorentz system, Table 4.3(b) shows that $\vec{E}, V, \vec{B}$, and Φ_B should be given the prefix "h" to match the Heaviside-Lorentz notation used in previous chapters; and when the three equations are written in the rationalized mks system, Table 4.3(a) shows that no prefixes are to be expected.

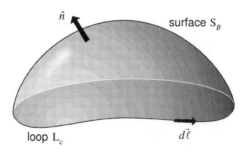

Figure 4.8 Loop $\mathbf{L}_c$ is the edge of (unclosed) surface S_B.

4.2.56 Maxwell's equations

The electric field $\vec{E}$, electric displacement field $\vec{D}$, magnetic field $\vec{H}$, and magnetic induction field $\vec{B}$ always obey the equations

$$\vec{\nabla} \cdot \vec{D} = \Pi \rho_Q, \qquad\qquad \vec{\nabla} \cdot \vec{B} = 0,$$

$$\vec{\nabla} \times \vec{H} = k_0 \left(\Pi \vec{J} + \frac{\partial \vec{D}}{\partial t} \right), \quad \vec{\nabla} \times \vec{E} + k_0 \frac{\partial \vec{B}}{\partial t} = 0.$$

Here t is the time coordinate, ρ_Q is the volume charge density, $\vec{J}$ is the volume current density, and the $\vec{\nabla}$ operator has the same meaning as in Appendix 2.B of Chapter 2. These four equations need to be combined with the formulas in Sections 4.2.26, 4.2.27, 4.2.28, 4.2.42, and 4.2.43 to specify the connections between the $\vec{E}$, $\vec{D}$, $\vec{H}$, and $\vec{B}$ fields. When Maxwell's equations are applied to a collection of fields inside a substance or material with a constant relative dielectric constant ε_r and a constant relative magnetic permeability μ_r, they can be written as

$$\nabla \cdot \vec{E} = \frac{\Pi}{\varepsilon_r \widetilde{\varepsilon}} \rho_Q,$$

$$\vec{\nabla} \cdot \vec{B} = 0,$$

$$\vec{\nabla} \times \vec{B} = (k_0 \Pi \mu_r \widetilde{\mu}) \vec{J} + (k_0 \mu_r \widetilde{\mu} \varepsilon_r \widetilde{\varepsilon}) \frac{\partial \vec{E}}{\partial t},$$

$$\vec{\nabla} \times \vec{E} + k_0 \frac{\partial \vec{B}}{\partial t} = 0;$$

or as

$$\vec{\nabla} \cdot \vec{D} = \Pi \rho_Q,$$

$$\vec{\nabla} \cdot \vec{H} = 0,$$

$$\vec{\nabla} \times \vec{H} = k_0 \left(\Pi \vec{J} + \frac{\partial \vec{D}}{\partial t} \right),$$

$$\vec{\nabla} \times \vec{D} + (k_0 \mu_r \widetilde{\mu} \varepsilon_r \widetilde{\varepsilon}) \frac{\partial \vec{H}}{\partial t} = 0.$$

The values of $\widetilde{\varepsilon}$, $\widetilde{\mu}$, k_0, and Π in these 12 equations come from Table 4.2. When Maxwell's equations are written in the rationalized mks system, Table 4.3(a) shows that $\vec{D}$, $\vec{H}$, $\widetilde{\mu} = \mu_0$, and $\widetilde{\varepsilon} = \varepsilon_0$ should be given the prefix "f" to match the rationalized mks notation used in the previous chapters; and when Maxwell's equations are written in the Heaviside-Lorentz system, Table 4.3(b) shows that $\vec{E}$, $\vec{D}$, $\vec{B}$, $\vec{H}$, $\vec{J}$, and ρ_Q should be given the prefix "h" to match the Heaviside-Lorentz notation used in the previous chapters. In Sections 4.2.26, 4.2.27, and 4.2.28, we read that in empty space, where $\mu_r = \varepsilon_r = 1$, both the $\vec{E}, \vec{D}$ fields and the $\vec{B}, \vec{H}$ fields become identical when working in electromagnetic systems where $\widetilde{\mu} = \widetilde{\varepsilon} = 1$. The above three sets of Maxwell's equations obey this rule by requiring the $\vec{E}$ and $\vec{D}$ fields and the $\vec{B}$ and $\vec{H}$ fields to solve the same equations in terms of $\vec{J}$ and ρ_Q when $\mu_r = \varepsilon_r = \widetilde{\mu} = \widetilde{\varepsilon} = 1$.

4.2.57 THE WAVE EQUATIONS FOR THE ELECTRIC AND MAGNETIC INDUCTION FIELDS

The electric field $\vec{E}$ and magnetic induction field $\vec{B}$ satisfy the wave equations

$$\nabla^2 \vec{E} - \Pi k_0^2 (\mu_r \widetilde{\mu}) \sigma \frac{\partial \vec{E}}{\partial t} - k_0^2 (\mu_r \widetilde{\mu})(\varepsilon_r \widetilde{\varepsilon}) \frac{\partial^2 \vec{E}}{\partial t^2} = 0$$

and

$$\nabla^2 \vec{B} - \Pi k_0^2 (\mu_r \widetilde{\mu}) \sigma \frac{\partial \vec{B}}{\partial t} - k_0^2 (\mu_r \widetilde{\mu})(\varepsilon_r \widetilde{\varepsilon}) \frac{\partial^2 \vec{B}}{\partial t^2} = 0,$$

when travelling through a substance or material characterized by a constant conductivity σ, a constant relative magnetic permeability μ_r, and a constant relative dielectric constant ε_r. Here, t is the time coordinate, and the Laplacian operator ∇^2 is explained at the end of Appendix 2.B of Chapter 2. When the wave equations are written in the rationalized mks system, Table 4.3(a) shows that $\widetilde{\varepsilon} = \varepsilon_0$ and $\widetilde{\mu} = \mu_0$ should be given the prefix "f" to match the rationalized mks notation used in the previous chapters; and when the wave equations are written in the Heaviside-Lorentz system, Table 4.3(b) shows that $\vec{E}, \vec{B}$, and σ should be given the prefix "h" to match the Heaviside-Lorentz notation used in the previous chapters.

4.2.58 THE POYNTING VECTOR AND ENERGY DENSITY OF AN ELECTROMAGNETIC FIELD

The formula for the energy density $\mathcal{U}$, in units of energy per unit volume, of an electromagnetic field at any field point is

$$\mathcal{U} = \frac{1}{2\Pi}(\vec{E} \cdot \vec{D} + \vec{B} \cdot \vec{H}),$$

where $\vec{E}, \vec{D}, \vec{H}, \vec{B}$ are, respectively, the electric field, the electric displacement field, the magnetic field, and the magnetic induction field at the field point. The formula for $\mathcal{U}$ in empty space can be written as

$$\mathcal{U} = \frac{1}{2\Pi}\left(\widetilde{\varepsilon}|\vec{E}|^2 + \widetilde{\mu}^{-1}|\vec{B}|^2\right).$$

The formula for the Poynting vector $\vec{S}$ at the field point is

$$\vec{S} = \frac{1}{\Pi k_0}\left(\vec{E} \times \vec{H}\right),$$

and in empty space this can be written as

$$\vec{S} = \frac{1}{\Pi\widetilde{\mu}k_0}\left(\vec{E} \times \vec{B}\right).$$

Vector $\vec{S}$ points in the direction of the energy flow carried by the electromagnetic field at the field point, and the magnitude of $\vec{S}$ is the energy per unit time per unit area carried by the electromagnetic field at the field point. When no other sources or sinks of energy are present, the equation for conservation of electromagnetic energy is

$$\frac{\partial \mathcal{U}}{\partial t} + \vec{\nabla} \cdot \vec{S} = 0.$$

The values of $\widetilde{\varepsilon}$, $\widetilde{\mu}$, k_0, and Π in the first four equations come from Table 4.2. The fifth equation has none of these free parameters, so it has the same form in all the electromagnetic systems of Table 4.2. When the first four equations are written in the rationalized mks system, Table 4.3(a) shows that $\vec{D}$, $\vec{H}$, $\widetilde{\varepsilon} = \varepsilon_0$, and $\widetilde{\mu} = \mu_0$ should be given the prefix "f" to match the rationalized mks notation used in the previous chapters; and when they are written in the Heaviside-Lorentz system, Table 4.3(b) shows that $\vec{E}$, $\vec{D}$, $\vec{H}$, and $\vec{B}$ should be given the prefix "h" to match the Heaviside-Lorentz notation used in the previous chapters. The fifth equation never needs an "f" or an "h" prefix because it only involves the nonelectromagnetic quantities $\mathcal{U}$ and $\vec{S}$.

4.2.59 ELECTRIC DIPOLE RADIATING IN A VACUUM

When an electric dipole $\vec{p}$ changes its magnitude or direction over time t, it emits electromagnetic radiation. The field point at which the electromagnetic radiation is evaluated is a distance r from dipole $\vec{p}$. This distance is much larger than the wavelength λ of the emitted radiation, and λ is itself much larger than the characteristic length of the dipole. The formula for the electric field $\vec{E}$ of the radiation coming from the dipole is

$$\vec{E}\Big|_{\text{at time } t} = \frac{\Pi k_0^2 \widetilde{\mu}}{4\pi r}\left\{\hat{e}_r \times \left(\hat{e}_r \times \frac{d^2\vec{p}}{dt^2}\right)\right\}\Bigg|_{\substack{\text{evaluated at the retarded} \\ \text{time } t'=t-(r/c)}},$$

where $\hat{e}_r$ is the dimensionless unit vector pointing from the dipole to the field point and c is the speed of light. The magnetic field $\vec{H}$ of the radiation coming from the dipole is, at that same field point,

$$\vec{H}\Big|_{\text{at time } t} = -\frac{\Pi k_0}{4\pi c r}\left\{\hat{e}_r \times \frac{d^2\vec{p}}{dt^2}\right\}\Bigg|_{\substack{\text{evaluated at the retarded} \\ \text{time } t'=t-(r/c)}},$$

and the magnetic induction field $\vec{B}$ of the radiation coming from the dipole is, at that field point,

$$\vec{B}\Big|_{\text{at time } t} = -\frac{\Pi k_0 \widetilde{\mu}}{4\pi c r}\left\{\hat{e}_r \times \frac{d^2\vec{p}}{dt^2}\right\}\Bigg|_{\substack{\text{evaluated at the retarded} \\ \text{time } t'=t-(r/c)}}.$$

The right-hand sides of the formulas for the $\vec{H}$ and $\vec{B}$ fields become the same when working in electromagnetic systems where $\widetilde{\mu} = 1$. This is just another example of how, when $\widetilde{\mu} = 1$, the $\vec{H}$ and $\vec{B}$ fields become identical in empty space (see Sections 4.2.27 and 4.2.28). If $\vec{p}$, whose characteristic length is much smaller than λ, is thought of as being at the center of a large sphere of radius r, which is much larger than λ, then the instantaneous electromagnetic energy flux W leaving the sphere at any point on the sphere is

$$W\Big|_{\text{at time } t} = \frac{\Pi k_0^2 \widetilde{\mu}}{16\pi^2 r^2 c}\left\{\left|\hat{e}_r \times \frac{d^2\vec{p}}{dt^2}\right|^2\right\}\Bigg|_{\substack{\text{evaluated at the retarded} \\ \text{time } t'=t-(r/c)}}.$$

In this formula, W has units of energy per unit time per unit area of the sphere, and $\hat{e}_r$ is the dimensionless unit vector pointing from the center of the sphere to the point on the sphere where W is being evaluated. For a sinusoidally oscillating dipole,

$$\vec{p}(t) = \vec{p}_0 \sin(\omega t),$$

with vector $\vec{p}_0$ the amplitude of the oscillation and ω its angular frequency in radians per unit time, the time-averaged energy flux W_{av} leaving the sphere at any point on the sphere is

$$W_{av} = \frac{\Pi k_0^2 \omega^4 \widetilde{\mu}}{32\pi^2 r^2 c}|\hat{e}_r \times \vec{p}_0|^2.$$

Again, $\hat{e}_r$ is the dimensionless unit vector pointing from the center of the sphere to the point on the sphere where the energy flux is being evaluated. An integration over the surface of the sphere shows that $W_{av}^{(tot)}$, the total and time-averaged radiant energy per unit time emitted by a sinusoidally oscillating dipole, is

$$W_{av}^{(tot)} = \frac{\Pi k_0^2 \omega^4 \widetilde{\mu}}{12\pi c}|\vec{p}_0|^2.$$

The values of $\widetilde{\mu}, k_0$, and Π come from Table 4.2. When these equations are written in the rationalized mks system, Table 4.3(a) shows that $\vec{H}$ and $\widetilde{\mu} = \mu_0$ should be given the prefix "f" to match the rationalized mks notation used in the previous chapters; and when they are written in the Heaviside-Lorentz system, Table 4.3(b) shows that $\vec{p}, \vec{p}_0, \vec{E}, \vec{B}$, and $\vec{H}$ should be given the prefix "h" to match the Heaviside-Lorentz notation used in the previous chapters.

4.2.60 Current-Loop Magnetic Dipole Radiating in a Vacuum

When a current-loop magnetic dipole $\vec{m}_I$ changes its magnitude or direction over time t, it emits electromagnetic radiation. The electromagnetic radiation is evaluated at a field point that is a distance r from $\vec{m}_I$. Distance r is much larger than the wavelength λ of the emitted radiation, and λ is itself much larger than the dipole's characteristic length. The electric field $\vec{E}$ of the radiation coming from the dipole is given by

$$\vec{E}\Big|_{\text{at time } t} = \frac{\Pi k_0 \widetilde{\mu}}{4\pi c r} \left\{ \hat{e}_r \times \frac{d^2 \vec{m}_I}{dt^2} \right\} \Bigg|_{\substack{\text{evaluated at the retarded} \\ \text{time } t' = t - (r/c)}},$$

where c is the speed of light and $\hat{e}_r$ is the dimensionless unit vector pointing from the dipole to the field point. At that same field point, the magnetic field $\vec{H}$ of the radiation coming from the dipole is

$$\vec{H}\Big|_{\text{at time } t} = \frac{\Pi}{4\pi c^2 r} \left\{ \hat{e}_r \times \left(\hat{e}_r \times \frac{d^2 \vec{m}_I}{dt^2} \right) \right\} \Bigg|_{\substack{\text{evaluated at the retarded} \\ \text{time } t' = t - (r/c)}},$$

and the magnetic induction field $\vec{B}$ is, at that field point,

$$\vec{B}\Big|_{\text{at time } t} = \frac{\Pi \widetilde{\mu}}{4\pi c^2 r} \left\{ \hat{e}_r \times \left(\hat{e}_r \times \frac{d^2 \vec{m}_I}{dt^2} \right) \right\} \Bigg|_{\substack{\text{evaluated at the retarded} \\ \text{time } t' = t - (r/c)}}.$$

These two formulas for the $\vec{H}$ and $\vec{B}$ fields have the same right-hand sides in electromagnetic systems where $\widetilde{\mu} = 1$. This just shows how, when $\widetilde{\mu} = 1$, the $\vec{H}$ and $\vec{B}$ fields become identical in empty space (see Sections 4.2.27 and 4.2.28). If the dipole, with a characteristic length that is much smaller than λ, is regarded as being at the center of a large sphere of radius r, where r is much larger than λ, then the electromagnetic energy flux W leaving the sphere at any point on the sphere and at any time t is

$$W\Big|_{\text{at time } t} = \frac{\Pi \widetilde{\mu}}{16\pi^2 r^2 c^3} \left\{ \left| \hat{e}_r \times \frac{d^2 \vec{m}_I}{dt^2} \right|^2 \right\} \Bigg|_{\substack{\text{evaluated at the retarded} \\ \text{time } t' = t - (r/c)}}.$$

Here, W has units of energy per unit time per unit area of the sphere, and $\hat{e}_r$ is the dimensionless unit vector pointing from the center of the sphere to the point on the sphere where W is being evaluated. We can represent a sinusoidally oscillating dipole by

$$\vec{m}_I(t) = \vec{m}_0 \sin(\omega t),$$

with vector $\vec{m}_0$ the amplitude of the oscillation and ω its angular frequency in radians per unit time. Now, at any point on the sphere, the time-averaged energy flux W_{av} leaving the sphere is

$$W_{av} = \frac{\Pi \omega^4 \widetilde{\mu}}{32 \pi^2 r^2 c^3} |\hat{e}_r \times \vec{m}_0|^2.$$

In this formula, $\hat{e}_r$ is again the dimensionless unit vector pointing from the center of the sphere to the point on the sphere where the energy flux is being evaluated. Integrating over the surface of the sphere, we find that $W_{av}^{(tot)}$, the total and time-averaged radiant energy per unit time emitted by a sinusoidally oscillating dipole, is

$$W_{av}^{(tot)} = \frac{\Pi \omega^4 \widetilde{\mu}}{12 \pi c^3} |\vec{m}_0|^2.$$

The free-parameter values $\widetilde{\mu}$, k_0, and Π come from Table 4.2. When we write these equations in the rationalized mks system, Table 4.3(a) shows that $\vec{H}$ and $\widetilde{\mu} = \mu_0$ should be given the prefix "f" to match the rationalized mks notation used in the previous chapters; and when we write them in the Heaviside-Lorentz system, Table 4.3(b) shows that $\vec{m}_I, \vec{m}_0, \vec{E}, \vec{B}$, and $\vec{H}$ should be given the prefix "h" to match the Heaviside-Lorentz notation used in the previous chapters.

4.3 UNDERSTANDING THE SUBSTITUTION TABLES

Before using the free-parameter method to transform an expression from one electromagnetic system to another, we must locate the expression in the previous section's list of equations and formulas—and if we cannot locate it, we must derive it from others that we can. Always there is a location step, perhaps even a derivation step, before the electromagnetic expression is transformed. Substitution tables avoid this inefficiency by working directly with the electromagnetic variables themselves. When using a substitution table, we only need to identify what physical quantity each variable represents—for example, this is the electric field, this is the charge, this is the magnetomotive force, etc.—rather than match the electromagnetic expression as a whole to a predetermined list of equations and formulas.

There is nothing particularly subtle about the way substitution tables work. We know from the work done in previous chapters that the transformation of physical quantities from one electromagnetic system to another can always, even when rationalization is involved, be represented by a multiplicative constant. As a reminder of how this works, we consider the transformation of a magnetic field H from the rationalized mks system to unrationalized Gaussian units. Following the notation of Section 3.8, we can write in rationalized mks that

$$\text{magnetic field in rationalized mks} = {}_f H_{\text{mks}} \frac{\text{amp}}{\text{m}}, \tag{4.12a}$$

where ${}_f H_{\text{mks}}$ is the numeric part of the magnetic field in rationalized mks, and amp/m are the units of the magnetic field in rationalized mks. Similarly, according to Section 3.2, we can write in Gaussian units that

$$\text{magnetic field in Gaussian units} = H_{\text{gs}} \frac{\text{abamp}}{\text{cm}}, \tag{4.12b}$$

with H_{gs} the numeric part of the magnetic field in Gaussian units and abamp/cm = oersted the corresponding magnetic-field units (see Table 3.1). The numeric parts ${}_f H_{\text{mks}}$ and H_{gs} of the magnetic field are numbers representing the magnetic field measurements coming from instruments calibrated in the rationalized mks and Gaussian systems respectively. If the strength of the magnetic field doubles, ${}_f H_{\text{mks}}$ and H_{gs} must both double; if the strength of the magnetic field triples, ${}_f H_{\text{mks}}$ and H_{gs} must both triple; and so on. This means the ratio of these two numbers must be another number which always has the same value,

$$\frac{{}_f H_{\text{mks}}}{H_{\text{gs}}} = \text{constant.}$$

The units amp, abamp, m, and cm are unchanging physical quantities, which makes any product or ratio involving them also an unchanging physical quantity. Consequently, the ratio

$$\frac{\text{magnetic field in rationalized mks}}{\text{magnetic field in Gaussian units}} = \left({}_f H_{\text{mks}} \frac{\text{amp}}{\text{m}} \right) \bigg/ \left(H_{\text{gs}} \frac{\text{abamp}}{\text{cm}} \right)$$

is an unchanging physical quantity. We call this ratio T_H and write

$$\left({}_f H_{\text{mks}} \frac{\text{amp}}{\text{m}} \right) \bigg/ \left(H_{\text{gs}} \frac{\text{abamp}}{\text{cm}} \right) = T_H$$

or

$$H_{\text{gs}} \frac{\text{abamp}}{\text{cm}} \cdot T_H = {}_f H_{\text{mks}} \frac{\text{amp}}{\text{m}}. \tag{4.13a}$$

Using the methods of Chapters 2 and 3 to solve for T_H, we get

$$T_H = \left(\frac{{}_f H_{\text{mks}}}{H_{\text{gs}}} \right) \cdot \left(\frac{\text{cm}}{\text{m}} \right) \cdot \left(\frac{\text{amp}}{\text{abamp}} \right).$$

From Table 3.17, we see that

$$\frac{_f H_{\text{mks}}}{H_{\text{gs}}} = \frac{10^3}{4\pi};$$

and from Table 2.3 and Table 3.8, we get

$$\left(\frac{\text{cm}}{\text{m}}\right) \cdot \left(\frac{\text{amp}}{\text{abamp}}\right) = \left(\frac{\text{cm}}{\text{m}}\right) \cdot \left(\frac{\text{coul}}{\text{sec}}\right) \cdot \left(\frac{\text{sec}}{\text{gm}^{1/2}\text{cm}^{1/2}}\right) = \frac{\text{coul} \cdot \text{cm}^{1/2}}{\text{gm}^{1/2} \cdot \text{m}}.$$

Hence, we can write

$$T_H = \sqrt{\frac{10^6}{(4\pi)^2}\left(\frac{\text{coul}^2 \cdot \text{cm}}{\text{gm} \cdot \text{m}^2}\right)} = \sqrt{\frac{10^6}{(4\pi)^2}\left(\frac{\text{coul}^2 \cdot 10^{-2}\text{m}}{10^{-3}\text{kg} \cdot \text{m}^2}\right)} = \sqrt{\frac{10^7}{(4\pi)^2}\frac{\text{coul}^2}{\text{kg} \cdot \text{m}}}.$$

Since $(\text{kg} \cdot \text{m})/\text{coul}^2 = \text{henry}/\text{m}$, we see that, according to Eq. (3.36e),

$$T_H = \sqrt{\frac{1}{(4\pi)^2\mu_0}}.$$

This can also be written as (see Table 3.14)

$$T_H = \sqrt{\frac{1}{4\pi\, _f\mu\,_0}}. \tag{4.13b}$$

We note that, as expected, T_H is not a dimensionless number but instead a dimensional physical quantity having the units $\sqrt{\text{m}/\text{henry}}$.

The reasoning applied to find T_H can always be used to construct a constant for the transformation of a physical quantity from one electromagnetic system to another. The transformation constant, for example, which takes electric charge from Gaussian units to the rationalized mks system, satisfies the equation (see Tables 3.1 and 3.9)

$$(Q_{\text{gs}}\text{statcoul})T_Q = Q_{\text{mks}}\text{coul} \tag{4.14a}$$

or

$$T_Q = \left(\frac{Q_{\text{mks}}}{Q_{\text{gs}}}\right) \cdot \left(\frac{\text{coul}}{\text{statcoul}}\right).$$

According to Table 2.3 and Table 3.17, this can be written as

$$T_Q = \frac{10}{c_{\text{cgs}}} \cdot \left(\frac{\text{coul} \cdot \text{sec}}{\text{gm}^{1/2}\text{cm}^{3/2}}\right) = \sqrt{\frac{10^2}{c_{\text{cgs}}^2} \cdot \left(\frac{\text{coul}^2}{10^{-3}\text{kg} \cdot 10^{-2}\text{m}}\right) \cdot \left(\frac{\text{sec}^2}{\text{cm}^2}\right)}.$$

Since $c_{cgs} \cdot (cm/sec) = c$, the speed of light, this becomes

$$T_Q = \sqrt{\frac{10^7}{c^2} \cdot \left(\frac{coul^2}{kg \cdot m}\right)} = \frac{1}{c}\sqrt{\frac{4\pi}{f\mu_0}}, \qquad (4.14b)$$

where in the last step we have again used $(kg \cdot m)/coul^2 =$ henry/m, Eq. (3.36e) and Table 3.14 to simplify the left-hand side of the formula. Equation (3.49e) shows that T_Q can also be written as

$$T_Q = \sqrt{4\pi f\varepsilon_0}. \qquad (4.14c)$$

Since T_Q in Eqs. (4.14b,c) is not the same transformation constant as T_H in Eq. (4.13b), this second example shows that we can expect to find different transformation constants for different electromagnetic physical quantities.

Having found T_Q for taking the electric charge from Gaussian units to the rationalized mks system, we can now, whenever we see Q in a rationalized mks equation, replace Q by $(T_Q \cdot Q)$ and start regarding Q as being in Gaussian units. The reason this works is easy to see: the charge Q in the original formula stands for

$$Q_{mks}coul,$$

and having replaced this by $(T_Q \cdot Q)$, with the new Q in Gaussian units standing for

$$Q_{gs}statcoul,$$

we know from the way T_Q behaves in Eq. (4.14a) that

$$T_Q \cdot Q = T_Q \cdot (Q_{gs}statcoul) = Q_{mks}coul.$$

Consequently, each "slot" for the electric charge in the original formula still contains

$$Q_{mks}coul,$$

which means the equation must still hold true. If there is also a slot in the original formula for the magnetic field, we can, using the rationalized mks notation of Section 3.8, replace $_fH$ everywhere by $(T_H \cdot H)$, with H taken to be the magnetic field in Gaussian units. According to Eq. (4.13a), all the magnetic-field slots still hold

$$_fH_{mks}\frac{amp}{m},$$

which means the equation again remains true.

We can clearly extend this reasoning to all of the electromagnetic physical quantities in the rationalized mks formula. For each electromagnetic physical quantity Y, we find the transformation constant T_Y taking Y from the Gaussian units to rationalized mks system. We then replace every Y in the original formula by $(T_Y \cdot Y)$, with Y now being thought of as the same physical quantity in Gaussian units. Just as was the case for the magnetic field and electric charge, each Y-slot still contains the appropriate rationalized mks quantity, which means the equation remains true even though each Y variable is now regarded as being in Gaussian units. Since both the original equation and the transformed equation state a relationship between the same groups of physical variables—that is, the transformation does not change an electric field into a magnetic field, or a resistance into an inductance, etc.—the transformed equation ought to be the Gaussian counterpart, possibly in disguised form, of the rationalized mks equation. This same logic obviously applies to any equation or formula for which we know all the T_Y values. In fact, it applies to any transformation of physical quantities Y from one electromagnetic system to another that can be represented by a multiplicative constant T_Y; that is, it applies to all the electromagnetic transformations discussed in this book. Consequently, extending the list of Y and T_Y to the entire set of physical quantities and electromagnetic systems discussed in the previous chapters prepares us to transform an enormous range of equations and formulas from one electromagnetic system to another. In practice, what we have done, as shown in the Appendix to this chapter, is to list substitution tables of $(T_Y \cdot Y)$ versus Y for going back and forth between all six electromagnetic systems listed in Table 4.2.

4.4 USING THE SUBSTITUTION TABLES

Substitution Tables 4.4(a)–(b) through 4.18(a)–(b) follow the same prefix conventions used in the previous chapters for the rationalized mks and Heaviside-Lorentz systems. These "f" and "h" prefixes, however, do not play a crucial role in the procedures described below. If, following the notation of almost all other authors, we omit the "f" and "h" prefixes, the substitution tables can still be used without any problems. In other words, the reader can mentally block out all the "f" and "h" prefixes, ignoring their presence both in the equations being transformed and in the substitution tables, and still get the right answers.

For the first example of how to use substitution tables, we convert the Lorentz force law in the rationalized mks system,

$$\vec{F} = Q\vec{E} + Q(\vec{v} \times \vec{B}), \tag{4.15a}$$

into Gaussian units. Here, $\vec{E}$ is the electric field experienced by a point charge Q moving at a velocity $\vec{v}$ in a magnetic induction field $\vec{B}$. From Table 4.4(a), we see that

$$Q \rightarrow \frac{Q}{c} \sqrt{\frac{4\pi}{f \mu_0}},$$

$$\vec{E} \rightarrow \vec{E} \cdot c \sqrt{\frac{f \mu_0}{4\pi}},$$

and

$$\vec{B} \rightarrow \vec{B} \cdot \sqrt{\frac{f \mu_0}{4\pi}}.$$

Putting the specified replacements into Eq. (4.15a) gives

$$\vec{F} = \left(\frac{Q}{c} \sqrt{\frac{4\pi}{f \mu_0}} \right) \left(\vec{E} \cdot c \sqrt{\frac{f \mu_0}{4\pi}} \right) + \left(\frac{Q}{c} \sqrt{\frac{4\pi}{f \mu_0}} \right) \left[\vec{v} \times \left(\vec{B} \sqrt{\frac{f \mu_0}{4\pi}} \right) \right]$$

or

$$\vec{F} = Q\vec{E} + \frac{1}{c} Q(\vec{v} \times \vec{B}). \tag{4.15b}$$

Checking back to Section 4.2.29, we see that this is the correct answer. There is often, as seen here, a large amount of "cancelling out" after making the replacements specified by the substitution tables.

The substitution tables can be applied to any electromagnetic equation, not just the formulas of classical electromagnetism that can be derived from Maxwell's equations and the Lorentz force law. From quantum field theory we have the formula for the dimensionless fine-structure constant in Gaussian units,

$$\alpha = \frac{e^2}{\hbar c}, \tag{4.16a}$$

where e is the charge on an electron in statcoul, $\hbar$ is Planck's constant divided by 2π, and c is the speed of light. Table 4.4(b) in the Appendix shows that charge transforms as

$$Q \rightarrow Q \cdot c \sqrt{\frac{f \mu_0}{4\pi}},$$

when going from Gaussian units to the rationalized mks system, so in rationalized mks

$$\alpha = \frac{\left(e \cdot c \sqrt{\frac{f \mu_0}{4\pi}} \right)^2}{\hbar c} = \frac{e^2 c f \mu_0}{4\pi \hbar}$$

or, using the formula $_f\varepsilon_0\,_f\mu_0 = c^{-2}$ [see Eq. (3.49e)] to replace $_f\mu_0$ by $_f\varepsilon_0$,

$$\alpha = \frac{e^2}{4\pi\,_f\varepsilon_0\hbar c}. \tag{4.16b}$$

In this rationalized mks formula, e is now the electron charge in coul.

Although we said at the beginning of this section that the substitution tables in the Appendix give the correct answer even when their "f" and "h" prefixes are ignored, this is true only as long as we do not examine what happens when transforming the same formula into different electromagnetic systems. Suppose, for example, Table 4.12(a) in the Appendix is now used to transform Eq. (4.16a) into the unrationalized mks system. We get

$$\alpha = \frac{(ec\sqrt{\mu_0})^2}{\hbar c} = \frac{e^2 c\mu_0}{\hbar}.$$

In unrationalized mks $\mu_0\varepsilon_0 = c^{-2}$ [see Eq. (3.49e)], so the formula for the fine-structure constant can also be written as

$$\alpha = \frac{e^2}{\varepsilon_0\hbar\,c}. \tag{4.16c}$$

Again, e is the electron charge in *coul*. When Eq. (4.16b) is compared to Eq. (4.16c), we see that ignoring the prefix "f" on ε_0 invites us to write

$$\frac{e^2}{4\pi\varepsilon_0\hbar c} = \frac{e^2}{\varepsilon_0\hbar c}.$$

On both sides of this "equation," e is the electron charge in coul. Consequently, if we are so unwise as to forget that ε_0 on the left-hand side is the permittivity of free space in the rationalized mks system whereas ε_0 on the right-hand side is the permittivity of free space in the unrationalized mks system, we might cancel out ε_0 on both sides to get

$$\frac{1}{4\pi} \overset{?}{=} 1.$$

Paying attention to the prefix "f" protects us from this obviously incorrect result; now when we write

$$\frac{e^2}{4\pi\,_f\varepsilon_0\hbar\,c} = \frac{e^2}{\varepsilon_0\hbar\,c}$$

we end up with the correct formula,

$$_f\varepsilon_0 = \frac{\varepsilon_0}{4\pi},$$

which relates the two values of the permittivity of free space in the rationalized and unrationalized mks systems.

Another example of how the prefixes in the substitution tables can be helpful occurs when Table 4.9(a) in the Appendix is used to convert formula (4.16a) to the Heaviside-Lorentz system. For the charge conversion we have

$$Q \to \frac{{}_h Q}{\sqrt{4\pi}},$$

which suggests that Eq. (4.16a) in the Heaviside-Lorentz system should be written as

$$\alpha = \left(\frac{{}_h e}{\sqrt{4\pi}} \right)^2 \cdot \frac{1}{\hbar c} = \frac{({}_h e)^2}{4\pi \hbar c}.$$

Here, the prefix "h" reminds us that ${}_h e$ is now the electron charge in the Heaviside-Lorentz system, which is larger by a factor of $\sqrt{4\pi}$ than the electron charge in Gaussian units because charge is a physical quantity that is rescaled by a factor of $\sqrt{4\pi}$ when going from the Gaussian to the Heaviside-Lorentz system (see Section 3.3). Ignoring the "h" prefix leads us to write

$$\alpha = \frac{e^2}{4\pi \hbar c}$$

as the formula for the fine-structure constant in the Heaviside-Lorentz system. This formula is, of course, correct; but without the "h" prefix it invites us to overlook that "e" is now the electron charge in the Heaviside-Lorentz system of rescaled electromagnetic physical quantities rather than the electron charge in Gaussian units. In both the Gaussian and Heaviside-Lorentz systems, the electron charge has the same units, making it easy to forget that the numeric part of the electron charge is not the same.

The formula for the strength H of a magnetic field at a distance r from a long wire carrying a current I is

$$H = \frac{2I}{r} \tag{4.17a}$$

in esu units [see Eq. (2.6) or Section 4.2.47). Using Table 4.16(a) in the Appendix to convert to emu units gives

$$(cH) = \frac{2(cI)}{r} \quad \text{or} \quad H = \frac{2I}{r}. \tag{4.17b}$$

Here the formula remains unchanged because the transformation constants cancel out. Using Table 4.17(a) to go from esu units to the unrationalized mks system gives

$$\left(H \cdot c\sqrt{\mu_0}\right) = \frac{2(I \cdot c\sqrt{\mu_0})}{r}$$

$$\text{or} \quad H = \frac{2I}{r} \text{ again.} \tag{4.17c}$$

We see that twice now a complicated set of transformation constants have cancelled out, leaving the formula unchanged. In this particular case, the equation changes form when Table 4.10(b) is used to transform Eq. (4.17a) to Gaussian units, giving

$$(cH) = \frac{2I}{r} \quad \text{which can be written as} \quad H = \frac{2I}{rc}; \tag{4.17d}$$

or when Table 4.13(b) is used to transform to the Heaviside-Lorentz system, giving

$$\left({}_hHc\sqrt{4\pi}\right) = \frac{2({}_hI/\sqrt{4\pi})}{r},$$

which can be written as

$${}_hH = \frac{{}_hI}{2\pi rc}; \tag{4.17e}$$

or when Table 4.6(b) is used to transform to rationalized mks, giving

$${}_fH\left(c\sqrt{4\pi {}_f\mu_0}\right) = \frac{2\left(I \cdot c\sqrt{\dfrac{{}_f\mu_0}{4\pi}}\right)}{r},$$

which can be written as

$${}_fH = \frac{I}{2\pi r}. \tag{4.17f}$$

It is easy to show that the substitution tables handle Maxwell's equations correctly. Starting off with the Heaviside-Lorentz version of these equations in empty space, with ${}_h\vec{E}$ the electric field, ${}_h\vec{B}$ the magnetic induction, ${}_h\rho_Q$ the volume charge density, ${}_h\vec{J}$ the volume current density, and c the speed of light, we have

$$\vec{\nabla} \cdot {}_h\vec{E} = {}_h\rho_Q, \tag{4.18a}$$

$$\vec{\nabla} \cdot {}_h\vec{B} = 0, \tag{4.18b}$$

$$\vec{\nabla} \times {}_h\vec{E} + \frac{1}{c}\frac{\partial {}_h\vec{B}}{\partial t} = 0, \tag{4.18c}$$

$$\vec{\nabla} \times {}_h\vec{B} = \frac{1}{c} {}_h\vec{J} + \frac{1}{c} \frac{\partial_h \vec{E}}{\partial t}. \tag{4.18d}$$

Table 4.5(b) can be used to convert to the rationalized mks system, giving

$$\vec{\nabla} \cdot \left(\frac{\vec{E}}{c\sqrt{f\mu_0}} \right) = \rho_Q \cdot (c\sqrt{f\mu_0}),$$

which simplifies to

$$\vec{\nabla} \cdot \vec{E} = (c^2 {}_f\mu_0)\rho_Q; \tag{4.19a}$$

$$\vec{\nabla} \cdot \left(\frac{\vec{B}}{\sqrt{f\mu_0}} \right) = 0,$$

which simplifies to

$$\vec{\nabla} \cdot \vec{B} = 0; \tag{4.19b}$$

$$\vec{\nabla} \times \left(\frac{\vec{E}}{c\sqrt{f\mu_0}} \right) + \frac{1}{c} \frac{\partial}{\partial t} \left(\frac{\vec{B}}{\sqrt{f\mu_0}} \right) = 0,$$

which simplifies to

$$\vec{\nabla} \times \vec{E} + \frac{\partial \vec{B}}{\partial t} = 0; \tag{4.19c}$$

and

$$\vec{\nabla} \times \left(\frac{\vec{B}}{\sqrt{f\mu_0}} \right) = \frac{1}{c}\vec{J} \cdot (c\sqrt{f\mu_0}) + \frac{1}{c} \frac{\partial}{\partial t} \left(\frac{\vec{E}}{c\sqrt{f\mu_0}} \right),$$

which simplifies to

$$\vec{\nabla} \times \vec{B} = {}_f\mu_0 \vec{J} + \frac{1}{c^2} \frac{\partial \vec{E}}{\partial t}. \tag{4.19d}$$

The relationship ${}_f\mu_0 {}_f\varepsilon_0 = c^{-2}$ from Eq. (3.49e) can be used to put Eqs. (4.19a) and (4.19d) into the form

$$\vec{\nabla} \cdot \vec{E} = {}_f\varepsilon_0^{-1} \rho_Q \tag{4.19e}$$

and

$$\vec{\nabla} \times \vec{B} = {}_f\mu_0 \vec{J} + {}_f\mu_0 {}_f \varepsilon_0 \frac{\partial \vec{E}}{\partial t}, \tag{4.19f}$$

which is perhaps the more conventional way to write these equations in the ratio-nalized mks system. Note that the Eqs. (4.19a) and (4.19d) coming directly from the substitution table are not wrong; they are just written in an unconventional manner.

Table 4.9b shows how to put Eqs. (4.18a–d) into Gaussian units,

$$\vec{\nabla} \cdot \left(\frac{\vec{E}}{\sqrt{4\pi}} \right) = (\rho_Q \sqrt{4\pi}),$$

which simplifies to

$$\vec{\nabla} \cdot \vec{E} = 4\pi\rho_Q; \tag{4.20a}$$

$$\vec{\nabla} \cdot \left(\frac{\vec{B}}{\sqrt{4\pi}} \right) = 0,$$

which simplifies to

$$\vec{\nabla} \cdot \vec{B} = 0; \tag{4.20b}$$

$$\vec{\nabla} \times \left(\frac{\vec{E}}{\sqrt{4\pi}} \right) + \frac{1}{c} \frac{\partial}{\partial t} \left(\frac{\vec{B}}{\sqrt{4\pi}} \right) = 0,$$

which simplifies to

$$\vec{\nabla} \times \vec{E} + \frac{1}{c} \frac{\partial \vec{B}}{\partial t} = 0; \tag{4.20c}$$

and

$$\vec{\nabla} \times \left(\frac{\vec{B}}{\sqrt{4\pi}} \right) = \frac{1}{c} \vec{J} \cdot (\sqrt{4\pi}) + \frac{1}{c} \frac{\partial}{\partial t} \left(\frac{\vec{E}}{\sqrt{4\pi}} \right),$$

which simplifies to

$$\vec{\nabla} \times \vec{B} = \frac{4\pi}{c} \vec{J} + \frac{1}{c} \frac{\partial \vec{E}}{\partial t}. \tag{4.20d}$$

To put Eqs. (4.18a–d) into esu units, we consult Table 4.13(a) to get

$$\vec{\nabla} \cdot \left(\frac{\vec{E}}{\sqrt{4\pi}} \right) = \rho_Q \cdot \sqrt{4\pi},$$

which simplifies to

$$\vec{\nabla} \cdot \vec{E} = 4\pi \rho_Q; \qquad (4.21\text{a})$$

$$\vec{\nabla} \cdot \left[\vec{B} \cdot \left(\frac{c}{\sqrt{4\pi}} \right) \right] = 0,$$

which simplifies to

$$\vec{\nabla} \cdot \vec{B} = 0; \qquad (4.21\text{b})$$

$$\vec{\nabla} \times \left(\frac{\vec{E}}{\sqrt{4\pi}} \right) + \frac{1}{c} \frac{\partial}{\partial t} \left[\vec{B} \cdot \left(\frac{c}{\sqrt{4\pi}} \right) \right] = 0,$$

which simplifies to

$$\vec{\nabla} \times \vec{E} + \frac{\partial \vec{B}}{\partial t} = 0; \qquad (4.21\text{c})$$

and

$$\vec{\nabla} \times \left[\vec{B} \cdot \left(\frac{c}{\sqrt{4\pi}} \right) \right] = \frac{1}{c} \vec{J} \cdot (\sqrt{4\pi}) + \frac{1}{c} \frac{\partial}{\partial t} \left(\frac{\vec{E}}{\sqrt{4\pi}} \right),$$

which simplifies to

$$\vec{\nabla} \times \vec{B} = \frac{1}{c^2} \left(4\pi \vec{J} + \frac{\partial \vec{E}}{\partial t} \right). \qquad (4.21\text{d})$$

The free-parameter formulas in Section 4.2.56 above show that Eqs. (4.20a–d) and (4.21a–d) are the correct version of Maxwell's equations for empty space in Gaussian and esu units.

Table 4.14(a) shows how to convert Eqs. (4.18a–d) to emu units, giving

$$\vec{\nabla} \cdot \left(\frac{\vec{E}}{c\sqrt{4\pi}} \right) = \rho_Q (c\sqrt{4\pi}),$$

which simplifies to

$$\vec{\nabla} \cdot \vec{E} = 4\pi c^2 \rho_Q; \tag{4.22a}$$

$$\vec{\nabla} \cdot \left(\frac{\vec{B}}{\sqrt{4\pi}} \right) = 0,$$

which simplifies to

$$\vec{\nabla} \cdot \vec{B} = 0; \tag{4.22b}$$

$$\vec{\nabla} \times \left(\frac{\vec{E}}{c\sqrt{4\pi}} \right) + \frac{1}{c} \frac{\partial}{\partial t} \left(\frac{\vec{B}}{\sqrt{4\pi}} \right) = 0,$$

which simplifies to

$$\vec{\nabla} \times \vec{E} + \frac{\partial \vec{B}}{\partial t} = 0; \tag{4.22c}$$

and

$$\vec{\nabla} \times \left(\frac{\vec{B}}{\sqrt{4\pi}} \right) = \frac{1}{c} (\vec{J} \cdot c\sqrt{4\pi}) + \frac{1}{c} \frac{\partial}{\partial t} \left(\frac{\vec{E}}{c\sqrt{4\pi}} \right),$$

which simplifies to

$$\vec{\nabla} \times \vec{B} = 4\pi \vec{J} + \frac{1}{c^2} \frac{\partial \vec{E}}{\partial t}. \tag{4.22d}$$

Equation (2.12b) states that the permittivity of free space is $\varepsilon_0 = c^{-2}$ in emu units, which means that Eqs. (4.22a) and (4.22d) can also be written as

$$\vec{\nabla} \cdot \vec{E} = \frac{4\pi \rho_Q}{\varepsilon_0} \tag{4.22e}$$

and

$$\vec{\nabla} \times \vec{B} = 4\pi \vec{J} + \varepsilon_0 \frac{\partial \vec{E}}{\partial t}. \tag{4.22f}$$

To get Maxwell's equations for empty space in the unrationalized mks system, we apply Table 4.15(a) in the Appendix to Eqs. (4.18a–d) to get

$$\vec{\nabla} \cdot \left(\frac{\vec{E}}{c\sqrt{4\pi \mu_0}} \right) = \rho_Q (c\sqrt{4\pi \mu_0}),$$

which simplifies to

$$\vec{\nabla} \cdot \vec{E} = 4\pi \mu_0 c^2 \rho_Q; \qquad (4.23\text{a})$$

$$\vec{\nabla} \cdot \left(\frac{\vec{B}}{\sqrt{4\pi \mu_0}} \right) = 0,$$

which simplifies to

$$\vec{\nabla} \cdot \vec{B} = 0; \qquad (4.23\text{b})$$

$$\vec{\nabla} \times \left(\frac{\vec{E}}{c\sqrt{4\pi \mu_0}} \right) + \frac{1}{c} \frac{\partial}{\partial t} \left(\frac{\vec{B}}{\sqrt{4\pi \mu_0}} \right) = 0,$$

which simplifies to

$$\vec{\nabla} \times \vec{E} + \frac{\partial \vec{B}}{\partial t} = 0; \qquad (4.23\text{c})$$

and

$$\vec{\nabla} \times \left(\frac{\vec{B}}{\sqrt{4\pi \mu_0}} \right) = \frac{1}{c} \left(\vec{J} \cdot c\sqrt{4\pi \mu_0} \right) + \frac{1}{c} \frac{\partial}{\partial t} \left(\frac{\vec{E}}{c\sqrt{4\pi \mu_0}} \right),$$

which simplifies to

$$\vec{\nabla} \times \vec{B} = 4\pi \mu_0 \vec{J} + \frac{1}{c^2} \frac{\partial \vec{E}}{\partial t} \qquad (4.23\text{d})$$

Again we can put the first and last equations, (4.23a) and (4.23d), into a more conventional form, this time using $\mu_0 \varepsilon_0 = c^{-2}$ from Eq. (3.49e), to get

$$\vec{\nabla} \cdot \vec{E} = \frac{4\pi \rho_Q}{\varepsilon_0} \qquad (4.23\text{e})$$

and

$$\vec{\nabla} \times \vec{B} = 4\pi \mu_0 \vec{J} + \mu_0 \varepsilon_0 \frac{\partial \vec{E}}{\partial t}. \qquad (4.23\text{f})$$

All electromagnetic equations have the same form in the esuq and emuq units introduced in Chapter 2 as they do in the unrationalized mks system, so Eqs. (4.23a–f) can also be thought of as Maxwell's equations for empty space in the esuq and emuq systems of units. Any substitution table in the Appendix that is labeled as transforming equations to or from the unrationalized mks system can also be regarded as transforming equations to or from the esuq and emuq systems of units.

Substitution tables require an intelligent understanding of the physical quantities involved, especially when working with magnetic fields or magnetic dipoles. The formula for the magnetic induction field $\vec{B}$ radiated by a magnetic dipole in empty space is, for example, in Gaussian units,

$$\vec{B}\Big|_{\text{at time } t} = \frac{1}{c^2 r}\left\{ \hat{e}_r \times \left(\hat{e}_r \times \frac{d^2 \vec{m}_I}{dt^2} \right)\right\}\Bigg|_{\substack{\text{evaluated at the retarded} \\ \text{time } t'=t-(r/c)}} . \qquad (4.24a)$$

Here, r is the distance from the current-loop magnetic dipole $\vec{m}_I$ to the field point, $\hat{e}_r$ is the dimensionless unit vector pointing from the magnetic dipole to the field point, and c is the speed of light. Now, as explained in Section 4.2.13, a current-loop magnetic dipole is the same thing as a permanent-magnet dipole when working in Gaussian units; and, as explained in Section 4.2.27 or 4.2.28, the magnetic field and magnetic-induction field are the same in empty space when working in Gaussian units. Hence, when the reader comes across Eq. (4.24a) in some textbook using Gaussian units, it is likely to be written as either

$$\vec{B}\Big|_{\text{at time } t} = \frac{1}{c^2 r}\left\{ \hat{e}_r \times \left(\hat{e}_r \times \frac{d^2 \vec{m}}{dt^2} \right)\right\}\Bigg|_{\substack{\text{evaluated at the retarded} \\ \text{time } t'=t-(r/c)}}$$

or

$$\vec{H}\Big|_{\text{at time } t} = \frac{1}{c^2 r}\left\{ \hat{e}_r \times \left(\hat{e}_r \times \frac{d^2 \vec{m}}{dt^2} \right)\right\}\Bigg|_{\substack{\text{evaluated at the retarded} \\ \text{time } t'=t-(r/c)}} ,$$

with no distinction being made between $\vec{B}$ and $\vec{H}$ in empty space. Before this sort of equation can be converted to an electromagnetic system that makes these distinctions, such as the rationalized mks system, we must decide whether $\vec{m}$ should be thought of as a current-loop magnetic dipole $\vec{m}_I$ or as a permanent-magnet dipole $\vec{m}_H$, and whether the field in empty space should be thought of as a $\vec{B}$ field or an $\vec{H}$ field. As is pointed out in the discussion following Eq. (2.68d), all we are really deciding is whether to represent the relationship between the physical quantities $\vec{B}, \vec{H}, \vec{m}_I, \vec{m}_H$ in terms of $\vec{B}$ and $\vec{m}_I$, $\vec{H}$ and $\vec{m}_I$, $\vec{B}$ and $\vec{m}_H$, or $\vec{H}$ and $\vec{m}_H$ in the new electromagnetic system.

The choice between $\vec{m}_I$ and $\vec{m}_H$ is relatively easy; almost all textbooks written in the last 50 years use $\vec{m}_I$ exclusively to represent their magnetic dipoles, so that is the choice made now. It is also conventional to choose $\vec{H}$ to represent the magnetic component of a radiating electromagnetic field, because it makes the end result slightly simpler; so that is also the choice made now. Hence, instead of Eq. (4.24a), we write

$$\vec{H}\Big|_{\text{at time } t} = \frac{1}{c^2 r} \left\{ \hat{e}_r \times \left(\hat{e}_r \times \frac{d^2 \vec{m}_I}{dt^2} \right) \right\}\Bigg|_{\substack{\text{evaluated at the retarded} \\ \text{time } t'=t-(r/c)}} \tag{4.24b}$$

in Gaussian units. Table 4.4(b) shows how to convert this to the rationalized mks system to get

$$_f \vec{H} \sqrt{4\pi \, _f \mu_0}\Big|_{\text{at time } t} = \frac{1}{c^2 r} \left\{ \hat{e}_r \times \left[\hat{e}_r \times \frac{d^2}{dt^2} \left(\vec{m}_I \sqrt{\frac{_f \mu_0}{4\pi}} \right) \right] \right\}\Bigg|_{\substack{\text{evaluated at the retarded} \\ \text{time } t'=t-(r/c)}}$$

or

$$_f \vec{H}\Big|_{\text{at time } t} = \frac{1}{4\pi c^2 r} \left\{ \hat{e}_r \times \left(\hat{e}_r \times \frac{d^2 \vec{m}_I}{dt^2} \right) \right\}\Bigg|_{\substack{\text{evaluated at the retarded} \\ \text{time } t'=t-(r/c)}}. \tag{4.24c}$$

Table 4.10(a) shows how to convert this to esu units, another electromagnetic system which always preserves the distinction between $\vec{m}_I, \vec{m}_H$ and between $\vec{B}, \vec{H}$:

$$\left(\frac{1}{c} \vec{H} \right)\Bigg|_{\text{at time } t} = \frac{1}{c^2 r} \left\{ \hat{e}_r \times \left[\hat{e}_r \times \frac{d^2}{dt^2} \left(\frac{1}{c} \vec{m}_I \right) \right] \right\}\Bigg|_{\substack{\text{evaluated at the retarded} \\ \text{time } t'=t-(r/c)}}$$

or

$$\vec{H}\Big|_{\text{at time } t} = \frac{1}{c^2 r} \left\{ \hat{e}_r \times \left(\hat{e}_r \times \frac{d^2 \vec{m}_I}{dt^2} \right) \right\}\Bigg|_{\substack{\text{evaluated at the retarded} \\ \text{time } t'=t-(r/c)}}. \tag{4.24d}$$

This is the same form as Eq. (4.24b). The final system represented in these tables that always preserves the distinction between $\vec{m}_I, \vec{m}_H$ and between $\vec{B}, \vec{H}$ is the unrationalized mks system. Table 4.12(a) shows that in this system, Eq. (4.24b) becomes

$$\left(\vec{H} \sqrt{\mu_0} \right)\Big|_{\text{at time } t} = \frac{1}{c^2 r} \left\{ \hat{e}_r \times \left[\hat{e}_r \times \frac{d^2}{dt^2} \left(\vec{m}_I \sqrt{\mu_0} \right) \right] \right\}\Bigg|_{\substack{\text{evaluated at the retarded} \\ \text{time } t'=t-(r/c)}}$$

or

$$\vec{H}\Big|_{\text{at time } t} = \frac{1}{c^2 r} \left\{ \hat{e}_r \times \left(\hat{e}_r \times \frac{d^2 \vec{m}_I}{dt^2} \right) \right\}\Bigg|_{\substack{\text{evaluated at the retarded} \\ \text{time } t'=t-(r/c)}}. \tag{4.24e}$$

This also ends up having the same form as Eq. (4.24b).

If we choose to work with Eq. (4.24a) instead of replacing $\vec{B}$ with $\vec{H}$ to get Eq. (4.24b), then Tables 4.4(b), 4.10(a), and 4.12(a) would give us

$$\vec{B}\Big|_{\text{at time } t} = \frac{f\mu_0}{4\pi c^2 r}\left\{\hat{e}_r \times \left(\hat{e}_r \times \frac{d^2\vec{m}_I}{dt^2}\right)\right\}\Bigg|_{\substack{\text{evaluated at the retarded} \\ \text{time } t'=t-(r/c)}} \qquad (4.25a)$$

for its rationalized mks counterpart,

$$\vec{B}\Big|_{\text{at time } t} = \frac{1}{c^4 r}\left\{\hat{e}_r \times \left(\hat{e}_r \times \frac{d^2\vec{m}_I}{dt^2}\right)\right\}\Bigg|_{\substack{\text{evaluated at the retarded} \\ \text{time } t'=t-(r/c)}} \qquad (4.25b)$$

for its esu counterpart, and

$$\vec{B}\Big|_{\text{at time } t} = \frac{\mu_0}{c^2 r}\left\{\hat{e}_r \times \left(\hat{e}_r \times \frac{d^2\vec{m}_I}{dt^2}\right)\right\}\Bigg|_{\substack{\text{evaluated at the retarded} \\ \text{time } t'=t-(r/c)}} \qquad (4.25c)$$

for its unrationalized mks counterpart. In the rationalized mks system

$$_f\vec{H} = \vec{B}/_f\mu_0,$$

in esu units*

$$\vec{H} = c^2\vec{B},$$

and in the unrationalized mks system

$$\vec{H} = \vec{B}/\mu_0.$$

These three formulas can be plugged into Eqs. (4.24c, d, e) to get Eqs. (4.25a, b, c) directly, showing that we can change back and forth between B and H without consulting the substitution tables.

One last point worth making is that when we convert from esu units to the unrationalized mks system in Table 4.17(a), the parameter μ_0 in esu units has a different meaning from the parameter μ_0 in the unrationalized mks system. In esu units, μ_0 is really just another name for c^{-2}, as explained in Eq. (2.11b). Consequently, we can write the relationship between H and B in esu units as

$$\vec{H} = \frac{\vec{B}}{\mu_0} - 4\pi\vec{M}_I$$

---

* The esu relationship between $\vec{H}$ and $\vec{B}$ can also be written as $\vec{H} = \vec{B}/\mu_0$ in empty space because $\mu_0 = c^{-2}$ in esu units—see Eq. (2.11b).

instead of as

$$\vec{H} = c^2 \vec{B} - 4\pi \vec{M}_I.$$

Applying Table 4.17(a) to either of these equations now gives

$$\vec{H} \cdot \left(c\sqrt{\mu_0}\right) = c^2 \left(\frac{\vec{B}}{c\sqrt{\mu_0}}\right) - 4\pi \vec{M}_I \cdot \left(c\sqrt{\mu_0}\right),$$

which reduces to the same relationship between H and B in the unrationalized mks system, namely

$$\vec{H} = \frac{\vec{B}}{\mu_0} - 4\pi \vec{M}_I.$$

The instruction in Table 4.17(a) to replace μ_0 by c^{-2} should not be applied to this final result; we are only allowed to substitute once for μ_0 when converting from esu units to the unrationalized mks system. After we have substituted once for all the μ_0's in the original equation, the μ_0's in the final unrationalized mks equation should be left alone because they stand for 10^{-7}henry/m, not c^{-2}.

4.5 PROBLEMS WITH THE FREE-PARAMETER METHOD AND SUBSTITUTION TABLES

Free-parameter methods and substitution tables work best when used by people who understand what it means to recognize—or not recognize—charge as a separate dimension. Consider, for example, the electromagnetic equation

$$_f\mu_0 \, _f\varepsilon_0 = \frac{1}{c^2}$$

in the rationalized mks system. In rationalized mks, both $_f\varepsilon_0$ and $_f\mu_0$ are physical quantities with units of farad/m and henry/m, respectively, but nowhere in Section 4.2 is there a free-parameter equation or formula resembling it. Deciding next to try the substitution tables, we use Table 4.4(a) in the Appendix to convert it to Gaussian units, getting

$$_f\mu_0 \cdot \left(_f\mu_0 c^2\right)^{-1} = \frac{1}{c^2} \quad \text{or} \quad \frac{1}{c^2} = \frac{1}{c^2}.$$

The same thing happens when Table 4.5(a) is used to convert to the Heaviside-Lorentz system, or when Tables 4.6(a) and 4.7(a) are used to convert to esu and emu units, respectively. Only when the equation is transformed to the unrationalized mks system is it preserved; Table 4.8(a) shows that it then becomes

$$(4\pi\mu_0) \cdot \left(\frac{\varepsilon_0}{4\pi}\right) = \frac{1}{c^2} \quad \text{or} \quad \mu_0\varepsilon_0 = \frac{1}{c^2}.$$

The reason things go wrong in the Gaussian, Heaviside-Lorentz, esu, and emu systems is that in these systems, charge is not recognized as a separate dimension— therefore no counterpart equation exists. The unrationalized mks system, on the other hand, imitates the rationalized mks system in recognizing charge as a separate dimension; consequently, in the unrationalized mks system there is a counterpart formula.

The only practical way to avoid conundrums of this sort is to understand the material in Chapters 1 through 3 well enough to see why not all electromagnetic equations have counterparts in every electromagnetic system discussed in this book. Equations that do not have counterparts in all the electromagnetic systems are very difficult to set up as sensible free-parameter expressions; and when transformed using substitution tables, they tend to end up as nonsense expressions like $c^{-2} = c^{-2}$. Naive users should approach free-parameter lists and substitution tables cautiously until they become familiar with the basic rules for transforming electromagnetic equations and formulas.

APPENDIX. SUBSTITUTION TABLES FOR THE RATIONALIZED MKS SYSTEM (ALSO CALLED SI UNITS), GAUSSIAN CGS UNITS, THE HEAVISIDE-LORENTZ CGS SYSTEM, ESU UNITS, EMU UNITS, AND THE UNRATIONALIZED MKS SYSTEM

This Appendix contains 30 substitution tables, which can be used to transform electromagnetic equations into any of the electromagnetic systems described in this book. If a table whose name ends in "a" gives the transformation from the "X" electromagnetic system to the "Y" electromagnetic system, then the table immediately following it whose name ends in "b" gives the reverse transformation, describing how to transform from the "Y" electromagnetic system to the "X" electromagnetic system. For example, since Table 4.9(a) transforms equations from Gaussian cgs units to the Heaviside-Lorentz cgs system, Table 4.9(b) must transform equations from the Heaviside-Lorentz cgs system to Gaussian cgs units. Similarly, since Table 4.18(b) transforms equations from the unrationalized mks system to emu units, Table 4.18(a) must transform equations from emu units to the unrationalized mks system. The substitution information for the rationalized mks system (or SI units) comes first, covered in Tables 4.4(a)–(b) to 4.8(a)–(b). Tables 4.9(a)–(b) to 4.12(a)–(b) for the Gaussian transformations come next. Since the transformation between Gaussian cgs units and the rationalized mks system has already been described in the first set of tables, there are two fewer tables than before. Next are Tables 4.13(a)–(b) to 4.15(a)–(b) for Heaviside-Lorentz transformations, followed by Tables 4.16(a)–(b) to 4.18(a)–(b) for the remaining esu, emu, and unrationalized mks transformations. According to the discussion at the beginning of Section 3.6, all electromagnetic equations have the same form in the esuq and emuq units of Chapter 2 as they do in the unrationalized mks system. This means that all tables in this appendix showing how to transform equations to and from the unrationalized mks system also show how to transform equations to and from the esuq and emuq systems of units.

REFERENCES

1. E. S. Shire, *Classical Electricity and Magnetism*, Cambridge University Press, New York, 1960.
2. J. D. Jackson, *Classical Electrodynamics*, 3rd edition, pp. 177–178, John Wiley & Sons, New York, 1999 for a discussion of how to convert the first double integral into the second double integral.

Table 4.4(a) To go from the rationalized mks system (also called SI units) to Gaussian cgs units.

magnetic vector potential A **Replace A by $A\sqrt{_f\mu_0/(4\pi)}$**	magnetic permeability $_f\mu$ **Replace $_f\mu$ by $\mu_r \cdot {_f\mu_0}$**
magnetic induction B **Replace B by $B\sqrt{_f\mu_0/(4\pi)}$**	relative magnetic permeability μ_r **Leave μ_r alone**
capacitance C **Replace C by $(4\pi C)/(_f\mu_0 c^2)$**	magnetic permeability of free space $_f\mu_0$ **Leave $_f\mu_0$ alone**
electric displacement $_fD$ **Replace $_fD$ by $D/(c\sqrt{4\pi\,_f\mu_0})$**	magnetic pole strength $_fp_H$ **Replace $_fp_H$ by $p_H\sqrt{4\pi\,_f\mu_0}$**
electric field E **Replace E by $cE\sqrt{_f\mu_0/(4\pi)}$**	electric dipole moment p **Replace p by $pc^{-1}\sqrt{(4\pi)/_f\mu_0}$**
dielectric constant $_f\varepsilon$ **Replace $_f\varepsilon$ by $\varepsilon_r/(_f\mu_0 c^2)$**	electric dipole density P **Replace P by $Pc^{-1}\sqrt{(4\pi)/_f\mu_0}$**
relative dielectric constant ε_r **Leave ε_r alone**	permeance $_f\mathcal{P}$ **Replace $_f\mathcal{P}$ by $\mathcal{P} \cdot {_f\mu_0}$**
permittivity of free space $_f\varepsilon_0$ **Replace $_f\varepsilon_0$ by $(_f\mu_0 c^2)^{-1}$**	charge Q **Replace Q by $Qc^{-1}\sqrt{(4\pi)/_f\mu_0}$**
magnetomotive force $_f\mathcal{F}$ **Replace $_f\mathcal{F}$ by $\mathcal{F}/\sqrt{4\pi\,_f\mu_0}$**	resistance R **Replace R by $(_f\mu_0 c^2 R)/(4\pi)$**
magnetic flux Φ_B **Replace Φ_B by $\Phi_B\sqrt{_f\mu_0/(4\pi)}$**	reluctance $_f\mathcal{R}$ **Replace $_f\mathcal{R}$ by $\mathcal{R}/_f\mu_0$**
conductance G **Replace G by $(4\pi G)/(_f\mu_0 c^2)$**	volume charge density ρ_Q **Replace ρ_Q by $\rho_Q c^{-1}\sqrt{(4\pi)/_f\mu_0}$**
magnetic field $_fH$ **Replace $_fH$ by $H/\sqrt{4\pi\,_f\mu_0}$**	resistivity ρ_R **Replace ρ_R by $(_f\mu_0 c^2 \rho_R)/(4\pi)$**
current I **Replace I by $Ic^{-1}\sqrt{(4\pi)/_f\mu_0}$**	elastance S **Replace S by $(_f\mu_0 c^2 S)/(4\pi)$**
volume current density J **Replace J by $Jc^{-1}\sqrt{(4\pi)/_f\mu_0}$**	surface charge density S_Q **Replace S_Q by $S_Q c^{-1}\sqrt{(4\pi)/_f\mu_0}$**
surface current density $\mathcal{J}_S$ **Replace $\mathcal{J}_S$ by $\mathcal{J}_S c^{-1}\sqrt{(4\pi)/_f\mu_0}$**	conductivity σ **Replace σ by $(4\pi\sigma)/(_f\mu_0 c^2)$**
inductance L **Replace L by $(_f\mu_0 c^2 L)/(4\pi)$**	electric potential V **Replace V by $cV\sqrt{_f\mu_0/(4\pi)}$**
permanent-magnet dipole moment $_fm_H$ **Replace $_fm_H$ by $m_H\sqrt{4\pi\,_f\mu_0}$**	magnetic scalar potential $_f\Omega_H$ **Replace $_f\Omega_H$ by $\Omega_H/\sqrt{4\pi\,_f\mu_0}$**
current-loop magnetic dipole moment m_I **Replace m_I by $m_I\sqrt{(4\pi)/_f\mu_0}$**	electric susceptibility $_f\chi_e$ **Replace $_f\chi_e$ by $4\pi\chi_e$**
permanent-magnet dipole density $_fM_H$ **Replace $_fM_H$ by $M_H\sqrt{4\pi\,_f\mu_0}$**	magnetic susceptibility $_f\chi_m$ **Replace $_f\chi_m$ by $4\pi\chi_m$**
current-loop magnetic dipole density M_I **Replace M_I by $M_I\sqrt{(4\pi)/_f\mu_0}$**	

Table 4.4(b) To go from Gaussian cgs units to the rationalized mks system (also called SI units).

magnetic vector potential A **Replace A by $A\sqrt{(4\pi)/{}_f\mu_0}$**	relative magnetic permeability μ_r **Leave μ_r alone**
magnetic induction B **Replace B by $B\sqrt{(4\pi)/{}_f\mu_0}$**	magnetic pole strength p_H **Replace p_H by ${}_f p_H/\sqrt{4\pi\,{}_f\mu_0}$**
capacitance C **Replace C by $({}_f\mu_0 c^2 C)/(4\pi)$**	electric dipole moment p **Replace p by $pc\sqrt{{}_f\mu_0/(4\pi)}$**
electric displacement D **Replace D by ${}_f D \cdot (c\sqrt{4\pi\,{}_f\mu_0})$**	electric dipole density P **Replace P by $Pc\sqrt{{}_f\mu_0/(4\pi)}$**
electric field E **Replace E by $Ec^{-1}\sqrt{(4\pi)/{}_f\mu_0}$**	permeance $\mathcal{P}$ **Replace $\mathcal{P}$ by ${}_f\mathcal{P}/{}_f\mu_0$**
relative dielectric constant ε_r **Leave ε_r alone**	charge Q **Replace Q by $Qc\sqrt{{}_f\mu_0/(4\pi)}$**
magnetomotive force $\mathcal{F}$ **Replace $\mathcal{F}$ by ${}_f\mathcal{F}\sqrt{4\pi\,{}_f\mu_0}$**	resistance R **Replace R by $(4\pi R)/({}_f\mu_0 c^2)$**
magnetic flux Φ_B **Replace Φ_B by $\Phi_B\sqrt{(4\pi)/{}_f\mu_0}$**	reluctance $\mathcal{R}$ **Replace $\mathcal{R}$ by ${}_f\mathcal{R} \cdot {}_f\mu_0$**
conductance G **Replace G by $({}_f\mu_0 c^2 G)/(4\pi)$**	volume charge density ρ_Q **Replace ρ_Q by $\rho_Q c\sqrt{{}_f\mu_0/(4\pi)}$**
magnetic field H **Replace H by ${}_f H\sqrt{4\pi\,{}_f\mu_0}$**	resistivity ρ_R **Replace ρ_R by $(4\pi\rho_R)/({}_f\mu_0 c^2)$**
current I **Replace I by $Ic\sqrt{{}_f\mu_0/(4\pi)}$**	elastance S **Replace S by $(4\pi S)/({}_f\mu_0 c^2)$**
volume current density J **Replace J by $Jc\sqrt{{}_f\mu_0/(4\pi)}$**	surface charge density S_Q **Replace S_Q by $S_Q c\sqrt{{}_f\mu_0/(4\pi)}$**
surface current density $\mathcal{J}_S$ **Replace $\mathcal{J}_S$ by $J_S c\sqrt{{}_f\mu_0/(4\pi)}$**	conductivity σ **Replace σ by $({}_f\mu_0 c^2\sigma)/(4\pi)$**
inductance L **Replace L by $(4\pi L)/({}_f\mu_0 c^2)$**	electric potential V **Replace V by $Vc^{-1}\sqrt{(4\pi)/{}_f\mu_0}$**
permanent-magnet dipole moment m_H **Replace m_H by ${}_f m_H/\sqrt{4\pi\,{}_f\mu_0}$**	magnetic scalar potential Ω_H **Replace Ω_H by ${}_f\Omega_H\sqrt{4\pi\,{}_f\mu_0}$**
current-loop magnetic dipole moment m_I **Replace m_I by $m_I\sqrt{{}_f\mu_0/(4\pi)}$**	electric susceptibility χ_e **Replace χ_e by ${}_f\chi_e/(4\pi)$**
permanent-magnet dipole density M_H **Replace M_H by ${}_f M_H/\sqrt{4\pi\,{}_f\mu_0}$**	magnetic susceptibility χ_m **Replace χ_m by ${}_f\chi_m/(4\pi)$**
current-loop magnetic dipole density M_I **Replace M_I by $M_I\sqrt{{}_f\mu_0/(4\pi)}$**	

Table 4.5(a) To go from the rationalized mks system (also called SI units) to the Heaviside-Lorentz cgs system.

magnetic vector potential A	magnetic permeability $_f\mu$
Replace A by $_hA\sqrt{_f\mu_0}$	**Replace $_f\mu$ by $\mu_r \cdot {_f\mu_0}$**
magnetic induction B	relative magnetic permeability μ_r
Replace B by $_hB\sqrt{_f\mu_0}$	**Leave μ_r alone**
capacitance C	magnetic permeability of free space $_f\mu_0$
Replace C by $_hC/(_f\mu_0c^2)$	**Leave $_f\mu_0$ alone**
electric displacement $_fD$	magnetic pole strength $_fp_H$
Replace $_fD$ by $_hD/(c\sqrt{_f\mu_0})$	**Replace $_fp_H$ by $_hp_H\sqrt{_f\mu_0}$**
electric field E	electric dipole moment p
Replace E by $_hEc\sqrt{_f\mu_0}$	**Replace p by $_hp(c\sqrt{_f\mu_0})^{-1}$**
dielectric constant $_f\varepsilon$	electric dipole density P
Replace $_f\varepsilon$ by $\varepsilon_r/(_f\mu_0c^2)$	**Replace P by $_hP(c\sqrt{_f\mu_0})^{-1}$**
relative dielectric constant ε_r	permeance $_f\mathcal{P}$
Leave ε_r alone	**Replace $_f\mathcal{P}$ by $\mathcal{P} \cdot {_f\mu_0}$**
permittivity of free space $_f\varepsilon_0$	charge Q
Replace $_f\varepsilon_0$ by $(_f\mu_0c^2)^{-1}$	**Replace Q by $_hQ(c\sqrt{_f\mu_0})^{-1}$**
magnetomotive force $_f\mathcal{F}$	resistance R
Replace $_f\mathcal{F}$ by $_h\mathcal{F}/\sqrt{_f\mu_0}$	**Replace R by $_hR \cdot (_f\mu_0c^2)$**
magnetic flux Φ_B	reluctance $_f\mathcal{R}$
Replace Φ_B by $_h\Phi_B\sqrt{_f\mu_0}$	**Replace $_f\mathcal{R}$ by $\mathcal{R}/_f\mu_0$**
conductance G	volume charge density ρ_Q
Replace G by $_hG/(_f\mu_0c^2)$	**Replace ρ_Q by $_h\rho_Q(c\sqrt{_f\mu_0})^{-1}$**
magnetic field $_fH$	resistivity ρ_R
Replace $_fH$ by $_hH/\sqrt{_f\mu_0}$	**Replace ρ_R by $_h\rho_R \cdot (_f\mu_0c^2)$**
current I	elastance S
Replace I by $_hI(c\sqrt{_f\mu_0})^{-1}$	**Replace S by $_hS \cdot (_f\mu_0c^2)$**
volume current density J	surface charge density S_Q
Replace J by $_hJ(c\sqrt{_f\mu_0})^{-1}$	**Replace S_Q by $_hS_Q(c\sqrt{_f\mu_0})^{-1}$**
surface current density $\mathcal{J}_S$	conductivity σ
Replace $\mathcal{J}_S$ by $_h\mathcal{J}_S(c\sqrt{_f\mu_0})^{-1}$	**Replace σ by $_h\sigma/(_f\mu_0c^2)$**
inductance L	electric potential V
Replace L by $_hL \cdot (_f\mu_0c^2)$	**Replace V by $_hVc\sqrt{_f\mu_0}$**
permanent-magnet dipole moment $_fm_H$	magnetic scalar potential $_f\Omega_H$
Replace $_fm_H$ by $_hm_H\sqrt{_f\mu_0}$	**Replace $_f\Omega_H$ by $_h\Omega_H/\sqrt{_f\mu_0}$**
current-loop magnetic dipole moment m_I	electric susceptibility $_f\chi_e$
Replace m_I by $_hm_I/\sqrt{_f\mu_0}$	**Replace $_f\chi_e$ by $_h\chi_e$**
permanent-magnet dipole density $_fM_H$	magnetic susceptibility $_f\chi_m$
Replace $_fM_H$ by $_hM_H\sqrt{_f\mu_0}$	**Replace $_f\chi_m$ by $_h\chi_m$**
current-loop magnetic dipole density M_I	
Replace M_I by $_hM_I/\sqrt{_f\mu_0}$	

Table 4.5(b) To go from the Heaviside-Lorentz cgs system to the rationalized mks system (also called SI units).

magnetic vector potential $_h A$ **Replace $_h A$ by $A/\sqrt{f\mu_0}$**	relative magnetic permeability μ_r **Leave μ_r alone**
magnetic induction $_h B$ **Replace $_h B$ by $B/\sqrt{f\mu_0}$**	magnetic pole strength $_h p_H$ **Replace $_h p_H$ by $_f p_H/\sqrt{f\mu_0}$**
capacitance $_h C$ **Replace $_h C$ by $C \cdot (_f\mu_0 c^2)$**	electric dipole moment $_h p$ **Replace $_h p$ by $p \cdot (c\sqrt{f\mu_0})$**
electric displacement $_h D$ **Replace $_h D$ by $_f D \cdot (c\sqrt{f\mu_0})$**	electric dipole density $_h P$ **Replace $_h P$ by $P \cdot (c\sqrt{f\mu_0})$**
electric field $_h E$ **Replace $_h E$ by $E/(c\sqrt{f\mu_0})$**	permeance $\mathcal{P}$ **Replace $\mathcal{P}$ by $_f\mathcal{P}/_f\mu_0$**
relative dielectric constant ε_r **Leave ε_r alone**	charge $_h Q$ **Replace $_h Q$ by $Q \cdot (c\sqrt{f\mu_0})$**
magnetomotive force $_h\mathcal{F}$ **Replace $_h\mathcal{F}$ by $_f\mathcal{F} \cdot \sqrt{f\mu_0}$**	resistance $_h R$ **Replace $_h R$ by $R/(_f\mu_0 c^2)$**
magnetic flux $_h\Phi_B$ **Replace $_h\Phi_B$ by $\Phi_B/\sqrt{f\mu_0}$**	reluctance $\mathcal{R}$ **Replace $\mathcal{R}$ by $_f\mathcal{R} \cdot _f\mu_0$**
conductance $_h G$ **Replace $_h G$ by $G \cdot (_f\mu_0 c^2)$**	volume charge density $_h\rho_Q$ **Replace $_h\rho_Q$ by $\rho_Q \cdot (c\sqrt{f\mu_0})$**
magnetic field $_h H$ **Replace $_h H$ by $_f H \cdot \sqrt{f\mu_0}$**	resistivity $_h\rho_R$ **Replace $_h\rho_R$ by $\rho_R/(_f\mu_0 c^2)$**
current $_h I$ **Replace $_h I$ by $I \cdot (c\sqrt{f\mu_0})$**	elastance $_h S$ **Replace $_h S$ by $S/(_f\mu_0 c^2)$**
volume current density $_h J$ **Replace $_h J$ by $J \cdot (c\sqrt{f\mu_0})$**	surface charge density $_h S_Q$ **Replace $_h S_Q$ by $S_Q \cdot (c\sqrt{f\mu_0})$**
surface current density $_h\mathcal{J}_S$ **Replace $_h\mathcal{J}_S$ by $\mathcal{J}_S \cdot (c\sqrt{f\mu_0})$**	conductivity $_h\sigma$ **Replace $_h\sigma$ by $\sigma \cdot (_f\mu_0 c^2)$**
inductance $_h L$ **Replace $_h L$ by $L/(_f\mu_0 c^2)$**	electric potential $_h V$ **Replace $_h V$ by $V/(c\sqrt{f\mu_0})$**
permanent-magnet dipole moment $_h m_H$ **Replace $_h m_H$ by $_f m_H/\sqrt{f\mu_0}$**	magnetic scalar potential $_h\Omega_H$ **Replace $_h\Omega_H$ by $_f\Omega_H \cdot \sqrt{f\mu_0}$**
current-loop magnetic dipole moment $_h m_I$ **Replace $_h m_I$ by $m_I \cdot \sqrt{f\mu_0}$**	electric susceptibility $_h\chi_e$ **Replace $_h\chi_e$ by $_f\chi_e$**
permanent-magnet dipole density $_h M_H$ **Replace $_h M_H$ by $_f M_H/\sqrt{f\mu_0}$**	magnetic susceptibility $_h\chi_m$ **Replace $_h\chi_m$ by $_f\chi_m$**
current-loop magnetic dipole density $_h M_I$ **Replace $_h M_I$ by $M_I \cdot \sqrt{f\mu_0}$**	

Table 4.6(a) To go from the rationalized mks system (also called SI units) to esu units.

magnetic vector potential A	magnetic permeability $_f\mu$
Replace A by $A \cdot (c\sqrt{_f\mu_0/(4\pi)})$	**Replace $_f\mu$ by $\mu_r \cdot _f\mu_0$**
magnetic induction B	relative magnetic permeability μ_r
Replace B by $B \cdot (c\sqrt{_f\mu_0/(4\pi)})$	**Leave μ_r alone**
capacitance C	magnetic permeability of free space $_f\mu_0$
Replace C by $(4\pi C)/(_f\mu_0 c^2)$	**Leave $_f\mu_0$ alone**
electric displacement $_f D$	magnetic pole strength $_f p_H$
Replace $_f D$ by $D/(c\sqrt{4\pi \,_f\mu_0})$	**Replace $_f p_H$ by $p_H(c\sqrt{4\pi \,_f\mu_0})$**
electric field E	electric dipole moment p
Replace E by $cE\sqrt{_f\mu_0/(4\pi)}$	**Replace p by $pc^{-1}\sqrt{(4\pi)/_f\mu_0}$**
dielectric constant $_f\varepsilon$	electric dipole density P
Replace $_f\varepsilon$ by $\varepsilon_r/(_f\mu_0 c^2)$	**Replace P by $Pc^{-1}\sqrt{(4\pi)/_f\mu_0}$**
relative dielectric constant ε_r	permeance $_f\mathcal{P}$
Leave ε_r alone	**Replace $_f\mathcal{P}$ by $\mathcal{P} \cdot (_f\mu_0 c^2)$**
permittivity of free space $_f\varepsilon_0$	charge Q
Replace $_f\varepsilon_0$ by $(_f\mu_0 c^2)^{-1}$	**Replace Q by $Qc^{-1}\sqrt{(4\pi)/_f\mu_0}$**
magnetomotive force $_f\mathcal{F}$	resistance R
Replace $_f\mathcal{F}$ by $\mathcal{F}/(c\sqrt{4\pi \,_f\mu_0})$	**Replace R by $(_f\mu_0 c^2 R)/(4\pi)$**
magnetic flux Φ_B	reluctance $_f\mathcal{R}$
Replace Φ_B by $\Phi_B(c\sqrt{_f\mu_0/(4\pi)})$	**Replace $_f\mathcal{R}$ by $\mathcal{R}/(_f\mu_0 c^2)$**
conductance G	volume charge density ρ_Q
Replace G by $(4\pi G)/(_f\mu_0 c^2)$	**Replace ρ_Q by $\rho_Q c^{-1}\sqrt{(4\pi)/_f\mu_0}$**
magnetic field $_f H$	resistivity ρ_R
Replace $_f H$ by $H/(c\sqrt{4\pi \,_f\mu_0})$	**Replace ρ_R by $(_f\mu_0 c^2 \rho_R)/(4\pi)$**
current I	elastance S
Replace I by $Ic^{-1}\sqrt{(4\pi)/_f\mu_0}$	**Replace S by $(_f\mu_0 c^2 S)/(4\pi)$**
volume current density J	surface charge density S_Q
Replace J by $Jc^{-1}\sqrt{(4\pi)/_f\mu_0}$	**Replace S_Q by $S_Q c^{-1}\sqrt{(4\pi)/_f\mu_0}$**
surface current density $\mathcal{J}_S$	conductivity σ
Replace $\mathcal{J}_S$ by $\mathcal{J}_S c^{-1}\sqrt{(4\pi)/_f\mu_0}$	**Replace σ by $(4\pi\sigma)/(_f\mu_0 c^2)$**
inductance L	electric potential V
Replace L by $(_f\mu_0 c^2 L)/(4\pi)$	**Replace V by $cV\sqrt{_f\mu_0/(4\pi)}$**
permanent-magnet dipole moment $_f m_H$	magnetic scalar potential $_f\Omega_H$
Replace $_f m_H$ by $m_H(c\sqrt{4\pi \,_f\mu_0})$	**Replace $_f\Omega_H$ by $\Omega_H/(c\sqrt{4\pi \,_f\mu_0})$**
current-loop magnetic dipole moment m_I	electric susceptibility $_f\chi_e$
Replace m_I by $m_I c^{-1}\sqrt{(4\pi)/_f\mu_0}$	**Replace $_f\chi_e$ by $4\pi\chi_e$**
permanent-magnet dipole density $_f M_H$	magnetic susceptibility $_f\chi_m$
Replace $_f M_H$ by $M_H(c\sqrt{4\pi \,_f\mu_0})$	**Replace $_f\chi_m$ by $4\pi\chi_m$**
current-loop magnetic dipole density M_I	
Replace M_I by $M_I c^{-1}\sqrt{(4\pi)/_f\mu_0}$	

Table 4.6(b) To go from esu units to the rationalized mks system (also called SI units).

magnetic vector potential A	relative magnetic permeability μ_r
Replace A by $Ac^{-1}\sqrt{(4\pi)/_f\mu_0}$	**Leave μ_r alone**
magnetic induction B	magnetic permeability of free space μ_0
Replace B by $Bc^{-1}\sqrt{(4\pi)/_f\mu_0}$	**Replace μ_0 by** c^{-2}
capacitance C	magnetic pole strength p_H
Replace C by $(_f\mu_0 c^2 C)/(4\pi)$	**Replace p_H by** $_f p_H/(c\sqrt{4\pi\,_f\mu_0})$
electric displacement D	electric dipole moment p
Replace D by $_f D \cdot (c\sqrt{4\pi\,_f\mu_0})$	**Replace p by** $pc\sqrt{_f\mu_0/(4\pi)}$
electric field E	electric dipole density P
Replace E by $Ec^{-1}\sqrt{(4\pi)/_f\mu_0}$	**Replace P by** $Pc\sqrt{_f\mu_0/(4\pi)}$
relative dielectric constant ε_r	permeance $\mathcal{P}$
Leave ε_r alone	**Replace $\mathcal{P}$ by** $_f\mathcal{P}/(_f\mu_0 c^2)$
magnetomotive force $\mathcal{F}$	charge Q
Replace $\mathcal{F}$ by $_f\mathcal{F} \cdot c\sqrt{4\pi\,_f\mu_0}$	**Replace Q by** $Q \cdot c\sqrt{_f\mu_0/(4\pi)}$
magnetic flux Φ_B	resistance R
Replace Φ_B by $\Phi_B c^{-1}\sqrt{(4\pi)/_f\mu_0}$	**Replace R by** $(4\pi R)/(_f\mu_0 c^2)$
conductance G	reluctance $\mathcal{R}$
Replace G by $(_f\mu_0 c^2 G)/(4\pi)$	**Replace $\mathcal{R}$ by** $_f\mathcal{R} \cdot (_f\mu_0 c^2)$
magnetic field H	volume charge density ρ_Q
Replace H by $_f H \cdot c\sqrt{4\pi\,_f\mu_0}$	**Replace ρ_Q by** $\rho_Q c\sqrt{_f\mu_0/(4\pi)}$
current I	resistivity ρ_R
Replace I by $Ic\sqrt{_f\mu_0/(4\pi)}$	**Replace ρ_R by** $(4\pi\rho_R)/(_f\mu_0 c^2)$
volume current density J	elastance S
Replace J by $Jc\sqrt{_f\mu_0/(4\pi)}$	**Replace S by** $(4\pi S)/(_f\mu_0 c^2)$
surface current density $\mathcal{J}_S$	surface charge density S_Q
Replace $\mathcal{J}_S$ by $\mathcal{J}_S c\sqrt{_f\mu_0/(4\pi)}$	**Replace S_Q by** $S_Q c\sqrt{_f\mu_0/(4\pi)}$
inductance L	conductivity σ
Replace L by $(4\pi L)/(_f\mu_0 c^2)$	**Replace σ by** $(_f\mu_0 c^2\sigma)/(4\pi)$
permanent-magnet dipole moment m_H	electric potential V
Replace m_H by $_f m_H/(c\sqrt{4\pi\,_f\mu_0})$	**Replace V by** $Vc^{-1}\sqrt{(4\pi)/_f\mu_0}$
current-loop magnetic dipole moment m_I	magnetic scalar potential Ω_H
Replace m_I by $m_I \cdot c\sqrt{_f\mu_0/(4\pi)}$	**Replace Ω_H by** $_f\Omega_H \cdot c\sqrt{4\pi\,_f\mu_0}$
permanent-magnet dipole density M_H	electric susceptibility χ_e
Replace M_H by $_f M_H/(c\sqrt{4\pi\,_f\mu_0})$	**Replace χ_e by** $_f\chi_e/(4\pi)$
current-loop magnetic dipole density M_I	magnetic susceptibility χ_m
Replace M_I by $M_I \cdot c\sqrt{_f\mu_0/(4\pi)}$	**Replace χ_m by** $_f\chi_m/(4\pi)$
magnetic permeability μ	
Replace μ by $\mu_r c^{-2}$	

Table 4.7(a) To go from the rationalized mks system (also called SI units) to emu units.

magnetic vector potential A	magnetic permeability $_f\mu$
Replace A by $A\sqrt{_f\mu_0/(4\pi)}$	*Replace $_f\mu$ by $\mu_r \cdot {}_f\mu_0$*
magnetic induction B	relative magnetic permeability μ_r
Replace B by $B\sqrt{_f\mu_0/(4\pi)}$	*Leave μ_r alone*
capacitance C	magnetic permeability of free space $_f\mu_0$
Replace C by $(4\pi C)/_f\mu_0$	*Leave $_f\mu_0$ alone*
electric displacement $_fD$	magnetic pole strength $_fp_H$
Replace $_fD$ by $D/\sqrt{4\pi\,_f\mu_0}$	*Replace $_fp_H$ by $p_H\sqrt{4\pi\,_f\mu_0}$*
electric field E	electric dipole moment p
Replace E by $E\sqrt{_f\mu_0/(4\pi)}$	*Replace p by $p\sqrt{(4\pi)/_f\mu_0}$*
dielectric constant $_f\varepsilon$	electric dipole density P
Replace $_f\varepsilon$ by $\varepsilon_r/(_f\mu_0c^2)$	*Replace P by $P\sqrt{(4\pi)/_f\mu_0}$*
relative dielectric constant ε_r	permeance $_f\mathcal{P}$
Leave ε_r alone	*Replace $_f\mathcal{P}$ by $\mathcal{P}\cdot{}_f\mu_0$*
permittivity of free space $_f\varepsilon_0$	charge Q
Replace $_f\varepsilon_0$ by $(_f\mu_0c^2)^{-1}$	*Replace Q by $Q\sqrt{(4\pi)/_f\mu_0}$*
magnetomotive force $_f\mathcal{F}$	resistance R
Replace $_f\mathcal{F}$ by $\mathcal{F}/\sqrt{4\pi\,_f\mu_0}$	*Replace R by $(_f\mu_0R)/(4\pi)$*
magnetic flux Φ_B	reluctance $_f\mathcal{R}$
Replace Φ_B by $\Phi_B\sqrt{_f\mu_0/(4\pi)}$	*Replace $_f\mathcal{R}$ by $\mathcal{R}/_f\mu_0$*
conductance G	volume charge density ρ_Q
Replace G by $(4\pi G)/_f\mu_0$	*Replace ρ_Q by $\rho_Q\sqrt{(4\pi)/_f\mu_0}$*
magnetic field $_fH$	resistivity ρ_R
Replace $_fH$ by $H/\sqrt{4\pi\,_f\mu_0}$	*Replace ρ_R by $(_f\mu_0\rho_R)/(4\pi)$*
current I	elastance S
Replace I by $I\sqrt{(4\pi)/_f\mu_0}$	*Replace S by $(_f\mu_0S)/(4\pi)$*
volume current density J	surface charge density S_Q
Replace J by $J\sqrt{(4\pi)/_f\mu_0}$	*Replace S_Q by $S_Q\sqrt{(4\pi)/_f\mu_0}$*
surface current density $\mathcal{J}_S$	conductivity σ
Replace $\mathcal{J}_S$ by $\mathcal{J}_S\sqrt{(4\pi)/_f\mu_0}$	*Replace σ by $(4\pi\sigma)/_f\mu_0$*
inductance L	electric potential V
Replace L by $(_f\mu_0L)/(4\pi)$	*Replace V by $V\sqrt{_f\mu_0/(4\pi)}$*
permanent-magnet dipole moment $_fm_H$	magnetic scalar potential $_f\Omega_H$
Replace $_fm_H$ by $m_H\sqrt{4\pi\,_f\mu_0}$	*Replace $_f\Omega_H$ by $\Omega_H/\sqrt{4\pi\,_f\mu_0}$*
current-loop magnetic dipole moment m_I	electric susceptibility $_f\chi_e$
Replace m_I by $m_I\sqrt{(4\pi)/_f\mu_0}$	*Replace $_f\chi_e$ by $4\pi\chi_e$*
permanent-magnet dipole density $_fM_H$	magnetic susceptibility $_f\chi_m$
Replace $_fM_H$ by $M_H\sqrt{4\pi\,_f\mu_0}$	*Replace $_f\chi_m$ by $4\pi\chi_m$*
current-loop magnetic dipole density M_I	
Replace M_I by $M_I\sqrt{(4\pi)/_f\mu_0}$	

Table 4.7(b) To go from emu units to the rationalized mks system (also called SI units).

magnetic vector potential A	current-loop magnetic dipole density M_I
Replace A by $A\sqrt{(4\pi)/_f\mu_0}$	**Replace M_I by $M_I\sqrt{_f\mu_0/(4\pi)}$**
magnetic induction B	relative magnetic permeability μ_r
Replace B by $B\sqrt{(4\pi)/_f\mu_0}$	**Leave μ_r alone**
capacitance C	magnetic pole strength p_H
Replace C by $(_f\mu_0 C)/(4\pi)$	**Replace p_H by $_f p_H/\sqrt{4\pi\,_f\mu_0}$**
electric displacement D	electric dipole moment p
Replace D by $_f D\sqrt{4\pi\,_f\mu_0}$	**Replace p by $p\sqrt{_f\mu_0/(4\pi)}$**
electric field E	electric dipole density P
Replace E by $E\sqrt{(4\pi)/_f\mu_0}$	**Replace P by $P\sqrt{_f\mu_0/(4\pi)}$**
dielectric constant ε	permeance $\mathcal{P}$
Replace ε by $\varepsilon_r c^{-2}$	**Replace $\mathcal{P}$ by $_f\mathcal{P}/_f\mu_0$**
relative dielectric constant ε_r	charge Q
Leave ε_r alone	**Replace Q by $Q\sqrt{_f\mu_0/(4\pi)}$**
permittivity of free space ε_0	resistance R
Replace ε_0 by c^{-2}	**Replace R by $(4\pi R)/_f\mu_0$**
magnetomotive force $\mathcal{F}$	reluctance $\mathcal{R}$
Replace $\mathcal{F}$ by $_f\mathcal{F}\sqrt{4\pi\,_f\mu_0}$	**Replace $\mathcal{R}$ by $_f\mathcal{R}\cdot\,_f\mu_0$**
magnetic flux Φ_B	volume charge density ρ_Q
Replace Φ_B by $\Phi_B\sqrt{(4\pi)/_f\mu_0}$	**Replace ρ_Q by $\rho_Q\sqrt{_f\mu_0/(4\pi)}$**
conductance G	resistivity ρ_R
Replace G by $(_f\mu_0 G)/(4\pi)$	**Replace ρ_R by $(4\pi\rho_R)/_f\mu_0$**
magnetic field H	elastance S
Replace H by $_f H\sqrt{4\pi\,_f\mu_0}$	**Replace S by $(4\pi S)/_f\mu_0$**
current I	surface charge density S_Q
Replace I by $I\sqrt{_f\mu_0/(4\pi)}$	**Replace S_Q by $S_Q\sqrt{_f\mu_0/(4\pi)}$**
volume current density J	conductivity σ
Replace J by $J\sqrt{_f\mu_0/(4\pi)}$	**Replace σ by $(_f\mu_0\sigma)/(4\pi)$**
surface current density $\mathcal{J}_S$	electric potential V
Replace $\mathcal{J}_S$ by $\mathcal{J}_S\sqrt{_f\mu_0/(4\pi)}$	**Replace V by $V\sqrt{(4\pi)/_f\mu_0}$**
inductance L	magnetic scalar potential Ω_H
Replace L by $(4\pi L)/_f\mu_0$	**Replace Ω_H by $_f\Omega_H\sqrt{4\pi\,_f\mu_0}$**
permanent-magnet dipole moment m_H	electric susceptibility χ_e
Replace m_H by $_f m_H/\sqrt{4\pi\,_f\mu_0}$	**Replace χ_e by $_f\chi_e/(4\pi)$**
current-loop magnetic dipole moment m_I	magnetic susceptibility χ_m
Replace m_I by $m_I\sqrt{_f\mu_0/(4\pi)}$	**Replace χ_m by $_f\chi_m/(4\pi)$**
permanent-magnet dipole density M_H	
Replace M_H by $_f M_H/\sqrt{4\pi\,_f\mu_0}$	

Table 4.8(a) To go from the rationalized mks system, also called SI units, to the unrationalized mks system. (This table also takes us from the rationalized mks system to the esuq and emuq units explained in Chapter 2.)

magnetic vector potential A	magnetic permeability $_f\mu$
Leave A alone	*Replace $_f\mu$ by $4\pi\mu$*
magnetic induction B	relative magnetic permeability μ_r
Leave B alone	*Leave μ_r alone*
capacitance C	magnetic permeability of free space $_f\mu_0$
Leave C alone	*Replace $_f\mu_0$ by $4\pi\mu_0$*
electric displacement $_fD$	magnetic pole strength $_fp_H$
Replace $_fD$ by $D/(4\pi)$	*Replace $_fp_H$ by $4\pi p_H$*
electric field E	electric dipole moment p
Leave E alone	*Leave p alone*
dielectric constant $_f\varepsilon$	electric dipole density P
Replace $_f\varepsilon$ by $\varepsilon/(4\pi)$	*Leave P alone*
relative dielectric constant ε_r	permeance $_f\mathcal{P}$
Leave ε_r alone	*Replace $_f\mathcal{P}$ by $4\pi\mathcal{P}$*
permittivity of free space $_f\varepsilon_0$	charge Q
Replace $_f\varepsilon_0$ by $\varepsilon_0/(4\pi)$	*Leave Q alone*
magnetomotive force $_f\mathcal{F}$	resistance R
Replace $_f\mathcal{F}$ by $\mathcal{F}/(4\pi)$	*Leave R alone*
magnetic flux Φ_B	reluctance $_f\mathcal{R}$
Leave Φ_B alone	*Replace $_f\mathcal{R}$ by $\mathcal{R}/(4\pi)$*
conductance G	volume charge density ρ_Q
Leave G alone	*Leave ρ_Q alone*
magnetic field $_fH$	resistivity ρ_R
Replace $_fH$ by $H/(4\pi)$	*Leave ρ_R alone*
current I	elastance S
Leave I alone	*Leave S alone*
volume current density J	surface charge density S_Q
Leave J alone	*Leave S_Q alone*
surface current density $\mathcal{J}_S$	conductivity σ
Leave $\mathcal{J}_S$ alone	*Leave σ alone*
inductance L	electric potential V
Leave L alone	*Leave V alone*
permanent-magnet dipole moment $_fm_H$	magnetic scalar potential $_f\Omega_H$
Replace $_fm_H$ by $4\pi m_H$	*Replace $_f\Omega_H$ by $\Omega_H/(4\pi)$*
current-loop magnetic dipole moment m_I	electric susceptibility $_f\chi_e$
Leave m_I alone	*Replace $_f\chi_e$ by $4\pi\chi_e$*
permanent-magnet dipole density $_fM_H$	magnetic susceptibility $_f\chi_m$
Replace $_fM_H$ by $4\pi M_H$	*Replace $_f\chi_m$ by $4\pi\chi_m$*
current-loop magnetic dipole density M_I	
Leave M_I alone	

Table 4.8(b) To go from the unrationalized mks system to the rationalized mks system (also called SI units). (This table also takes us from the esuq and emuq units explained in Chapter 2 to the rationalized mks system.)

magnetic vector potential A *Leave A alone*	magnetic permeability μ *Replace μ by $_f\mu/(4\pi)$*
magnetic induction B *Leave B alone*	relative magnetic permeability μ_r *Leave μ_r alone*
capacitance C *Leave C alone*	magnetic permeability of free space μ_0 *Replace μ_0 by $_f\mu_0/(4\pi)$*
electric displacement D *Replace D by $4\pi \cdot {}_fD$*	magnetic pole strength p_H *Replace p_H by $_fp_H/(4\pi)$*
electric field E *Leave E alone*	electric dipole moment p *Leave p alone*
dielectric constant ε *Replace ε by $4\pi \cdot {}_f\varepsilon$*	electric dipole density P *Leave P alone*
relative dielectric constant ε_r *Leave ε_r alone*	permeance $\mathcal{P}$ *Replace $\mathcal{P}$ by $_f\mathcal{P}/(4\pi)$*
permittivity of free space ε_0 *Replace ε_0 by $4\pi \cdot {}_f\varepsilon_0$*	charge Q *Leave Q alone*
magnetomotive force $\mathcal{F}$ *Replace $\mathcal{F}$ by $4\pi \cdot {}_f\mathcal{F}$*	resistance R *Leave R alone*
magnetic flux Φ_B *Leave Φ_B alone*	reluctance $\mathcal{R}$ *Replace $\mathcal{R}$ by $4\pi \cdot {}_f\mathcal{R}$*
conductance G *Leave G alone*	volume charge density ρ_Q *Leave ρ_Q alone*
magnetic field H *Replace H by $4\pi \cdot {}_fH$*	resistivity ρ_R *Leave ρ_R alone*
current I *Leave I alone*	elastance S *Leave S alone*
volume current density J *Leave J alone*	surface charge density S_Q *Leave S_Q alone*
surface current density $\mathcal{J}_S$ *Leave $\mathcal{J}_S$ alone*	conductivity σ *Leave σ alone*
inductance L *Leave L alone*	electric potential V *Leave V alone*
permanent-magnet dipole moment m_H *Replace m_H by $_fm_H/(4\pi)$*	magnetic scalar potential Ω_H *Replace Ω_H by $4\pi \cdot {}_f\Omega_H$*
current-loop magnetic dipole moment m_I *Leave m_I alone*	electric susceptibility χ_e *Replace χ_e by $_f\chi_e/(4\pi)$*
permanent-magnet dipole density M_H *Replace M_H by $_fM_H/(4\pi)$*	magnetic susceptibility χ_m *Replace χ_m by $_f\chi_m/(4\pi)$*
current-loop magnetic dipole density M_I *Leave M_I alone*	

Table 4.9(a) To go from Gaussian cgs units to the Heaviside-Lorentz cgs system.

magnetic vector potential A	relative magnetic permeability μ_r
Replace A by $_h A \sqrt{4\pi}$	**Leave μ_r alone**
magnetic induction B	magnetic pole strength p_H
Replace B by $_h B \sqrt{4\pi}$	**Replace p_H by $_h p_H / \sqrt{4\pi}$**
capacitance C	electric dipole moment p
Replace C by $_h C / (4\pi)$	**Replace p by $_h p / \sqrt{4\pi}$**
electric displacement D	electric dipole density P
Replace D by $_h D \sqrt{4\pi}$	**Replace P by $_h P / \sqrt{4\pi}$**
electric field E	permeance $\mathcal{P}$
Replace E by $_h E \sqrt{4\pi}$	**Leave $\mathcal{P}$ alone**
relative dielectric constant ε_r	charge Q
Leave ε_r alone	**Replace Q by $_h Q / \sqrt{4\pi}$**
magnetomotive force $\mathcal{F}$	resistanceh R
Replace $\mathcal{F}$ by $_h \mathcal{F} \sqrt{4\pi}$	**Replace R by $4\pi \cdot {_h R}$**
magnetic flux Φ_B	reluctance $\mathcal{R}$
Replace Φ_B by $_h \Phi_B \sqrt{4\pi}$	**Leave $\mathcal{R}$ alone**
conductance G	volume charge density ρ_Q
Replace G by $_h G / (4\pi)$	**Replace ρ_Q by $_h \rho_Q / \sqrt{4\pi}$**
magnetic field H	resistivity ρ_R
Replace H by $_h H \sqrt{4\pi}$	**Replace ρ_R by $4\pi \cdot {_h \rho_R}$**
current I	elastance S
Replace I by $_h I / \sqrt{4\pi}$	**Replace S by $4\pi \cdot {_h S}$**
volume current density J	surface charge density S_Q
Replace J by $_h J / \sqrt{4\pi}$	**Replace S_Q by $_h S_Q / \sqrt{4\pi}$**
surface current density $\mathcal{J}_S$	conductivity σ
Replace $\mathcal{J}_S$ by $\mathcal{J}_S / \sqrt{4\pi}$	**Replace σ by $_h \sigma / (4\pi)$**
inductance L	electric potential V
Replace L by $4\pi \cdot {_h L}$	**Replace V by $_h V \sqrt{4\pi}$**
permanent-magnet dipole moment m_H	magnetic scalar potential Ω_H
Replace m_H by $_h m_H / \sqrt{4\pi}$	**Replace Ω_H by $_h \Omega_H \sqrt{4\pi}$**
current-loop magnetic dipole moment m_I	electric susceptibility χ_e
Replace m_I by $_h m_I / \sqrt{4\pi}$	**Replace χ_e by $_h \chi_e / (4\pi)$**
permanent-magnet dipole density M_H	magnetic susceptibility χ_m
Replace M_H by $_h M_H / \sqrt{4\pi}$	**Replace χ_m by $_h \chi_m / (4\pi)$**
current-loop magnetic dipole density M_I	
Replace M_I by $_h M_I / \sqrt{4\pi}$	

Table 4.9(b) To go from the Heaviside-Lorentz cgs system to Gaussian cgs units.

magnetic vector potential $_hA$ **Replace $_hA$ by $A/\sqrt{4\pi}$**	relative magnetic permeability μ_r **Leave μ_r alone**
magnetic induction $_hB$ **Replace $_hB$ by $B/\sqrt{4\pi}$**	magnetic pole strength $_hp_H$ **Replace $_hp_H$ by $p_H\sqrt{4\pi}$**
capacitance $_hC$ **Replace $_hC$ by $4\pi C$**	electric dipole moment $_hp$ **Replace $_hp$ by $p\sqrt{4\pi}$**
electric displacement $_hD$ **Replace $_hD$ by $D/\sqrt{4\pi}$**	electric dipole density $_hP$ **Replace $_hP$ by $P\sqrt{4\pi}$**
electric field $_hE$ **Replace $_hE$ by $E/\sqrt{4\pi}$**	permanence $\mathcal{P}$ **Leave $\mathcal{P}$ alone**
relative dielectric constant ε_r **Leave ε_r alone**	charge $_hQ$ **Replace $_hQ$ by $Q\sqrt{4\pi}$**
magnetomotive force $_h\mathcal{F}$ **Replace $_h\mathcal{F}$ by $\mathcal{F}/\sqrt{4\pi}$**	resistance $_hR$ **Replace $_hR$ by $R/(4\pi)$**
magnetic flux $_h\Phi_B$ **Replace $_h\Phi_B$ by $\Phi_B/\sqrt{4\pi}$**	reluctance $\mathcal{R}$ **Leave $\mathcal{R}$ alone**
conductance $_hG$ **Replace $_hG$ by $4\pi G$**	volume charge density $_h\rho_Q$ **Replace $_h\rho_Q$ by $\rho_Q\sqrt{4\pi}$**
magnetic field $_hH$ **Replace $_hH$ by $H/\sqrt{4\pi}$**	resistivity $_h\rho_R$ **Replace $_h\rho_R$ by $\rho_R/(4\pi)$**
current $_hI$ **Replace $_hI$ by $I\sqrt{4\pi}$**	elastance $_hS$ **Replace $_hS$ by $S/(4\pi)$**
volume current density $_hJ$ **Replace $_hJ$ by $J\sqrt{4\pi}$**	surface charge density $_hS_Q$ **Replace $_hS_Q$ by $S_Q\sqrt{4\pi}$**
surface current density $_h\mathcal{J}_S$ **Replace $_h\mathcal{J}_S$ by $\mathcal{J}_S\sqrt{4\pi}$**	conductivity $_h\sigma$ **Replace $_h\sigma$ by $4\pi\sigma$**
inductance $_hL$ **Replace $_hL$ by $L/(4\pi)$**	electric potential $_hV$ **Replace $_hV$ by $V/\sqrt{4\pi}$**
permanent-magnet dipole moment $_hm_H$ **Replace $_hm_H$ by $m_H\sqrt{4\pi}$**	magnetic scalar potential $_h\Omega_H$ **Replace $_h\Omega_H$ by $\Omega_H/\sqrt{4\pi}$**
current-loop magnetic dipole moment $_hm_I$ **Replace $_hm_I$ by $m_I\sqrt{4\pi}$**	electric susceptibility $_h\chi_e$ **Replace $_h\chi_e$ by $4\pi\chi_e$**
permanent-magnet dipole density $_hM_H$ **Replace $_hM_H$ by $M_H\sqrt{4\pi}$**	magnetic susceptibility $_h\chi_m$ **Replace $_h\chi_m$ by $4\pi\chi_m$**
current-loop magnetic dipole density $_hM_I$ **Replace $_hM_I$ by $M_I\sqrt{4\pi}$**	

Table 4.10(a) To go from Gaussian cgs units to esu units.

magnetic vector potential A	relative magnetic permeability μ_r
Replace A by $c \cdot A$	**Leave μ_r alone**
magnetic induction B	magnetic pole strength p_H.
Replace B by $c \cdot B$	**Replace p_H by $c \cdot p_H$**
capacitance C	electric dipole moment p
Leave C alone	**Leave p alone**
electric displacement D	electric dipole density P
Leave D alone	**Leave P alone**
electric field E	permeance $\mathcal{P}$
Leave E alone	**Replace $\mathcal{P}$ by $c^2 \mathcal{P}$**
relative dielectric constant ε_r	charge Q
Leave ε_r alone	**Leave Q alone**
magnetomotive force $\mathcal{F}$	resistance R
Replace $\mathcal{F}$ by F/c	**Leave R alone**
magnetic flux Φ_B	reluctance $\mathcal{R}$
Replace Φ_B by $c \cdot \Phi_B$	**Replace $\mathcal{R}$ by $\mathcal{R}/c^2$**
conductance G	volume charge density ρ_Q
Leave G alone	**Leave ρ_Q alone**
magnetic field H	resistivity ρ_R
Replace H by H/c	**Leave ρ_R alone**
current I	elastance S
Leave 1 alone	**Leave S alone**
volume current density J	surface charge density S_Q
Leave J alone	**Leave S_Q alone**
surface current density $\mathcal{J}_S$	conductivity σ
Leave $\mathcal{J}_S$ alone	**Leave σ alone**
inductance L	electric potential V
Leave L alone	**Leave V alone**
permanent-magnet dipole moment m_H	magnetic scalar potential Ω_H
Replace m_H by $c \cdot m_H$	**Replace Ω_H by Ω_H/c**
current-loop magnetic dipole moment m_I	electric susceptibility χ_e
Replace m_I by m_I/c	**Leave χ_e alone**
permanent-magnet dipole density M_H	magnetic susceptibility χ_m
Replace M_H by $c \cdot M_H$	**Leave χ_m alone**
current-loop magnetic dipole density M_I	
Replace M_I by M_I/c	

Table 4.10(b) To go from esu units to Gaussian cgs units.

magnetic vector potential A	relative magnetic permeability μ_r
Replace A by A/c	*Leave μ_r alone*
magnetic induction B	magnetic permeability of free space μ_0
Replace B by B/c	*Replace μ_0 by c^{-2}*
capacitance C	magnetic pole strength p_H
Leave C alone	*Replace p_H by p_H/c*
electric displacement D	electric dipole moment p
Leave D alone	*Leave p alone*
electric field E	electric dipole density P
Leave E alone	*Leave P alone*
relative dielectric constant ε_r	permeance $\mathcal{P}$
Leave ε_r alone	*Replace $\mathcal{P}$ by $\mathcal{P}/c^2$*
magnetomotive force $\mathcal{F}$	charge Q
Replace $\mathcal{F}$ by $\mathcal{F} \cdot c$	*Leave Q alone*
magnetic flux Φ_B	resistance R
Replace Φ_B by Φ_B/c	*Leave R alone*
conductance G	reluctance $\mathcal{R}$
Leave G alone	*Replace $\mathcal{R}$ by $\mathcal{R} \cdot c^2$*
magnetic field H	volume charge density ρ_Q
Replace H by $H \cdot c$	*Leave ρ_Q alone*
current I	resistivity ρ_R
Leave I alone	*Leave ρ_R alone*
volume current density J	elastance S
Leave J alone	*Leave S alone*
surface current density $\mathcal{J}_S$	surface charge density S_Q
Leave $\mathcal{J}_S$ alone	*Leave S_Q alone*
inductance L	conductivity σ
Leave L alone	*Leave σ alone*
permanent-magnet dipole moment m_H	electric potential V
Replace m_H by m_H/c	*Leave V alone*
current-loop magnetic dipole moment m_I	magnetic scalar potential Ω_H
Replace m_I by $m_I \cdot c$	*Replace Ω_H by $\Omega_H \cdot c$*
permanent-magnet dipole density M_H	electric susceptibility χ_e
Replace M_H by M_H/c	*Leave χ_e alone*
current-loop magnetic dipole density M_I	magnetic susceptibility χ_m
Replace M_I by $M_I \cdot c$	*Leave χ_m alone*
magnetic permeability μ	
Replace μ by $\mu_r c^{-2}$	

Table 4.11(a) To go from Gaussian cgs units to emu units.

magnetic vector potential A	relative magnetic permeability μ_r
Leave A alone	**Leave μ_r alone**
magnetic induction B	magnetic pole strength p_H
Leave B alone	**Leave p_H alone**
capacitance C	electric dipole moment p
Replace C by $c^2 C$	**Replace p by $c \cdot p$**
electric displacement D	electric dipole density P
Replace D by $c \cdot D$	**Replace P by $c \cdot P$**
electric field E	permeance $\mathcal{P}$
Replace E by E/c	**Leave $\mathcal{P}$ alone**
relative dielectric constant ε_r	charge Q
Leave ε_r alone	**Replace Q by $c \cdot Q$**
magnetomotive force $\mathcal{F}$	resistance R
Leave $\mathcal{F}$ alone	**Replace R by R/c^2**
magnetic flux Φ_B	reluctance $\mathcal{R}$
Leave Φ_B alone	**Leave $\mathcal{R}$ alone**
conductance G	volume charge density ρ_Q
Replace G by $c^2 G$	**Replace ρ_Q by $c \cdot \rho_Q$**
magnetic field H	resistivity ρ_R
Leave H alone	**Replace ρ_R by ρ_R/c^2**
current I	elastance S
Replace I by $c \cdot I$	**Replace S by S/c^2**
volume current density J	surface charge density S_Q
Replace J by $c \cdot J$	**Replace S_Q by $c \cdot S_Q$**
surface current density $\mathcal{J}_S$	conductivity σ
Replace $\mathcal{J}_S$ by $c \cdot \mathcal{J}_S$	**Replace σ by $c^2 \sigma$**
inductance L	electric potential V
Replace L by L/c^2	**Replace V by V/c**
permanent-magnet dipole moment m_H	magnetic scalar potential Ω_H
Leave m_H alone	**Leave Ω_H alone**
current-loop magnetic dipole moment m_I	electric susceptibility χ_e
Leave m_I alone	**Leave χ_e alone**
permanent-magnet dipole density M_H	magnetic susceptibility χ_m
Leave M_H alone	**Leave χ_m alone**
current-loop magnetic dipole density M_I	
Leave M_I alone	

Table 4.11(b) To go from emu units to Gaussian cgs units.

magnetic vector potential A	current-loop magnetic dipole density M_I
Leave A alone	*Leave M_I alone*
magnetic induction B	relative magnetic permeability μ_r
Leave B alone	*Leave μ_r alone*
capacitance C	magnetic pole strength p_H
Replace C by C/c^2	*Leave p_H alone*
electric displacement D	electric dipole moment p
Replace D by D/c	*Replace p by p/c*
electric field E	electric dipole density P
Replace E by $E \cdot c$	*Replace P by P/c*
dielectric constant ε	permeance $\mathcal{P}$
Replace ε by ε_r/c^2	*Leave $\mathcal{P}$ alone*
relative dielectric constant ε_r	charge Q
Leave ε_r alone	*Replace Q by Q/c*
permittivity of free space ε_0	resistance R
Replace ε_0 by c^{-2}	*Replace R by c^2/R*
magnetomotive force $\mathcal{F}$	reluctance $\mathcal{R}$
Leave $\mathcal{F}$ alone	*Leave $\mathcal{R}$ alone*
magnetic flux Φ_B	volume charge density ρ_Q
Leave Φ_B alone	*Replace ρ_Q by ρ_Q/c*
conductance G	resistivity ρ_R
Replace G by G/c^2	*Replace ρ_R by $c^2\rho_R$*
magnetic field H	elastance S
Leave H alone	*Replace S by c^2/S*
current I	surface charge density S_Q
Replace I by I/c	*Replace S_Q by S_Q/c*
volume current density J	conductivity σ
Replace J by J/c	*Replace σ by σ/c^2*
surface current density $\mathcal{J}_S$	electric potential V
Replace $\mathcal{J}_S$ by $\mathcal{J}_S/c$	*Replace V by $V \cdot c$*
inductance L	magnetic scalar potential Ω_H
Replace L by $L \cdot c^2$	*Leave Ω_H alone*
permanent-magnet dipole moment m_H	electric susceptibility χ_e
Leave m_H alone	*Leave χ_e alone*
current-loop magnetic dipole moment m_I	magnetic susceptibility χ_m
Leave m_I alone	*Leave χ_m alone*
permanent-magnet dipole density M_H	
Leave M_H alone	

Table 4.12(a) To go from Gaussian cgs units to the unrationalized mks system. (This table also takes us from Gaussian cgs units to the esuq and emuq units explained in Chapter 2.)

magnetic vector potential A	relative magnetic permeability μ_r
Replace A by $A/\sqrt{\mu_0}$	*Leave μ_r alone*
magnetic induction B	magnetic pole strength p_H
Replace B by $B/\sqrt{\mu_0}$	*Replace p_H by $p_H/\sqrt{\mu_0}$*
capacitance C	electric dipole moment p
Replace C by $(\mu_0 c^2) \cdot C$	*Replace p by $p \cdot (c\sqrt{\mu_0})$*
electric displacement D	electric dipole density P
Replace D by $D \cdot (c\sqrt{\mu_0})$	*Replace P by $P \cdot (c\sqrt{\mu_0})$*
electric field E	permeance $\mathcal{P}$
Replace E by $E/(c\sqrt{\mu_0})$	*Replace $\mathcal{P}$ by $\mathcal{P}/\mu_0$*
relative dielectric constant ε_r	charge Q
Leave ε_r alone	*Replace Q by $Q \cdot (c\sqrt{\mu_0})$*
magnetomotive force $\mathcal{F}$	resistance R
Replace $\mathcal{F}$ by $\mathcal{F}\sqrt{\mu_0}$	*Replace R by $R/(\mu_0 c^2)$*
magnetic flux Φ_B	reluctance $\mathcal{R}$
Replace Φ_B by $\Phi_B/\sqrt{\mu_0}$	*Replace $\mathcal{R}$ by $\mathcal{R} \cdot \mu_0$*
conductance G	volume charge density ρ_Q
Replace G by $(\mu_0 c^2) \cdot G$	*Replace ρ_Q by $\rho_Q \cdot (c\sqrt{\mu_0})$*
magnetic field H	resistivity ρ_R
Replace H by $H\sqrt{\mu_0}$	*Replace ρ_R by $\rho_R/(\mu_0 c^2)$*
current I	elastance S
Replace I by $I \cdot (c\sqrt{\mu_0})$	*Replace S by $S/(\mu_0 c^2)$*
volume current density J	surface charge density S_Q
Replace J by $J \cdot (c\sqrt{\mu_0})$	*Replace S_Q by $S_Q \cdot (c\sqrt{\mu_0})$*
surface current density $\mathcal{J}_S$	conductivity σ
Replace $\mathcal{J}_S$ by $\mathcal{J}_S \cdot (c\sqrt{\mu_0})$	*Replace σ by $(\mu_0 c^2) \cdot \sigma$*
inductance L	electric potential V
Replace L by $L/(\mu_0 c^2)$	*Replace V by $V/(c\sqrt{\mu_0})$*
permanent-magnet dipole moment m_H	magnetic scalar potential Ω_H
Replace m_H by $m_H/\sqrt{\mu_0}$	*Replace Ω_H by $\Omega_H\sqrt{\mu_0}$*
current-loop magnetic dipole moment m_I	electric susceptibility χ_e
Replace m_I by $m_I\sqrt{\mu_0}$	*Leave χ_e alone*
permanent-magnet dipole density M_H	magnetic susceptibility χ_m
Replace M_H by $M_H/\sqrt{\mu_0}$	*Leave χ_m alone*
current-loop magnetic dipole density M_I	
Replace M_I by $M_I\sqrt{\mu_0}$	

Table 4.12(b) To go from the unrationalized mks system to Gaussian cgs units. (This table also takes us from the esuq and emuq units explained in Chapter 2 to Gaussian cgs units.)

magnetic vector potential A *Replace A by $A\sqrt{\mu_0}$*	magnetic permeability μ *Replace μ by $\mu_r \cdot \mu_0$*
magnetic induction B *Replace B by $B\sqrt{\mu_0}$*	relative magnetic permeability μ_r *Leave μ_r alone*
capacitance C *Replace C by $C/(\mu_0 c^2)$*	magnetic permeability of free space μ_0 *Leave μ_0 alone*
electric displacement D *Replace D by $D/(c\sqrt{\mu_0})$*	magnetic pole strength p_H *Replace p_H by $p_H \cdot \sqrt{\mu_0}$*
electric field E *Replace E by $E \cdot (c\sqrt{\mu_0})$*	electric dipole moment p *Replace p by $p/(c\sqrt{\mu_0})$*
dielectric constant ε *Replace ε by $\varepsilon_r/(\mu_0 c^2)$*	electric dipole density P *Replace P by $P/(c\sqrt{\mu_0})$*
relative dielectric constant ε_r *Leave ε_r alone*	permeance $\mathcal{P}$ *Replace $\mathcal{P}$ by $\mu_0 \cdot \mathcal{P}$*
permittivity of free space ε_0 *Replace ε_0 by $(\mu_0 c^2)^{-1}$*	charge Q *Replace Q by $Q/(c\sqrt{\mu_0})$*
magnetomotive force $\mathcal{F}$ *Replace $\mathcal{F}$ by $\mathcal{F}/\sqrt{\mu_0}$*	resistance R *Replace R by $R \cdot (\mu_0 c^2)$*
magnetic flux Φ_B *Replace Φ_B by $\Phi_B \cdot \sqrt{\mu_0}$*	reluctance $\mathcal{R}$ *Replace $\mathcal{R}$ by $\mathcal{R}/\mu_0$*
conductance G *Replace G by $G/(\mu_0 c^2)$*	volume charge density ρ_Q *Replace ρ_Q by $\rho_Q/(c\sqrt{\mu_0})$*
magnetic field H *Replace H by $H/\sqrt{\mu_0}$*	resistivity ρ_R *Replace ρ_R by $\rho_R \cdot (\mu_0 c^2)$*
current I *Replace I by $I/(c\sqrt{\mu_0})$*	elastance S *Replace S by $S \cdot (\mu_0 c^2)$*
volume current density J *Replace J by $J/(c\sqrt{\mu_0})$*	surface charge density S_Q *Replace S_Q by $S_Q/(c\sqrt{\mu_0})$*
surface current density $\mathcal{J}_S$ *Replace $\mathcal{J}_S$ by $\mathcal{J}_S/(c\sqrt{\mu_0})$*	conductivity σ *Replace σ by $\sigma/(\mu_0 c^2)$*
inductance L *Replace L by $L \cdot (\mu_0 c^2)$*	electric potential V *Replace V by $V \cdot (c\sqrt{\mu_0})$*
permanent-magnet dipole moment m_H *Replace m_H by $m_H\sqrt{\mu_0}$*	magnetic scalar potential Ω_H *Replace Ω_H by $\Omega_H/\sqrt{\mu_0}$*
current-loop magnetic dipole moment m_I *Replace m_I by $m_I/\sqrt{\mu_0}$*	electric susceptibility χ_e *Leave χ_e alone*
permanent-magnet dipole density M_H *Replace M_H by $M_H\sqrt{\mu_0}$*	magnetic susceptibility χ_m *Leave χ_m alone*
current-loop magnetic dipole density M_I *Replace M_I by $M_I/\sqrt{\mu_0}$*	

Table 4.13(a) To go from the Heaviside-Lorentz cgs system to esu units.

magnetic vector potential $_hA$ ***Replace $_hA$ by $A \cdot (c/\sqrt{4\pi})$***	relative magnetic permeability μ_r ***Leave μ_r alone***
magnetic induction $_hB$ ***Replace $_hB$ by $B \cdot (c/\sqrt{4\pi})$***	magnetic pole strength $_hp_H$ ***Replace $_hp_H$ by $p_H \cdot (c\sqrt{4\pi})$***
capacitance $_hC$ ***Replace $_hC$ by $4\pi C$***	electric dipole moment $_hp$ ***Replace $_hp$ by $p\sqrt{4\pi}$***
electric displacement $_hD$ ***Replace $_hD$ by $D/\sqrt{4\pi}$***	electric dipole density $_hP$ ***Replace $_hP$ by $P\sqrt{4\pi}$***
electric field $_hE$ ***Replace $_hE$ by $E/\sqrt{4\pi}$***	permeance $\mathcal{P}$ ***Replace $\mathcal{P}$ by $c^2 \cdot \mathcal{P}$***
relative dielectric constant ε_r ***Leave ε_r alone***	charge $_hQ$ ***Replace $_hQ$ by $Q\sqrt{4\pi}$***
magnetomotive force $_h\mathcal{F}$ ***Replace $_h\mathcal{F}$ by $\mathcal{F}/(c\sqrt{4\pi})$***	resistance $_hR$ ***Replace $_hR$ by $R/(4\pi)$***
magnetic flux $_h\Phi_B$ ***Replace $_h\Phi_B$ by $\Phi_B \cdot (c/\sqrt{4\pi})$***	reluctance $\mathcal{R}$ ***Replace $\mathcal{R}$ by $\mathcal{R}/c^2$***
conductance $_hG$ ***Replace $_hG$ by $4\pi G$***	volume charge density $_h\rho_Q$ ***Replace $_h\rho_Q$ by $\rho_Q\sqrt{4\pi}$***
magnetic field $_hH$ ***Replace $_hH$ by $H/(c\sqrt{4\pi})$***	resistivity $_h\rho_R$ ***Replace $_h\rho_R$ by $\rho_R/(4\pi)$***
current $_hI$ ***Replace $_hI$ by $I\sqrt{4\pi}$***	elastance $_hS$ ***Replace $_hS$ by $S/(4\pi)$***
volume current density $_hJ$ ***Replace $_hJ$ by $J\sqrt{4\pi}$***	surface charge density $_hS_Q$ ***Replace $_hS_Q$ by $S_Q\sqrt{4\pi}$***
surface current density $_h\mathcal{J}_S$ ***Replace $_h\mathcal{J}_S$ by $\mathcal{J}_S\sqrt{4\pi}$***	conductivity $_h\sigma$ ***Replace $_h\sigma$ by $4\pi\sigma$***
inductance $_hL$ ***Replace $_hL$ by $L/(4\pi)$***	electric potential $_hV$ ***Replace $_hV$ by $V/\sqrt{4\pi}$***
permanent-magnet dipole moment $_hm_H$ ***Replace $_hm_H$ by $m_H \cdot (c\sqrt{4\pi})$***	magnetic scalar potential $_h\Omega_H$ ***Replace $_h\Omega_H$ by $\Omega_H/(c\sqrt{4\pi})$***
current-loop magnetic dipole moment $_hm_I$ ***Replace $_hm_I$ by $m_I \cdot (\sqrt{4\pi}/c)$***	electric susceptibility $_h\chi_e$ ***Replace $_h\chi_e$ by $4\pi\chi_e$***
permanent-magnet dipole density $_hM_H$ ***Replace $_hM_H$ by $M_H \cdot (c\sqrt{4\pi})$***	magnetic susceptibility $_h\chi_m$ ***Replace $_h\chi_m$ by $4\pi\chi_m$***
current-loop magnetic dipole density $_hM_I$ ***Replace $_hM_I$ by $M_I \cdot (\sqrt{4\pi}/c)$***	

Table 4.13(b) To go from esu units to the Heaviside-Lorentz cgs system.

magnetic vector potential A	relative magnetic permeability μ_r
Replace A by $_h A \cdot (\sqrt{4\pi}/c)$	**Leave μ_r alone**
magnetic induction B	magnetic permeability of free space μ_0
Replace B by $_h B \cdot (\sqrt{4\pi}/c)$	**Replace μ_0 by c^{-2}**
capacitance C	magnetic pole strength p_H
Replace C by $_h C/(4\pi)$	**Replace p_H by $_h p_H/(c\sqrt{4\pi})$**
electric displacement D	electric dipole moment p
Replace D by $_h D\sqrt{4\pi}$	**Replace p by $_h p/\sqrt{4\pi}$**
electric field E	electric dipole density P
Replace E by $_h E\sqrt{4\pi}$	**Replace P by $_h P/\sqrt{4\pi}$**
relative dielectric constant ε_r	permeance $\mathcal{P}$
Leave ε_r alone	**Replace $\mathcal{P}$ by $\mathcal{P}/c^2$**
magnetomotive force $\mathcal{F}$	charge Q
Replace $\mathcal{F}$ by $_h \mathcal{F} \cdot (c\sqrt{4\pi})$	**Replace Q by $_h Q/\sqrt{4\pi}$**
magnetic flux Φ_B	resistance R
Replace Φ_B by $_h \Phi_B \cdot (\sqrt{4\pi}/c)$	**Replace R by $4\pi \cdot {_h}R$**
conductance G	reluctance $\mathcal{R}$
Replace G by $_h G/(4\pi)$	**Replace $\mathcal{R}$ by $\mathcal{R} \cdot c^2$**
magnetic field H	volume charge density ρ_Q
Replace H by $_h H \cdot (c\sqrt{4\pi})$	**Replace ρ_Q by $_h\rho_Q/\sqrt{4\pi}$**
current I	resistivity ρ_R
Replace I by $_h I/\sqrt{4\pi}$	**Replace ρ_R by $4\pi \cdot {_h}\rho_R$**
volume current density J	elastance S
Replace J by $_h J/\sqrt{4\pi}$	**Replace S by $4\pi \cdot {_h}S$**
surface current density $\mathcal{J}_S$	surface charge density S_Q
Replace $\mathcal{J}_S$ by $_h\mathcal{J}_S/\sqrt{4\pi}$	**Replace S_Q by $_h S_Q/\sqrt{4\pi}$**
inductance L	conductivity σ
Replace L by $4\pi \cdot {_h}L$	**Replace σ by $_h\sigma/(4\pi)$**
permanent-magnet dipole moment m_H	electric potential V
Replace m_H by $_h m_H/(c\sqrt{4\pi})$	**Replace V by $_h V\sqrt{4\pi}$**
current-loop magnetic dipole moment m_I	magnetic scalar potential Ω_H
Replace m_I by $_h m_I \cdot (c/\sqrt{4\pi})$	**Replace Ω_H by $_h\Omega_H \cdot (c\sqrt{4\pi})$**
permanent-magnet dipole density M_H	electric susceptibility χ_e
Replace M_H by $_h M_H/(c\sqrt{4\pi})$	**Replace χ_e by $_h\chi_e/(4\pi)$**
current-loop magnetic dipole density M_I	magnetic susceptibility χ_m
Replace M_I by $_h M_I \cdot (c/\sqrt{4\pi})$	**Replace χ_m by $_h\chi_m/(4\pi)$**
magnetic permeability μ	
Replace μ by $\mu_r \cdot c^{-2}$	

Table 4.14(a) To go from the Heaviside-Lorentz cgs system to emu units.

magnetic vector potential $_hA$	relative magnetic permeability μ_r
Replace $_hA$ by $A/\sqrt{4\pi}$	**Leave μ_r alone**
magnetic induction $_hB$	magnetic pole strength $_hp_H$
Replace $_hB$ by $B/\sqrt{4\pi}$	**Replace $_hp_H$ by $p_H \cdot \sqrt{4\pi}$**
capacitance $_hC$	electric dipole moment $_hp$
Replace $_hC$ by $(4\pi c^2) \cdot C$	**Replace $_hp$ by $p \cdot (c\sqrt{4\pi})$**
electric displacement $_hD$	electric dipole density $_hP$
Replace $_hD$ by $D \cdot (c/\sqrt{4\pi})$	**Replace $_hP$ by $P \cdot (c\sqrt{4\pi})$**
electric field $_hE$	permeance $\mathcal{P}$
Replace $_hE$ by $E/(c\sqrt{4\pi})$	**Leave $\mathcal{P}$ alone**
relative dielectric constant ε_r	charge $_hQ$
Leave ε_r alone	**Replace $_hQ$ by $Q \cdot (c\sqrt{4\pi})$**
magnetomotive force $_h\mathcal{F}$	resistance $_hR$
Replace $_h\mathcal{F}$ by $\mathcal{F}/\sqrt{4\pi}$	**Replace $_hR$ by $R/(4\pi c^2)$**
magnetic flux $_h\Phi_B$	reluctance $\mathcal{R}$
Replace $_h\Phi_B$ by $\Phi_B/\sqrt{4\pi}$	**Leave $\mathcal{R}$ alone**
conductance $_hG$	volume charge density $_h\rho_Q$
Replace $_hG$ by $(4\pi c^2) \cdot G$	**Replace $_h\rho_Q$ by $\rho_Q \cdot (c\sqrt{4\pi})$**
magnetic field $_hH$	resistivity $_h\rho_R$
Replace $_hH$ by $H/\sqrt{4\pi}$	**Replace $_h\rho_R$ by $\rho_R/(4\pi c^2)$**
current $_hI$	elastance $_hS$
Replace $_hI$ by $I \cdot (c\sqrt{4\pi})$	**Replace $_hS$ by $S/(4\pi c^2)$**
volume current density $_hJ$	surface charge density $_hS_Q$
Replace $_hJ$ by $J \cdot (c\sqrt{4\pi})$	**Replace $_hS_Q$ by $S_Q \cdot (c\sqrt{4\pi})$**
surface current density $_h\mathcal{J}_S$	conductivity $_h\sigma$
Replace $_h\mathcal{J}_S$ by $\mathcal{J}_S \cdot (c\sqrt{4\pi})$	**Replace $_h\sigma$ by $(4\pi c^2) \cdot \sigma$**
inductance $_hL$	electric potential $_hV$
Replace $_hL$ by $L/(4\pi c^2)$	**Replace $_hV$ by $V/(c\sqrt{4\pi})$**
permanent-magnet dipole moment $_hm_H$	magnetic scalar potential $_h\Omega_H$
Replace $_hm_H$ by $m_H \cdot \sqrt{4\pi}$	**Replace $_h\Omega_H$ by $\Omega_H/\sqrt{4\pi}$**
current-loop magnetic dipole moment $_hm_I$	electric susceptibility $_h\chi_e$
Replace $_hm_I$ by $m_I \cdot \sqrt{4\pi}$	**Replace $_h\chi_e$ by $4\pi\chi_e$**
permanent-magnet dipole density $_hM_H$	magnetic susceptibility $_h\chi_m$
Replace $_hM_H$ by $M_H \cdot \sqrt{4\pi}$	**Replace $_h\chi_m$ by $4\pi\chi_m$**
current-loop magnetic dipole density $_hM_I$	
Replace $_hM_I$ by $M_I \cdot \sqrt{4\pi}$	

Table 4.14(b) To go from emu units to the Heaviside-Lorentz cgs system.

magnetic vector potential A	current-loop magnetic dipole density M_I
Replace A by $_hA \cdot \sqrt{4\pi}$	***Replace M_I by $_hM_I/\sqrt{4\pi}$***
magnetic induction B	relative magnetic permeability μ_r
Replace B by $_hB \cdot \sqrt{4\pi}$	***Leave μ_r alone***
capacitance C	magnetic pole strength p_H
Replace C by $_hC/(4\pi c^2)$	***Replace p_H by $_hp_H/\sqrt{4\pi}$***
electric displacement D	electric dipole moment p
Replace D by $_hD \cdot (\sqrt{4\pi}/c)$	***Replace p by $_hp/(c\sqrt{4\pi})$***
electric field E	electric dipole density P
Replace E by $_hE \cdot (c\sqrt{4\pi})$	***Replace P by $_hP/(c\sqrt{4\pi})$***
dielectric constant ε	permeance $\mathcal{P}$
Replace ε by $\varepsilon_r c^{-2}$	***Leave $\mathcal{P}$ alone***
relative dielectric constant ε_r	charge Q
Leave ε_r alone	***Replace Q by $_hQ/(c\sqrt{4\pi})$***
permittivity of free space ε_0	resistance R
Replace ε_0 by c^{-2}	***Replace R by $(4\pi c^2) \cdot {}_hR$***
magnetomotive force $\mathcal{F}$	reluctance $\mathcal{R}$
Replace $\mathcal{F}$ by $_h\mathcal{F} \cdot \sqrt{4\pi}$	***Leave $\mathcal{R}$ alone***
magnetic flux Φ_B	volume charge density ρ_Q
Replace Φ_B by $_h\Phi_B \cdot \sqrt{4\pi}$	***Replace ρ_Q by $_h\rho_Q/(c\sqrt{4\pi})$***
conductance G	resistivity ρ_R
Replace G by $_hG/(4\pi c^2)$	***Replace ρ_R by $(4\pi c^2) \cdot {}_h\rho_R$***
magnetic field H	elastance S
Replace H by $_hH \cdot \sqrt{4\pi}$	***Replace S by $(4\pi c^2) \cdot {}_hS$***
current I	surface charge density S_Q
Replace I by $_hI/(c\sqrt{4\pi})$	***Replace S_Q by $_hS_Q/(c\sqrt{4\pi})$***
volume current density J	conductivity σ
Replace J by $_hJ/(c\sqrt{4\pi})$	***Replace σ by $_h\sigma/(4\pi c^2)$***
surface current density $\mathcal{J}_S$	electric potential V
Replace $\mathcal{J}_S$ by $_h\mathcal{J}_S/(c\sqrt{4\pi})$	***Replace V by $_hV \cdot (c\sqrt{4\pi})$***
inductance L	magnetic scalar potential Ω_H
Replace L by $(4\pi c^2) \cdot {}_hL$	***Replace Ω_H by $_h\Omega_H \cdot \sqrt{4\pi}$***
permanent-magnet dipole moment m_H	electric susceptibility χ_e
Replace m_H by $_hm_H/\sqrt{4\pi}$	***Replace χ_e by $_h\chi_e/(4\pi)$***
current-loop magnetic dipole moment m_I	magnetic susceptibility χ_m
Replace m_I by $_hm_I/\sqrt{4\pi}$	***Replace χ_m by $_h\chi_m/(4\pi)$***
permanent-magnet dipole density M_H	
Replace M_H by $_hM_H/\sqrt{4\pi}$	

Table 4.15(a) To go from the Heaviside-Lorentz cgs system to the unrationalized mks system. (This table also takes us from the Heaviside-Lorentz cgs system to the esuq and emuq units explained in Chapter 2.)

magnetic vector potential $_hA$ ***Replace*** $_hA$ ***by*** $A/\sqrt{4\pi\mu_0}$	relative magnetic permeability μ_r ***Leave*** μ_r ***alone***
magnetic induction $_hB$ ***Replace*** $_hB$ ***by*** $B/\sqrt{4\pi\mu_0}$	magnetic pole strength $_hp_H$ ***Replace*** $_hp_H$ ***by*** $p_H \cdot \sqrt{4\pi/\mu_0}$
capacitance $_hC$ ***Replace*** $_hC$ ***by*** $(4\pi\mu_0c^2) \cdot C$	electric dipole moment $_hp$ ***Replace*** $_hp$ ***by*** $p \cdot (c\sqrt{4\pi\mu_0})$
electric displacement $_hD$ ***Replace*** $_hD$ ***by*** $D \cdot c\sqrt{\mu_0/(4\pi)}$	electric dipole density $_hP$ ***Replace*** $_hP$ ***by*** $P \cdot (c\sqrt{4\pi\mu_0})$
electric field $_hE$ ***Replace*** $_hE$ ***by*** $E/(c\sqrt{4\pi\mu_0})$	permeance $\mathcal{P}$ ***Replace*** $\mathcal{P}$ ***by*** $\mathcal{P}/\mu_0$
relative dielectric constant ε_r ***Leave*** ε_r ***alone***	charge $_hQ$ ***Replace*** $_hQ$ ***by*** $Q \cdot (c\sqrt{4\pi\mu_0})$
magnetomotive force $_h\mathcal{F}$ ***Replace*** $_h\mathcal{F}$ ***by*** $\mathcal{F} \cdot \sqrt{\mu_0/(4\pi)}$	resistance $_hR$ ***Replace*** $_hR$ ***by*** $R/(4\pi\mu_0c^2)$
magnetic flux $_h\Phi_B$ ***Replace*** $_h\Phi_B\sqrt{f\mu_0}$ ***by*** $\Phi_B/\sqrt{4\pi\mu_0}$	reluctance $\mathcal{R}$ ***Replace*** $\mathcal{R}$ ***by*** $\mu_0 \cdot \mathcal{R}$
conductance $_hG$ ***Replace*** $_hG$ ***by*** $(4\pi\mu_0c^2) \cdot G$	volume charge density $_h\rho_Q$ ***Replace*** $_h\rho_Q$ ***by*** $\rho_Q \cdot (c\sqrt{4\pi\mu_0})$
magnetic field $_hH$ ***Replace*** $_hH$ ***by*** $H \cdot \sqrt{\mu_0/(4\pi)}$	resistivity $_h\rho_R$ ***Replace*** $_h\rho_R$ ***by*** $\rho_R/(4\pi\mu_0c^2)$
current $_hI$ ***Replace*** $_hI$ ***by*** $I \cdot (c\sqrt{4\pi\mu_0})$	elastance $_hS$ ***Replace*** $_hS$ ***by*** $S/(4\pi\mu_0c^2)$
volume current density $_hJ$ ***Replace*** $_hJ$ ***by*** $J \cdot (c\sqrt{4\pi\mu_0})$	surface charge density $_hS_Q$ ***Replace*** $_hS_Q$ ***by*** $S_Q \cdot (c\sqrt{4\pi\mu_0})$
surface current density $_h\mathcal{J}_S$ ***Replace*** $_h\mathcal{J}_S$ ***by*** $\mathcal{J}_S \cdot (c\sqrt{4\pi\mu_0})$	conductivity $_h\sigma$ ***Replace*** $_h\sigma$ ***by*** $(4\pi\mu_0c^2) \cdot \sigma$
inductance $_hL$ ***Replace*** $_hL$ ***by*** $L/(4\pi\mu_0c^2)$	electric potential $_hV$ ***Replace*** $_hV$ ***by*** $V/(c\sqrt{4\pi\mu_0})$
permanent-magnet dipole moment $_hm_H$ ***Replace*** $_hm_H$ ***by*** $m_H \cdot \sqrt{4\pi/\mu_0}$	magnetic scalar potential $_h\Omega_H$ ***Replace*** $_h\Omega_H$ ***by*** $\Omega_H \cdot \sqrt{\mu_0/(4\pi)}$
current-loop magnetic dipole moment $_hm_I$ ***Replace*** $_hm_I$ ***by*** $m_I \cdot \sqrt{4\pi\mu_0}$	electric susceptibility $_h\chi_e$ ***Replace*** $_h\chi_e$ ***by*** $4\pi\chi_e$
permanent-magnet dipole density $_hM_H$ ***Replace*** $_hM_H$ ***by*** $M_H \cdot \sqrt{4\pi/\mu_0}$	magnetic susceptibility $_h\chi_m$ ***Replace*** $_h\chi_m$ ***by*** $4\pi\chi_m$
current-loop magnetic dipole density $_hM_I$ ***Replace*** $_hM_I$ ***by*** $M_I \cdot \sqrt{4\pi\mu_0}$	

Table 4.15(b) To go from the unrationalized mks system to the Heaviside-Lorentz cgs system. (This table also takes us from the esuq and emuq units explained in Chapter 2 to the Heaviside-Lorentz cgs system.)

magnetic vector potential A	magnetic permeability μ
Replace A by $_h A \sqrt{4\pi\mu_0}$	*Replace μ by $\mu_r \cdot \mu_0$*
magnetic induction B	relative magnetic permeability μ_r
Replace B by $_h B \sqrt{4\pi\mu_0}$	*Leave μ_r alone*
capacitance C	magnetic permeability of free space μ_0
Replace C by $_h C/(4\pi\mu_0 c^2)$	*Leave μ_0 alone*
electric displacement D	magnetic pole strength p_H
Replace D by $_h D c^{-1}\sqrt{4\pi/\mu_0}$	*Replace p_H by $_h p_H \sqrt{\mu_0/(4\pi)}$*
electric field E	electric dipole moment p
Replace E by $_h E c \sqrt{4\pi\mu_0}$	*Replace p by $_h p (c\sqrt{4\pi\mu_0})^{-1}$*
dielectric constant ε	electric dipole density P
Replace ε by $\varepsilon_r/(\mu_0 c^2)$	*Replace P by $_h P (c\sqrt{4\pi\mu_0})^{-1}$*
relative dielectric constant ε_r	permeance $\mathcal{P}$
Leave ε_r alone	*Replace $\mathcal{P}$ by $\mathcal{P} \cdot \mu_0$*
permittivity of free space ε_0	charge Q
Replace ε_0 by $(\mu_0 c^2)^{-1}$	*Replace Q by $_h Q (c\sqrt{4\pi\mu_0})^{-1}$*
magnetomotive force $\mathcal{F}$	resistance R
Replace $\mathcal{F}$ by $_h \mathcal{F}\sqrt{4\pi/\mu_0}$	*Replace R by $_h R \cdot (4\pi\mu_0 c^2)$*
magnetic flux Φ_B	reluctance $\mathcal{R}$
Replace Φ_B by $_h \Phi_B \sqrt{4\pi\mu_0}$	*Replace $\mathcal{R}$ by $\mathcal{R}/\mu_0$*
conductance G	volume charge density ρ_Q
Replace G by $_h G/(4\pi\mu_0 c^2)$	*Replace ρ_Q by $_h \rho_Q (c\sqrt{4\pi\mu_0})^{-1}$*
magnetic field H	resistivity ρ_R
Replace H by $_h H \sqrt{4\pi/\mu_0}$	*Replace ρ_R by $_h \rho_R \cdot (4\pi\mu_0 c^2)$*
current I	elastance S
Replace I by $_h I (c\sqrt{4\pi\mu_0})^{-1}$	*Replace S by $_h S \cdot (4\pi\mu_0 c^2)$*
volume current density J	surface charge density S_Q
Replace J by $_h J (c\sqrt{4\pi\mu_0})^{-1}$	*Replace S_Q by $_h S_Q (c\sqrt{4\pi\mu_0})^{-1}$*
surface current density $\mathcal{J}_S$	conductivity σ
Replace $\mathcal{J}_S$ by $_h \mathcal{J}_S (c\sqrt{4\pi\mu_0})^{-1}$	*Replace σ by $_h \sigma/(4\pi\mu_0 c^2)$*
inductance L	electric potential V
Replace L by $_h L \cdot (4\pi\mu_0 c^2)$	*Replace V by $_h V c \sqrt{4\pi\mu_0}$*
permanent-magnet dipole moment m_H	magnetic scalar potential Ω_H
Replace m_H by $_h m_H \sqrt{\mu_0/(4\pi)}$	*Replace Ω_H by $_h \Omega_H \sqrt{4\pi/\mu_0}$*
current-loop magnetic dipole moment m_I	electric susceptibility χ_e
Replace m_I by $_h m_I/\sqrt{4\pi\mu_0}$	*Replace χ_e by $_h \chi_e/(4\pi)$*
permanent-magnet dipole density M_H	magnetic susceptibility χ_m
Replace M_H by $_h M_H \sqrt{\mu_0/(4\pi)}$	*Replace χ_m by $_h \chi_m/(4\pi)$*
current-loop magnetic dipole density M_I	
Replace M_I by $_h M_I/\sqrt{4\pi\mu_0}$	

Table 4.16(a) To go from esu units to emu units.

magnetic vector potential A	relative magnetic permeability μ_r
Replace A by A/c	***Leave μ_r alone***
magnetic induction B	magnetic permeability of free space μ_0
Replace B by B/c	***Replace μ_0 by c^{-2}***
capacitance C	magnetic pole strength p_H
Replace C by $c^2 C$	***Replace p_H by p_H/c***
electric displacement D	electric dipole moment p
Replace D by cD	***Replace p by $c \cdot p$***
electric field E	electric dipole density P
Replace E by E/c	***Replace P by $c \cdot P$***
relative dielectric constant ε_r	permeance $\mathcal{P}$
Leave ε_r alone	***Replace $\mathcal{P}$ by $\mathcal{P}/c^2$***
magnetomotive force $\mathcal{F}$	charge Q
Replace $\mathcal{F}$ by $\mathcal{F} \cdot c$	***Replace Q by $c \cdot Q$***
magnetic flux Φ_B	resistance R
Replace Φ_B by Φ_B/c	***Replace R by R/c^2***
conductance G	reluctance $\mathcal{R}$
Replace G by $c^2 G$	***Replace $\mathcal{R}$ by $\mathcal{R} \cdot c^2$***
magnetic field H	volume charge density ρ_Q
Replace H by $H \cdot c$	***Replace ρ_Q by $c \cdot \rho_Q$***
current I	resistivity ρ_R
Replace I by $c \cdot I$	***Replace ρ_R by ρ_R/c^2***
volume current density J	elastance S
Replace J by $c \cdot J$	***Replace S by S/c^2***
surface current density $\mathcal{J}_S$	surface charge density S_Q
Replace $\mathcal{J}_S$ by $c \cdot \mathcal{J}_S$	***Replace S_Q by $c \cdot S_Q$***
inductance L	conductivity σ
Replace L by L/c^2	***Replace σ by $c^2 \sigma$***
permanent-magnet dipole moment m_H	electric potential V
Replace m_H by m_H/c	***Replace V by V/c***
current-loop magnetic dipole moment m_I	magnetic scalar potential Ω_H
Replace m_I by $m_I \cdot c$	***Replace Ω_H by $\Omega_H \cdot c$***
permanent-magnet dipole density M_H	electric susceptibility χ_e
Replace M_H by M_H/c	***Leave χ_e alone***
current-loop magnetic dipole density M_I	magnetic susceptibility χ_m
Replace M_I by $M_I \cdot c$	***Leave χ_m alone***
magnetic permeability μ	
Replace μ by $\mu_r c^{-2}$	

Table 4.16(b) To go from emu units to esu units.

magnetic vector potential A	current-loop magnetic dipole density M_I
Replace A by $c \cdot A$	**Replace M_I by M_I/c**
magnetic induction B	relative magnetic permeability μ_r
Replace B by $c \cdot B$	**Leave μ_r alone**
capacitance C	magnetic pole strength p_H
Replace C by C/c^2	**Replace p_H by $p_H \cdot c$**
electric displacement D	electric dipole moment p
Replace D by D/c	**Replace p by p/c**
electric field E	electric dipole density P
Replace E by $E \cdot c$	**Replace P by P/c**
dielectric constant ε	permeance $\mathcal{P}$
Replace ε by ε_r/c^2	**Replace $\mathcal{P}$ by $c^2\mathcal{P}$**
relative dielectric constant ε_r	charge Q
Leave ε_r alone	**Replace Q by Q/c**
permittivity of free space ε_0	resistance R
Replace ε_0 by c^{-2}	**Replace R by c^2R**
magnetomotive force $\mathcal{F}$	reluctance $\mathcal{R}$
Replace $\mathcal{F}$ by $\mathcal{F}/c$	**Replace $\mathcal{R}$ by $\mathcal{R}/c^2$**
magnetic flux Φ_B	volume charge density ρ_Q
Replace Φ_B by $\Phi_B \cdot c$	**Replace ρ_Q by ρ_Q/c**
conductance G	resistivity ρ_R
Replace G by G/c^2	**Replace ρ_R by $c^2\rho_R$**
magnetic field H	elastance S
Replace H by H/c	**Replace S by c^2S**
current I	surface charge density S_Q
Replace I by I/c	**Replace S_Q by S_Q/c**
volume current density J	conductivity σ
Replace J by J/c	**Replace σ by σ/c^2**
surface current density $\mathcal{J}_S$	electric potential V
Replace $\mathcal{J}_S$ by $\mathcal{J}_S/c$	**Replace V by $V \cdot c$**
inductance L	magnetic scalar potential Ω_H
Replace L by $L \cdot c^2$	**Replace Ω_H by Ω_H/c**
permanent-magnet dipole moment m_H	electric susceptibility χ_e
Replace m_H by $m_H \cdot c$	**Leave χ_e alone**
current-loop magnetic dipole moment m_I	magnetic susceptibility χ_m
Replace m_I by m_I/c	**Leave χ_m alone**
permanent-magnet dipole density M_H	
Replace M_H by $M_H \cdot c$	

Table 4.17(a) To go from esu units to the unrationalized mks system. (This table also takes us from esu units to the esuq and emuq units explained in Chapter 2.)

magnetic vector potential A **Replace A by $A/(c\sqrt{\mu_0})$**	relative magnetic permeability μ_r **Leave μ_r alone**
magnetic induction B **Replace B by $B/(c\sqrt{\mu_0})$**	magnetic permeability of free space μ_0 **Replace μ_0 by c^{-2}**
capacitance C **Replace C by $\mu_0 c^2 C$**	magnetic pole strength p_H **Replace p_H by $p_H/(c\sqrt{\mu_0})$**
electric displacement D **Replace D by $(c\sqrt{\mu_0})D$**	electric dipole moment p **Replace p by $(c\sqrt{\mu_0}) \cdot p$**
electric field E **Replace E by $E/(c\sqrt{\mu_0})$**	electric dipole density P **Replace P by $(c\sqrt{\mu_0}) \cdot P$**
relative dielectric constant ε_r **Leave ε_r alone**	permeance $\mathcal{P}$ **Replace $\mathcal{P}$ by $\mathcal{P}/(\mu_0 c^2)$**
magnetomotive force $\mathcal{F}$ **Replace $\mathcal{F}$ by $\mathcal{F} \cdot (c\sqrt{\mu_0})$**	charge Q **Replace Q by $(c\sqrt{\mu_0}) \cdot Q$**
magnetic flux Φ_B **Replace Φ_B by $\Phi_B/(c\sqrt{\mu_0})$**	resistance R **Replace R by $R/(\mu_0 c^2)$**
conductance G **Replace G by $\mu_0 c^2 G$**	reluctance $\mathcal{R}$ **Replace $\mathcal{R}$ by $\mathcal{R} \cdot (\mu_0 c^2)$**
magnetic field H **Replace H by $H \cdot (c\sqrt{\mu_0})$**	volume charge density ρ_Q **Replace ρ_Q by $(c\sqrt{\mu_0}) \cdot \rho_Q$**
current I **Replace I by $(c\sqrt{\mu_0}) \cdot I$**	resistivity ρ_R **Replace ρ_R by $\rho_R/(\mu_0 c^2)$**
volume current density J **Replace J by $(c\sqrt{\mu_0}) \cdot J$**	elastance S **Replace S by $S/(\mu_0 c^2)$**
surface current density $\mathcal{J}_S$ **Replace $\mathcal{J}_S$ by $(c\sqrt{\mu_0}) \cdot \mathcal{J}_S$**	surface charge density S_Q **Replace S_Q by $(c\sqrt{\mu_0}) \cdot S_Q$**
inductance L **Replace L by $L/(\mu_0 c^2)$**	conductivity σ **Replace σ by $\mu_0 c^2 \sigma$**
permanent-magnet dipole moment m_H **Replace m_H by $m_H/(c\sqrt{\mu_0})$**	electric potential V **Replace V by $V/(c\sqrt{\mu_0})$**
current-loop magnetic dipole moment m_I **Replace m_I by $m_I \cdot (c\sqrt{\mu_0})$**	magnetic scalar potential Ω_H **Replace Ω_H by $\Omega_H \cdot (c\sqrt{\mu_0})$**
permanent-magnet dipole density M_H **Replace M_H by $M_H/(c\sqrt{\mu_0})$**	electric susceptibility χ_e **Leave χ_e alone**
current-loop magnetic dipole density M_I **Replace M_I by $M_I \cdot (c\sqrt{\mu_0})$**	magnetic susceptibility χ_m **Leave χ_m alone**
magnetic permeability μ **Replace μ by $\mu_r c^{-2}$**	

Table 4.17(b) To go from the unrationalized mks system to esu units. (This table also takes us from the esuq and emuq units explained in Chapter 2 to esu units.)

magnetic vector potential A	magnetic permeability μ
Replace A by $A \cdot (c\sqrt{\mu_0})$	*Replace μ by $\mu_r \cdot \mu_0$*
magnetic induction B	relative magnetic permeability μ_r
Replace B by $B \cdot (c\sqrt{\mu_0})$	*Leave μ_r alone*
capacitance C	magnetic permeability of free space μ_0
Replace C by $C/(\mu_0 c^2)$	*Leave μ_0 alone*
electric displacement D	magnetic pole strength p_H
Replace D by $D/(c\sqrt{\mu_0})$	*Replace p_H by $p_H \cdot (c\sqrt{\mu_0})$*
electric field E	electric dipole moment p
Replace E by $E \cdot (c\sqrt{\mu_0})$	*Replace p by $p/(c\sqrt{\mu_0})$*
dielectric constant ε	electric dipole density P
Replace ε by $\varepsilon_r/(\mu_0 c^2)$	*Replace P by $P/(c\sqrt{\mu_0})$*
relative dielectric constant ε_r	permeance $\mathcal{P}$
Leave ε_r alone	*Replace $\mathcal{P}$ by $(\mu_0 c^2) \cdot \mathcal{P}$*
permittivity of free space ε_0	charge Q
Replace ε_0 by $(\mu_0 c^2)^{-1}$	*Replace Q by $Q/(c\sqrt{\mu_0})$*
magnetomotive force $\mathcal{F}$	resistance R
Replace $\mathcal{F}$ by $\mathcal{F}/(c\sqrt{\mu_0})$	*Replace R by $R \cdot (\mu_0 c^2)$*
magnetic flux Φ_B	reluctance $\mathcal{R}$
Replace Φ_B by $\Phi_B \cdot (c\sqrt{\mu_0})$	*Replace $\mathcal{R}$ by $\mathcal{R}/(\mu_0 c^2)$*
conductance G	volume charge density ρ_Q
Replace G by $G/(\mu_0 c^2)$	*Replace ρ_Q by $\rho_Q/(c\sqrt{\mu_0})$*
magnetic field H	resistivity ρ_R
Replace H by $H/(c\sqrt{\mu_0})$	*Replace ρ_R by $\rho_R \cdot (\mu_0 c^2)$*
current I	elastance S
Replace I by $I/(c\sqrt{\mu_0})$	*Replace S by $S \cdot (\mu_0 c^2)$*
volume current density J	surface charge density S_Q
Replace J by $J/(c\sqrt{\mu_0})$	*Replace S_Q by $S_Q/(c\sqrt{\mu_0})$*
surface current density $\mathcal{J}_S$	conductivity σ
Replace $\mathcal{J}_S$ by $\mathcal{J}_S/(c\sqrt{\mu_0})$	*Replace σ by $\sigma/(\mu_0 c^2)$*
inductance L	electric potential V
Replace L by $L \cdot (\mu_0 c^2)$	*Replace V by $V \cdot (c\sqrt{\mu_0})$*
permanent-magnet dipole moment m_H	magnetic scalar potential Ω_H
Replace m_H by $m_H \cdot (c\sqrt{\mu_0})$	*Replace Ω_H by $\Omega_H/(c\sqrt{\mu_0})$*
current-loop magnetic dipole moment m_I	electric susceptibility χ_e
Replace m_I by $m_I/(c\sqrt{\mu_0})$	*Leave χ_e alone*
permanent-magnet dipole density M_H	magnetic susceptibility χ_m
Replace M_H by $M_H \cdot (c\sqrt{\mu_0})$	*Leave χ_m alone*
current-loop magnetic dipole density M_I	
Replace M_I by $M_I/(c\sqrt{\mu_0})$	

Table 4.18(a) To go from emu units to the unrationalized mks system. (This table also takes us from emu units to the esuq and emuq units explained in Chapter 2.)

magnetic vector potential A	current-loop magnetic dipole density M_I
Replace A by $A/\sqrt{\mu_0}$	*Replace M_I by $M_I \cdot \sqrt{\mu_0}$*
magnetic induction B	relative magnetic permeability μ_r
Replace B by $B/\sqrt{\mu_0}$	*Leave μ_r alone*
capacitance C	magnetic pole strength p_H
Replace C by $\mu_0 \cdot C$	*Replace p_H by $p_H/\sqrt{\mu_0}$*
electric displacement D	electric dipole moment p
Replace D by $D \cdot \sqrt{\mu_0}$	*Replace p by $p \cdot \sqrt{\mu_0}$*
electric field E	electric dipole density P
Replace E by $E/\sqrt{\mu_0}$	*Replace P by $P \cdot \sqrt{\mu_0}$*
dielectric constant ε	permeance $\mathcal{P}$
Replace ε by ε_r/c^2	*Replace $\mathcal{P}$ by $\mathcal{P}/\mu_0$*
relative dielectric constant ε_r	charge Q
Leave ε_r alone	*Replace Q by $Q \cdot \sqrt{\mu_0}$*
permittivity of free space ε_0	resistance R
Replace ε_0 by c^{-2}	*Replace R by R/μ_0*
magnetomotive force $\mathcal{F}$	reluctance $\mathcal{R}$
Replace $\mathcal{F}$ by $\mathcal{F} \cdot \sqrt{\mu_0}$	*Replace $\mathcal{R}$ by $\mathcal{R} \cdot \mu_0$*
magnetic flux Φ_B	volume charge density ρ_Q
Replace Φ_B by $\Phi_B/\sqrt{\mu_0}$	*Replace ρ_Q by $\rho_Q \cdot \sqrt{\mu_0}$*
conductance G	resistivity ρ_R
Replace G by $G \cdot \mu_0$	*Replace ρ_R by ρ_R/μ_0*
magnetic field H	elastance S
Replace H by $H \cdot \sqrt{\mu_0}$	*Replace S by S/μ_0*
current I	surface charge density S_Q
Replace I by $I \cdot \sqrt{\mu_0}$	*Replace S_Q by $S_Q \cdot \sqrt{\mu_0}$*
volume current density J	conductivity σ
Replace J by $J \cdot \sqrt{\mu_0}$	*Replace σ by $\mu_0 \cdot \sigma$*
surface current density $\mathcal{J}_S$	electric potential V
Replace $\mathcal{J}_S$ by $\mathcal{J}_S \cdot \sqrt{\mu_0}$	*Replace V by $V/\sqrt{\mu_0}$*
inductance L	magnetic scalar potential Ω_H
Replace L by L/μ_0	*Replace Ω_H by $\Omega_H \cdot \sqrt{\mu_0}$*
permanent-magnet dipole moment m_H	electric susceptibility χ_e
Replace m_H by $m_H/\sqrt{\mu_0}$	*Leave χ_e alone*
current-loop magnetic dipole moment m_I	magnetic susceptibility χ_m
Replace m_I by $m_I \cdot \sqrt{\mu_0}$	*Leave χ_m alone*
permanent-magnet dipole density M_H	
Replace M_H by $M_H/\sqrt{\mu_0}$	

Table 4.18(b) To go from the unrationalized mks system to emu units. (This table also takes us from the esuq and emuq units explained in Chapter 2 to emu units.)

magnetic vector potential A	magnetic permeability μ
Replace A by $A\sqrt{\mu_0}$	*Replace μ by $\mu_r \cdot \mu_0$*
magnetic induction B	relative magnetic permeability μ_r
Replace B by $B\sqrt{\mu_0}$	*Leave μ_r alone*
capacitance C	magnetic permeability of free space μ_0
Replace C by C/μ_0	*Leave μ_0 alone*
electric displacement D	magnetic pole strength p_H
Replace D by $D/\sqrt{\mu_0}$	*Replace p_H by $p_H \cdot \sqrt{\mu_0}$*
electric field E	electric dipole moment p
Replace E by $E\sqrt{\mu_0}$	*Replace p by $p/\sqrt{\mu_0}$*
dielectric constant ε	electric dipole density P
Replace ε by $\varepsilon_r/(\mu_0 c^2)$	*Replace P by $P/\sqrt{\mu_0}$*
relative dielectric constant ε_r	permeance $\mathcal{P}$
Leave ε_r alone	*Replace $\mathcal{P}$ by $\mu_0 \cdot \mathcal{P}$*
permittivity of free space ε_0	charge Q
Replace ε_0 by $(\mu_0 c^2)^{-1}$	*Replace Q by $Q/\sqrt{\mu_0}$*
magnetomotive force $\mathcal{F}$	resistance R
Replace $\mathcal{F}$ by $\mathcal{F}/\sqrt{\mu_0}$	*Replace R by $R \cdot \mu_0$*
magnetic flux Φ_B	reluctance $\mathcal{R}$
Replace Φ_B by $\Phi_B \cdot \sqrt{\mu_0}$	*Replace $\mathcal{R}$ by $\mathcal{R}/\mu_0$*
conductance G	volume charge density ρ_Q
Replace G by G/μ_0	*Replace ρ_Q by $\rho_Q/\sqrt{\mu_0}$*
magnetic field H	resistivity ρ_R
Replace H by $H/\sqrt{\mu_0}$	*Replace ρ_R by $\rho_R \cdot \mu_0$*
current I	elastance S
Replace I by $I/\sqrt{\mu_0}$	*Replace S by $S \cdot \mu_0$*
volume current density J	surface charge density S_Q
Replace J by $J/\sqrt{\mu_0}$	*Replace S_Q by $S_Q/\sqrt{\mu_0}$*
surface current density $\mathcal{J}_S$	conductivity σ
Replace $\mathcal{J}_S$ by $\mathcal{J}_S/\sqrt{\mu_0}$	*Replace σ by σ/μ_0*
inductance L	electric potential V
Replace L by $L \cdot \mu_0$	*Replace V by $V\sqrt{\mu_0}$*
permanent-magnet dipole moment m_H	magnetic scalar potential Ω_H
Replace m_H by $m_H\sqrt{\mu_0}$	*Replace Ω_H by $\Omega_H/\sqrt{\mu_0}$*
current-loop magnetic dipole moment m_I	electric susceptibility χ_e
Replace m_I by $m_I/\sqrt{\mu_0}$	*Leave χ_e alone*
permanent-magnet dipole density M_H	magnetic susceptibility χ_m
Replace M_H by $M_H\sqrt{\mu_0}$	*Leave χ_m alone*
current-loop magnetic dipole density M_I	
Replace M_I by $M_I/\sqrt{\mu_0}$	

Bibliography

ARTICLES

[1]   Birge, R. T., "On Electric and Magnetic Units and Dimensions," *The American Physics Teacher*, Vol. 2, No. 2 (May 1934) pp. 41–48.

[2]   Birge, R. T., "On the Establishment of Fundamental and Derived Units, with Special Reference to Electric Units. Part I," *The American Physics Teacher*, Vol. 3, No. 3 (1935) pp. 102–109.

[3]   Birge, R. T., "On the Establishment of Fundamental and Derived Units, with Special Reference to Electric Units. Part II," *The American Physics Teacher*, Vol. 3, No. 4 (1935) pp. 171–179.

[4]   Campbell, N. R. and L. Hartshorn, "The Experimental Basis of Electromagnetism: The Direct Current Circuit," *Proceedings of the Physical Society* (London), Vol. 58 (1946), pp. 634–653.

[5]   Campbell, N. R. and L. Hartshorn, "The Experimental Basis of Electromagnetism: Part II—Electrostatics," *Proceedings of the Physical Society* (London), Vol. 60 (1948), pp. 27–52.

[6]   Campbell, N. R. and L. Hartshorn, "The Experimental Basis of Electromagnetism—Part III: The Magnetic Field," *Proceedings of the Physical Society* (London), Vol. 62 (1950) pp. 422–429.

[7]   Cornelius, P., W. De Groot, and R. Vermeulen, "Quantity Equations, Rationalization and Change of Number of Fundamental Quantities I," *Applied Scientific Research, Section B Electrophysics, Acoustics, Optics, Mathematical Methods*, Vol. 12, No. 1 (1965) pp. 1–17.

[8]   Cornelius, P., W. de Groot, and R. Vermeulen, "Quantity Equations, Rationalization and Change of Number of Fundamental Quantities II," *Applied Scientific Research, Section B Electrophysics, Acoustics, Optics, Mathematical Methods*, Vol. 12, No. 4 (1965) pp. 235–247.

[9]   Cornelius, P., W. de Groot, and R. Vermeulen, "Quantity Equations, Rationalization and Change of Number of Fundamental Quantities III," *Applied Scientific Research, Section B Electrophysics, Acoustics, Optics, Mathematical Methods*, Vol. 12, No. 4 (1965) pp. 248–265.

[10]   Hartshorn, L., "The Experimental Basis of Electromagnetism—Part IV: Magnetic Materials," *Proceedings of the Physical Society* (London), Vol. 62 (1950) pp. 429–445.

[11]   Kennely, A. E., "Historical Outline of the Electrical Units," *Journal of Engineering Education*, Vol. 19 (1928) pp. 229–275.

[12]   Kennelly, A. E., "Magnetic Formulae Expressed in M. K. S. System of Units," *Proceedings of the American Philosophical Society Held at Philadelphia for Promoting Useful Knowledge*, Vol. 76, No. 3 (1936) pp. 343–377.

[13]   Page, C. H., "Physical Entities and Mathematical Representation," *Journal of Research of the National Bureau of Standards—B. Mathematics and Mathematical Physics*, Vol. 65B, No. 4 (Oct.-Dec. 1961) pp. 227–235.

[14]   Page, C. H., "The Mathematical Representation of Physical Entities," *IEEE Transactions on Education*, Vol. E-10, No. 2 (June 1967) pp. 70–74.

[15]   Silsbee, F. B., "Does Rationalization Change Units?" *Electrical Engineering*, Vol. 76, (April 1957) pp. 296–299.

[16]   Silsbee, F. B., "Systems of Electrical Units," *Journal of Research of the National Bureau of Standards—C: Engineering and Instrumentation*, Vol. 66C, No. 2 (April-June 1962) pp. 137–178.

BOOKS

[1]   Atkinson, E. (translator), *Elementary Treatise on Physics: Experimental and Applied*, translated from Ganot's *Elements de Physique*, A. W. Reinold (ed.), William Wood and Company, New York (1910).

[2]   Badt, F. B., and H. S. Carhart, *Derivation of Practical Electrical Units*, 2nd edition, Electrician Publishing Company, Chicago, IL (1893).

[3]   Bradshow, E., *Electrical Units with Special Reference to the M.K.S. System*, Chapman & Hall, Ltd., London (1952).

[4]   Bridgman, P. W., *Dimensional Analysis*, Yale University Press, New Haven (1931).

[5]   Chute, H. N., *Physical Laboratory Manual*, D. C. Heath & Co., Publishers, Boston (1907).

[6]   Cullwick, E. G., *The Fundamentals of Electromagnetism*, The Macmillan Company, New York (1939).

[7]   Daniell, V. V., *Dielectric Relaxation*, Academic Press, New York (1967).

[8]   Duff, A. W. (ed.), *A Textbook of Physics*, 2nd edition revised, P. Blakiston's Son & Co., Philadelphia (1909).

[9]   Electronics training staff of the cruft laboratory, Harvard University, *Electronic Circuits and Tubes*, McGraw-Hill Book Company, Inc. (H. E. Clifford and A. H. Wing, eds.) (1947).

[10]   Everett, J. D., *Illustrations of the Centimeter-Gramme-Second (C.G.S.) System of Units*, Physical Society of London, London (1875).

[11]   Eyges, L., *The Classical Electromagnetic Field*, Dover Publications, Inc., New York (1972).

[12]   Feather, N., *Electricity and Matter*, Aldine Publishing Company, Chicago (1968).

[13] Fink, D. G. (editor-in-chief) and J. M. Carroll (associate editor), *Standard Handbook for Electrical Engineers*, McGraw-Hill Book Company, New York (1969).

[14] Griffiths, D. J., *Introduction to Electrodynamics*, 2nd edition, Prentice Hall, Englewood Cliffs, New Jersey (1989).

[15] Harnwell, G. P., *Principles of Electricity and Magnetism*, McGraw-Hill Book Company, Inc., New York (1938).

[16] Harris, F. K., *Electrical Measurements*, John Wiley & Sons, Inc., New York (1952).

[17] Huang, K., *Statistical Mechanics*, John Wiley & Sons, Inc., New York (1963).

[18] Huntley, H. E., *Dimensional Analysis*, Dover Publications, Inc., New York (1967).

[19] Hvistendahl, H. S., *Engineering Units and Physical Quantities*, Macmillan & Co. Ltd., London (1964).

[20] Ipsen, D. C., *Units, Dimensions, and Dimensionless Numbers*, McGraw-Hill Book Company, Inc., New York (1960).

[21] Jackson, J. D., *Classical Electrodynamics*, John Wiley & Sons, Inc., New York (1962).

[22] Jackson, J. D., *Classical Electrodynamics*, 3rd edition, John Wiley & Sons, Inc., New York (1999).

[23] Jauncey, G. E. M., and A. S. Langsdorf, *M. K. S. and Dimensions and a Proposed M. K. O. S. System*, The Macmillan Company, New York (1940).

[24] Langhaar, H. L., *Dimensional Analysis and Theory of Models*, John Wiley & Sons, Inc., New York (1951).

[25] Lewis, R., *Engineering Quantities and Systems of Units*, John Wiley & Sons, Inc., New York (1972).

[26] Mahan, B. H., *University Chemistry*, 2nd edition, Addison-Wesley Publishing Company, Inc., Reading, MA (1969).

[27] Mason, Max, and Weaver, Warren, *The Electromagnetic Field*, Dover Publications, Inc., New York (1929).

[28] Maxwell, J. C., *A Treatise on Electricity and Magnetism*, Vols. 1 and 2, Dover Publications, Inc., New York (Copy of 1891 third edition, republished 1954).

[29] Members of the staff of the dept. of electrical engineering, MIT, *Applied Electronics: A First Course in Electronics, Electron Tubes, and Associated Circuits*, John Wiley & Sons, Inc., New York (1943).

[30] Menzel, D. H. (ed.), *Fundamental Formulas of Physics*, Vols. 1 and 2, Dover Publications, Inc., New York (1960).

[31] Page, L. and N. I. Adams, Jr., *Principles of Electricity: An Intermediate Text in Electricity and Magnetism*, D. Van Nostrand Company, Inc., Princeton, New Jersey (1958).

[32] Panofsky, W. K. H., and M. Phillips, *Classical Electricity and Magnetism*, 2nd edition, Addison-Wesley Publishing Company, Inc., Reading, MA (1962).

[33]   Papas, C. H., *Theory of Electromagnetic Wave Propagation*, Dover Publications, Inc., New York (1988).

[34]   Pender, H. and W. A. Del Mar (eds.), *Electrical Engineer's Handbook: Electric Power*, 4th edition, Wiley Engineering Handbook Series, John Wiley & Sons, Inc., New York (1949).

[35]   Purcell, E. M., *Electricity and Magnetism, Berkeley Physics Course— Volume 2*, published by McGraw-Hill Book Company, Inc., (Copyright 1965 by Education Development Center, Newton, MA).

[36]   Ramo, S., and J. R. Whinnery, *Fields and Waves in Modern Radio*, John Wiley & Sons, Inc., New York (1953).

[37]   Sas, R. K., and F. B. Pidduck, *The Meter-Kilogram-Second System of Electrical Units*, Chemical Publishing Co. of N. Y. Inc., New York (1947).

[38]   Shadowitz, A., *The Electromagnetic Field*, Dover Publications, Inc., New York (1975).

[39]   Shire, E. S., *Classical Electricity and Magnetism*, Cambridge University Press, London (1960).

[40]   Skilling, H. H., *Fundamentals of Electric Waves*, John Wiley & Sons, Inc., New York (1942).

[41]   Slater, J. C., and N. H. Frank, *Electromagnetism*, Dover Publications, Inc., New York (1947).

[42]   Smythe, W. R., *Static and Dynamic Electricity*, McGraw-Hill Book Company, New York (1939).

[43]   Smythe, W. R., *Static and Dynamic Electricity*, 3rd edition, McGraw-Hill Book Company, New York (1968).

[44]   Spangenberg, K. R., *Vacuum Tubes*, McGraw-Hill Book Company, Inc., New York (1948).

[45]   Starling, S. G., *Electricity and Magnetism For Advanced Students*, Longmans, Green, and Co., London (1921).

[46]   Stratton, J. A., *Electromagnetic Theory*, McGraw-Hill Book Company, Inc., New York (1941).

[47]   Taylor, E. F., and J. A. Wheeler, *Spacetime Physics*, W. H. Freeman and Company, San Francisco (1966).

[48]   Taylor, B. N., *Guide for the Use of the International System of Units*, NIST Special Publication 811, 1995 edition (published by the Dept. of Commerce, April 1995, Washington, DC).

[49]   Tunbridge, P., *Lord Kelvin: His Influence on Electrical Measurements and Units*, Peter Peregrinus, Ltd. (1992).

[50]   Varner, W. R., *The Fourteen Systems of Units*, Vantage Press, New York (1961).

[51]   Vigoureux, P., *Units and Standards for Electromagnetism*, Springer-Verlag New York Inc., New York (1971).

Index